MOLECULAR
BIOLOGY
INTELLIGENCE
UNIT

MECHANISMS OF DNA DAMAGE RECOGNITION IN MAMMALIAN CELLS

Hanspeter Naegeli, D.V.M.
Institute of Pharmacology and Toxicology
University of Zürich-Tierspital
Zürich, Switzerland

CHAPMAN & HALL
I⟨T⟩P An International Thomson Publishing Company

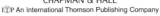

New York • Albany • Bonn • Boston • Cincinnati • Detroit • London • Madrid • Melbourne •
Mexico City • Pacific Grove • Paris • San Francisco • Singapore • Tokyo • Toronto • Washington

LANDES
BIOSCIENCE

AUSTIN, TEXAS
U.S.A.

MOLECULAR BIOLOGY INTELLIGENCE UNIT

MECHANISMS OF DNA DAMAGE RECOGNITION
IN MAMMALIAN CELLS
R.G. LANDES COMPANY
Austin, Texas, U.S.A.

U.S. and Canada Copyright © 1997 R.G. Landes Company and Chapman & Hall

Please address all inquiries to the Publishers:
R.G. Landes Company, 810 South Church Street, Georgetown, Texas, U.S.A. 78626
Phone: 512/ 863 7762; FAX: 512/ 863 0081

North American distributor:

Chapman & Hall, 115 Fifth Avenue, New York, New York, U.S.A. 10003

CHAPMAN & HALL

U.S. and Canada ISBN: 0-412-13311-3

While the authors, editors and publisher believe that drug selection and dosage and the specifications and usage of equipment and devices, as set forth in this book, are in accord with current recommendations and practice at the time of publication, they make no warranty, expressed or implied, with respect to material described in this book. In view of the ongoing research, equipment development, changes in governmental regulations and the rapid accumulation of information relating to the biomedical sciences, the reader is urged to carefully review and evaluate the information provided herein.

Library of Congress Cataloging-in-Publication Data

CIP applied for; not received at time of publication

Publisher's Note

R.G. Landes Bioscience Publishers produces books in six Intelligence Unit series: *Medical, Molecular Biology, Neuroscience, Tissue Engineering, Biotechnology* and *Environmental.* The authors of our books are acknowledged leaders in their fields. Topics are unique; almost without exception, no similar books exist on these topics.

Our goal is to publish books in important and rapidly changing areas of bioscience for sophisticated researchers and clinicians. To achieve this goal, we have accelerated our publishing program to conform to the fast pace at which information grows in bioscience. Most of our books are published within 90 to 120 days of receipt of the manuscript. We would like to thank our readers for their continuing interest and welcome any comments or suggestions they may have for future books.

Shyamali Ghosh
Publications Director
R.G. Landes Company

DEDICATION

To Regi and Pamela

CONTENTS

PREFACE

The maintenance of genetic integrity and continuity was once thought to be solely based on the inherent stability of DNA, the molecule of heredity, but it is now clear that DNA is intrinsically unstable and subject to continuous decay.[1] In a typical mammalian cell, spontaneous chemical reactions disrupt thousands of DNA nucleotides every day.[2] DNA is also constantly damaged by endogenous agents such as free radicals generated during oxidative processes,[3] or by environmental insults including various types of radiation and a large number of electrophilic chemicals.[1,4]

In response to these genotoxic reactions, all forms of life ranging from bacteria to mammals possess multiple systems for either repairing damage inflicted on DNA or, alternatively, increasing their tolerance to it.[1,5-8] To be effective, these damage processing mechanisms must be selectively targeted to sites of damage in cellular DNA. Thus, detection of DNA lesions by discriminating between normal and abnormal nucleotide constituents is a crucial step in all DNA damage response processes. To perform this molecular recognition function, cells are endowed with specialized mechanisms that search for and locate DNA damage in the genome.

The goal of the present review is to summarize existing knowledge on how mammalian cells monitor the structural and functional integrity of their DNA. The problem of DNA damage recognition is common to all organisms, but is further aggravated in higher eukaryotes and mammals by the size of their genomes and the resulting necessity to detect trace amounts of DNA lesions among a very large excess of nondamaged DNA. The biological and medical relevance of DNA damage recognition is evidenced by clinical syndromes characterized by various defects in the capacity to detect DNA damage. Chapter 1 describes inherited human diseases caused by molecular deficiencies in DNA damage recognition and shows that, in most cases, these defects cause a dramatic phenotype of increased susceptibility to mutagenesis and carcinogenesis. Chapter 2 reviews spontaneous or transgenic animal models that are being used to study the biochemical and clinical intricacies associated with these cancer-prone human diseases.

Chapter 3 briefly summarizes the basic chemistry of DNA and the predominant types of damage that disrupt its structure and function. To handle the wide spectrum of lesions occurring in the bases, sugars or phosphates which make up DNA, living cells have evolved several distinct molecular strategies to detect and repair DNA damage. The following chapters summarize these multiple strategies of damage recognition. Chapter 4 illustrates those damage recognition mechanisms (DNA damage reversal, base excision repair) that rely on a simple scheme of one enzyme-one substrate, each enzyme being able to recognize just one type or a narrow range of DNA lesions. Chapter 5 outlines the basic mechanisms involved in a more general mode of DNA damage recognition, in which a set of gene products recognizes and processes multiple types of DNA damage. This system, designated nucleotide excision repair, is characterized by its ability to accommodate a large number of chemically diverse DNA adducts, while ignoring the structural fluctuations of normal DNA.

The next chapters elaborate on the topic of DNA damage recognition in mammalian nucleotide excision repair. In particular, chapter 6 reviews the known factors involved in mammalian nucleotide excision repair and discusses possible biochemical mechanisms implicated in the damage recognition step of this pathway. Chapter 7 summarizes in more detail current experimental approaches and problems in trying to identify molecular determinants that attract mammalian nucleotide excision repair proteins to sites of damage. Molecular crosstalks between DNA excision repair and other nuclear transactions are the topic of chapter 8. In fact, damage recognition is effected in the context of various metabolic processes that occur simultaneously on the same DNA template, and several proteins that are seemingly unrelated to excision repair are able to bind to sites containing DNA adducts. These factors may potentially modulate DNA repair by either exposing or shielding damaged DNA constituents. Chapter 8 will also elaborate on the concept of protein hijacking that has been proposed as a possible mechanism of efficient anticancer therapy. Chapter 9 deals with the recognition of DNA damage by RNA polymerases and discusses its consequences with regard to the phenomenon of transcription-repair coupling.

Damage recognition is not restricted to DNA repair, as cells possess many other molecular mechanisms to increase their ability to survive in the presence of genotoxic agents. To provide an attractive example for such damage processing pathways, chapter 10 describes how DNA polymerases interact with mutagenic base adducts and summarizes experimental evidence in support of the hypothesis that mammalian cells possess specialized mechanisms to increase their tolerance to damaged templates during DNA replication.

As discussed in chapter 11, the question of how mammalian cells monitor and safeguard the integrity of their genome is highly relevant in many disciplines of modern toxicology and medicine. For example, knowledge of the mechanisms underlying DNA damage recognition should be useful to understand the molecular pathogenesis of cancer and develop appropriate prevention programs. Also, knowledge on DNA damage recognition should be useful in designing new therapeutic strategies or in assessing the low-dose carcinogenic potential of genotoxic agents. The molecular details of many DNA damage recognition pathways are still being unveiled and more research is required to fully appreciate the significance of these fascinating and important processes in mammalian systems.

REFERENCES

1. Friedberg EC, Walker GC, Siede W. DNA Repair and Mutagenesis. Washington, D.C.: American Society for Microbiology, 1995.
2. Lindahl T. Instability and decay of the primary structure of DNA. Nature 1993; 362:709-715.
3. Ames BN, Shinegaga MK, Hagen TM. Oxidants, antioxidants, and the degenerative diseases of ageing. Proc Natl Acad Sci USA 1993; 90:7915-7922.
4. Albert RE, Burns FJ. Carcinogenic atmospheric pollutants and the nature of low-level risks. In: Hiatt HH, Watson JD, Winston JA, eds. Origins of Human Cancer. Cold Spring Harbor: Cold Spring Harbor Laboratory, 1977:289-292.
5. Sancar A. Mechanisms of DNA excision repair. Science 1994; 266:1954-1956.
6. Hanawalt PC. Transcription-coupled repair and human disease. Science 1994; 266:1957-1958.
7. Tanaka K, Wood RD. Xeroderma pigmentosum and nucleotide excision repair of DNA. Trends Biochem Sci 1994; 19:83-86.
8. Hoeijmakers JHJ. Human nucleotide excision repair syndromes: molecular clues to unexpected intricacies. Eur J Cancer 1994; 30A:1912-1921.

ACKNOWLEDGMENTS

Research support for the author's laboratory comes from the Swiss National Science Foundation, the Wolfermann-Naegeli Stiftung and the Kanton of Zürich.

MEDICAL BACKGROUND: HUMAN DNA DAMAGE RECOGNITION AND PROCESSING DISORDERS

If left uncorrected, DNA damage poses multiple threats to the proper functioning of DNA. First, many DNA lesions cause cytotoxic cell death by interfering with essential transactions such as transcription or DNA replication.[1,2] Second, DNA lesions induce lethality by triggering pathways of programmed cell death (apoptosis).[3,4] Third, cells that survive are subject to permanent changes in the nucleotide sequence, or genetic code, of DNA. This mutagenic response is due to the increased probability that errors occur upon replication of damaged templates.[5,6] Multiple DNA damage processing pathways have been identified in mammalian cells, all of which are able to modulate the cytotoxic, apoptotic or mutagenic effects of DNA lesions. Among these pathways, DNA repair acts as a key line of defense to remove injuries to DNA and increase cell survival by allowing resumption of transcription or replication. DNA repair processes also minimize the mutagenic consequences of DNA damage, thereby preventing the accumulation of mutations at critical positions in DNA and reducing the incidence of cancer or inherited diseases.[1,7,8]

THE CELLULAR DNA DAMAGE RESPONSE NETWORK

There are several inherited human disorders caused by malfunctioning DNA damage processing mechanisms. To understand the pathophysiology and clinical manifestations of these diseases, the metabolic basis of a particular syndrome should be evaluated within the broader context of an integrated cellular DNA damage response system. In fact, as shown in Figure 1.1, damage processing pathways are organized as an intricate biochemical network. Inactivation of one pathway results in the compensatory channeling of the offending lesion into alternative pathways which, depending on their activities, mitigate or potentiate the biological endpoints deriving from DNA damage. A brief review of the cellular reactions to cyclobutane pyrimidine dimers, the quantitatively major DNA lesion induced by the ultraviolet (UV) component of solar radiation, may illustrate the complexity of such damage response networks.

Historically, photoreactivation was the first DNA repair reaction described in the literature. In 1949, Kelner[9] found that visible light enhances the ability of the microorganism *Streptomyces griseus* to recover from irradiation with UV light. About 10 years later it was

Mechanisms of DNA Damage Recognition in Mammalian Cells, by Hanspeter Naegeli.
© 1997 R.G. Landes Company.

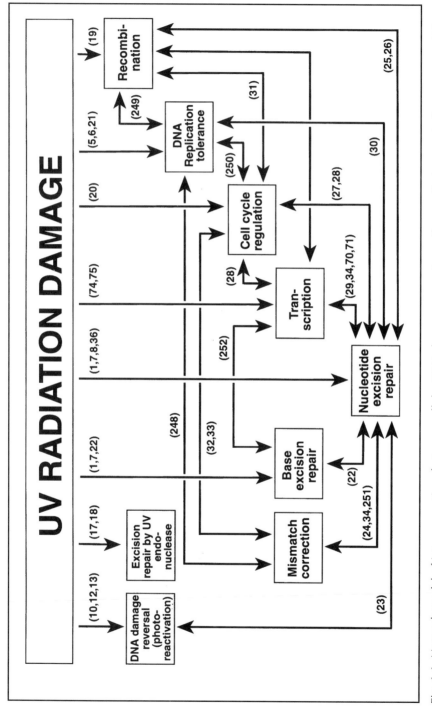

Fig. 1.1. Network model of interconnections between cellular DNA damage processing pathways. Each arrow indicates a synergistic or antagonistic interaction between different pathways responding to the presence of UV radiation damage, a predominant DNA lesion inflicted by the environment. The numbers in parenthesis indicate relevant references.

discovered that this photoreactivation phenomenon was promoted by a DNA repair enzyme, designated DNA photolyase.[10] This enzyme catalyzes a unique reaction in which the energy of visible light is utilized to break cyclobutane rings in DNA and convert cyclobutane pyrimidine dimers into monomeric pyrimidine bases. In a fascinating account of these early discoveries, Hanawalt indicated that an interesting observation possibly related to photoreactivation was already recorded much earlier.[11] In 1933, Hausser and von Oehmcke reported that the UV-induced browning of banana skin could be prevented by subsequent irradiation at longer (visible) wavelength. Retrospectively, this observation suggests the involvement of a photoreactivating enzyme in preserving the natural color of bananas.[11] Photoreactivation involves neither removal nor replacement of bases or nucleotides, but rather provides an example of DNA repair by which a covalent damage is simply reversed. The possible role of DNA photolyase in humans is a controversial issue.[12,13] The reversal of alkylation damage from DNA by O^6-methyl-guanine-DNA methyltransferase, on the other hand, is of unequivocal importance in all mammals and humans.[14]

During the early 1960s, Setlow, Carrier,[15] Boyce and Howard-Flanders[16] reported on the existence of an excision repair reaction that removes cyclobutane pyrimidine dimers from bacterial genomes. The repair scheme they discovered was termed nucleotide excision repair because it excises DNA damage as oligomeric DNA segments consisting of 12-13 nucleotides in prokaryotes and 27-29 nucleotides in eukaryotes. Figure 1.2 illustrates that this oligonucleotide excision activity is initiated by dual incision of damaged DNA strands. A prominent feature of nucleotide excision repair is its versatility, as the same multiprotein system is able to process a wide range of chemical DNA adducts in addition to cyclobutane pyrimidine dimers or other UV photoproducts. In a few known cases, for example in the bacterium *Micrococcus luteus* and in T4 phage-infected *Escherichia coli*, removal of cyclobutane pyrimidine dimers is also initiated by specific pyrimidine dimer-DNA glycosylases.[1,7] As indicated in Figure 1.2, this class of enzymes releases damaged bases directly, thereby generating abasic sites as reaction intermediates. An alternative mechanism for excision of UV radiation products (including cyclobutane pyrimidine dimers) depends on the action of novel UV-specific endonucleases that incise DNA immediately 5' to the lesion (Fig. 1.2). Such enzymes have been recently discovered in the yeast *Schizosaccharomyces pombe* and the filamentous fungus *Neurospora crassa*, but may also exist in other species.[17,18]

These examples involving the repair of cyclobutane pyrimidine dimers either by photolyase, nucleotide excision repair, base excision repair or UV-specific endonucleases show that the same DNA lesion may serve as a substrate for multiple repair systems. If DNA damage reversal or excision repair processes fail, DNA lesions may be channeled into a pathway of DNA recombination, although this system is mostly utilized to repair broken chromosomes.[19] If DNA damage persists despite the presence of all these repair reactions, cell cycle checkpoint systems are normally activated to provide more time for DNA repair mechanisms before progressing into the next stages of cell cycle.[20] If a particular type of DNA damage is completely intractable to DNA repair processes, the mammalian DNA replication machinery displays a remarkable ability to bypass DNA damage by exploiting specialized tolerance pathways that may be either error-free (nonmutagenic) or error-prone (mutagenic).[21]

Figure 1.1 shows that these different processes are intimately connected with each other forming an intricate network, and recent discoveries have revealed unanticipated relationships with other fundamental DNA transactions. For example, nucleotide excision repair acts as a backup system for lesions that are not processed by DNA damage reversal or base excision repair.[22] The same nucleotide excision repair system is stimulated by DNA photolyase[23] and its activity may be further modulated by proteins that are normally involved in base ex-

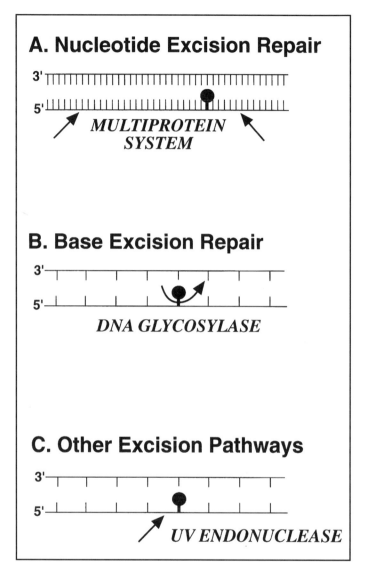

Fig. 1.2. Scheme illustrating the basic differences in the initiation of excision repair mechanisms: A, nucleotide excision repair; B, base excision repair; C, DNA incision by UV damage-specific endonucleases. Nucleotide excision repair is performed by dual incision near sites of carcinogen-DNA adducts. The first step in base excision repair is the removal of damaged or inappropriate bases by cleavage of N-glycosylic bonds, generating abasic sites. UV-specific endonucleases incise DNA on the 5' side of UV radiation products.

cision repair, mismatch correction,[24] recombination[25,26] or the cell cycle regulatory circuit (Fig. 1.1).[27,28] In addition, nucleotide excision repair is directly coupled to RNA polymerase II-dependent transcription[29] and to DNA replication,[30] as these pathways share common factors (Fig. 1.1). There are many additional connections between individual pathways of the DNA damage response network. For example, at least one component (p53) of the cell cycle regulatory system interacts with both transcription[28] and recombination subunits.[31] There is also evidence for a possible involvement of the mismatch correction system in cell cycle regulation.[32,33] Another intriguing level of complexity in the damage response network is indicated by the finding that coupling of nucleotide excision repair to transcription depends on mismatch repair proficiency.[34]

The first two chapters of this review are intended to summarize human diseases and animal models that illustrate the clinical, biological and toxicological significance of the mammalian DNA damage processing network. Table 1.1 shows human diseases associated with deficiencies in different pathways of this network (reviewed in chapter 1) and the corresponding animal models which, in most cases, have been established by transgenic approaches in mice (reviewed in chapter 2).

DNA REPAIR DISORDERS

Spontaneous or experimentally induced DNA repair mutants illustrate the disastrous biological effects of malfunctioning DNA repair systems. A large number of repair-defective bacteria or yeast strains, as well as repair-defective mammalian cell lines have been identified, isolated and characterized.[1,7,8] As one might expect, a common feature of all these DNA repair mutants is hypersensitivity to specific classes of DNA damaging agents. In most cases, DNA repair mutants also respond to treatment with these genotoxic agents by increased frequencies of mutagenesis.[35] As a consequence,

human DNA repair disorders are often associated with a greatly elevated risk of cancer.

XERODERMA PIGMENTOSUM

The phenotypic consequences of a failure in DNA repair is most vividly demonstrated by the occurrence of human DNA repair disorders, of which xeroderma pigmentosum has long been the prototypical example. Nearly 30 years ago, Cleaver published his discovery that cells of patients suffering from xeroderma pigmentosum carry a DNA excision repair deficiency.[36] Individuals afflicted by this rare recessively inherited disorder are defective in the nucleotide excision repair process that normally removes DNA damage induced by exposure to UV radiation (see Fig. 1.2). These patients present with extreme hypersensitivity to sunlight and develop dry skin (xeroderma), strong pigmentation abnormalities and malignancies at the exposed areas of the skin, mainly basal cell carcinomas, squamous cell carcinomas and melanomas (Table 1.2).[37,38] Compared to the normal population, the estimated incidence of xeroderma pigmentosum-associated skin cancer under the age of 19 is 4,900 times higher for squamous cell and basal cell carcinoma, and 8,000 times higher for malignant melanoma (Table 1.2).[39-41] These skin malignancies are further characterized by their early appearance, the median age for development of the first skin cancers in xeroderma pigmentosum individuals being 8 years. This is approximately 50 years younger than the median age of skin cancer in the general population.[42] In a large number of cases, the eyelids, conjunctiva and cornea are severely affected as these anterior ocular tissues are also exposed to UV radiation. Of about 1,000 xeroderma pigmentosum patients reported worldwide, approximately 20% exhibit accelerated neurodegeneration associated with neuronal death in the central and peripheral nervous system. This clinical finding suggests that the high metabolic rate combined with the long life span of neurons could make them especially dependent on proficient DNA repair mechanisms.[43]

Table 1.1. Summary of human diseases and animal models associated with defective processing of DNA damage

Biochemical pathway	Human diseases (chapter 1)	Animal models (chapter 2)
Nucleotide excision repair	Xeroderma pigmentosum, trichothiodystrophy, Cockayne syndrome	XPA-, XPC- and ERCC1-deficient mice, *Drosophila haywire* mutant
Base excision repair	None	Lethality of polymerase β- and AP endonuclease-deficient mouse embryos
Mismatch correction	Hereditary non-polyposis colorectal cancer	PMS2- and MSH2-deficient mice
DNA damage reversal	None (?)	MGMT-deficient mice
Recombination	Uncertain (see ref. 246)	SCID in mice and Arabian foals, Ku86-deficient mice, lethality of Rad51-deficient mouse embryos
Cell cycle checkpoint regulation	Li-Fraumeni syndrome, ataxia telangiectasia	p53-deficient mice, ATM-deficient mice
DNA replication tolerance	Xeroderma pigmentosum variant	*Drosophila mus* strains
DNA ligation	Patient 46BR	DNA ligase I-deficient mice (ref. 247)
Other	Fanconi's anemia, Bloom's syndrome	In progress

At the biochemical level, skin or blood cells obtained from xeroderma pigmentosum patients are defective or completely deficient in nucleotide excision repair of UV radiation products and bulky base adducts. Because of their incapability to efficiently eliminate bulky DNA damage, xeroderma pigmentosum cells are hypersensitive to UV radiation and many chemical carcinogens, and are hypermutable in response to these DNA damaging agents.[44-47] Characteristic mutations found in skin tumors of xeroderma pigmentosum patients are tandem CC→TT transitions which, at least in the skin, are thought to be uniquely induced by UV irradiation.[48,49]

Xeroderma pigmentosum occurs at a frequency of about 1 in 250,000 in North America and Europe, and 1 in 40,000 in Japan and North Africa. Individuals afflicted by this inherited disorder have been assigned to seven different genetic complementation groups (XP-A through XP-G, each carrying a mutation in a different gene).[37,39] An eighth form of the disease, called xeroderma pigmentosum variant, displays anomalies in DNA replication (see chapter 10).[50] These complementation groups were identified by measuring DNA repair synthesis, frequently referred to as unscheduled DNA synthesis, in multinucleate cells formed by fusing

Table 1.2. *Cancer frequency in a collection of xeroderma pigmentosum patients and comparison with the general population in the United States (adapted from ref. 40)*

Location and type of cancer	Age group (years)	Total number of patients	Number of patients with cancer	Observed cancer frequency	Expected cancer frequency (United States)	Ratio
Skin: basal cell and squamous cell carcinomas	0-19	77	49	0.64	0.00013	4,900
	0-39	123	52	0.42	0.0011	400
Skin: melanomas	0-19	77	8	0.10	0.000013	8,000
	0-39	123	14	0.11	0.00018	600
Eye: mainly non-melanoma cancers of conjunctiva or cornea	0-19	77	4	0.052	0.000052	1,000
	0-39	123	5	0.041	0.000059	700
Tongue: mainly squamous cell carcinomas of the anterior tongue	0-19	77	3	0.039	0.00000039	100,000
	0-39	123	3	0.024	0.0000080	3,000
Brain tumors	0-19	77	1	0.013	0.00026	50
	0-39	123	2	0.016	0.0004	40

fibroblasts or lymphoblasts from two different patients. If the two patients have mutations in separate genes, each cell is able to complement the deficiency of its fusion partner and the multinucleate cells acquire normal unscheduled DNA synthesis. Complementation groups A, C, D and the variant form are relatively more frequent and together comprise approximately 90% of all characterized xeroderma pigmentosum patients.[37,39] The remaining 10% is composed of xeroderma pigmentosum group B, F, E, G. The genes mutated in these different complementation groups, the biochemical properties of their protein products and the mechanisms involved in mammalian nucleotide excision repair will be discussed in detail in chapter 6.

Trichothiodystrophy

Among the xeroderma pigmentosum complementation groups, XP-D is the most heterogeneous. Mutations in the *XPD* gene not only cause xeroderma pigmentosum but can also give rise to phenotypically different syndromes such as trichothiodystrophy and Cockayne syndrome.[51-56] Table 1.3 compares the clinical and cellular manifestations of these three related disorders. A common hallmark of all three syndromes is the occurrence of hypersensitivity to the UV component of solar light.

Trichothiodystrophy is a rare autosomal recessive disorder that is mainly characterized by ichthyosis (fish-like scaliness of the skin), sulfur-deficient brittle hair, intellectual impairment, decreased fertility and short stature.[57] In contrast to xeroderma pigmentosum, neurological dysfunctions are primarily related to neuron dysmyelination. In at least one clinical case, the characteristic trichothiodystrophy symptoms were accompanied by a remarkable phenotype suggestive of a temperature-sensitive mutation in the affected *XPD* gene. This patient repeatedly showed a striking loss of scalp hair during infection periods that were associated with elevated body temperature.[58]

Trichothiodystrophy patients are frequently photosensitive and, at the biochemical level, deficient in nucleotide excision repair. Cells obtained from these trichothiodystrophy individuals manifest hypersensitivity to killing by UV light and reduced unscheduled DNA synthesis similar to cells from xeroderma pigmentosum patients.[56,59,60] Using a simian virus 40-based shuttle vector it was shown that trichothiodystrophy cells develop almost identical frequencies of UV-induced mutagenesis as classical xeroderma pigmentosum cells.[61,62] In striking contrast with xeroderma pigmentosum patients, however, individuals with trichothiodystrophy do not develop early skin cancer.[63] These observations indicate that cancer induction is not directly linked to the observed mutation rate. Thus, other factors such as the qualitative pattern of mutations or, alternatively, reduced catalase activity or low immune surveillance in xeroderma pigmentosum individuals may play a decisive role in cancer pathogenesis.[64-67] Experimental studies in mice demonstrated that UV radiation causes profound systemic immunosuppression.[68] Interestingly, this negative effect on the immune system was largely prevented by topical application to UV-irradiated mice of liposomes containing active cyclobutane pyrimidine dimer-DNA glycosylase.[68] In another report, some heavily sun-exposed Egyptian xeroderma pigmentosum patients showed abnormally weak reactions in tests of immune function.[66] Collectively, these observations suggest that cyclobutane dimers persisting in the DNA of UV-exposed skin cells are responsible for initiating cytokine-mediated immunosuppression[69] which, in turn, may contribute to the high incidence of skin cancer in xeroderma pigmentosum.

Trichothiodystrophy can result from mutations in one of three genes (*XPD*, *XPB* or *TTDA*).[1,8] The products of *XPB* and *XPD*, and perhaps also *TTDA*, are subunits of TFIIH, a complex multiprotein factor required for nucleotide excision repair and transcription by RNA polymerase II.[70] The finding that a number of proteins are shared

Table 1.3. Comparison of clinical and cellular features in xeroderma pigmentosum, trichothiodystrophy and Cockayne syndrome (adapted from ref. 8)

Abnormality	Xeroderma pigmentosum	Trichothiodystrophy	Cockayne syndrome
UV sensitivity	Yes	Yes	Yes
Freckle-like pigmentation abnormalities	Yes	No	No
Ichthyosis	No	Yes	No
Brittle hair with low sulphur	No	Yes	No
Skin cancer	Yes	No	No
Mental retardation	Some patients	Yes	Yes
Brain pathology	Atrophy	Dysmyelination	Dysmyelination
Sexual development	Immature in some patients	Decreased fertility	Decreased fertility
Dwarfism	No	Most patients	Most patients
Unscheduled DNA synthesis	Reduced (except in the variant form)	Reduced	Normal

between nucleotide excision repair and transcription suggests that at least some clinical features of trichothiodystrophy are the result of a transcription defect rather than the consequence of a DNA excision repair insufficiency. For example, the characteristic skin, nail and hair abnormalities observed in trichothiodystrophy patients may relate to defective transcription of genes encoding sulfur proteins of epidermal tissues. Similarly, neurodysmyelination may involve reduced transcription of genes encoding factors that are important for myelin sheet formation. This possible association between defective transcription and various phenotypic characteristics of trichothiodystrophy led to the suggestion that this disease defines, in part, a "transcription syndrome".[8,70]

An elegant approach was used to test this transcription hypothesis. Guzder and co-workers[71] expressed mutated human *XPD* genes in yeast *Saccharomyces cerevisiae* cells in which *RAD3* had been deleted. *RAD3* is the highly conserved yeast homolog of *XPD* and, like *XPD*, is required for nucleotide excision repair and, additionally, has an essential role in RNA polymerase II-dependent transcription. Due to this transcription function, yeast *rad3* null mutants are inviable, but the human *XPD* gene is able to rescue these *rad3* deletion mutations by heterologous complementation. However, it was found that *XPD* alleles containing precisely those mutations that were previously detected in trichothiodystrophy patients failed to rescue the *rad3* null mutation, providing additional evidence in favor of the hypothesis that trichothiodystrophy mutations impair the ability of XPD protein to function normally in basal transcription by RNA polymerase II. Thus, trichothiodystrophy may involve transcriptional defects in addition to malfunctioning of nucleotide excision repair. The transcription syndrome hypothesis also provides an alternative explanation for the missing cancer predisposition in trichothiodystrophy patients. Conceivably, reduced transcription of critical genes may exert antitumorigenic effects by inhibiting cell proliferation and limiting the propensity to undergo transformation.[55,72]

COCKAYNE SYNDROME

Patients suffering from Cockayne syndrome carry defects in a specific excision repair subpathway that results in preferential repair of the transcribed strand of active genes.[73-75] Cockayne syndrome is a rare autosomal recessive disease characterized primarily by growth failure and progressive neurological dysfunction involving microcephaly, hydrocephalus and calcification of basal ganglia (Table 1.3).[76,77] The neurological effects are related to neurodysmyelination, as in trichothiodystrophy. Body weight of Cockayne syndrome individuals tends to be affected more severely than body length, leading to a characteristic appearance of so-called cachectic dwarfism.[78,79] Patients suffering from Cockayne syndrome have faces with large sunken eyes, temporal wasting and a thin prominent nose. Characteristic signs of Cockayne syndrome also include sun hypersensitivity,[80-84] ophthalmic changes including pigmentary degeneration of the retina and cataracts, deafness, impaired sexual development, dental caries, and premature death. The mean age of death is about 12 years. In the occasional patients who have survived to adulthood, Cockayne syndrome was not associated with a significant increase in cancer frequency.[63,79]

Lymphoblastoid and fibroblast cell lines from Cockayne syndrome patients suffering from the classical form of the disease have normal overall nucleotide excision re-

pair,[74,75,85] but display abnormally decreased levels of RNA and DNA synthesis following exposure to UV radiation or bulky carcinogens.[86,87] An early suggestion[86,88] that the Cockayne syndrome defect may reside in the repair of transcribed genes was confirmed by the finding eight years later that Cockayne syndrome fibroblasts have lost the preferential repair of active genes but remain proficient in overall genome repair.[73,75] It was concluded that the sun hypersensitivity of Cockayne syndrome individuals is caused by a defect in the biochemical mechanism that couples transcription to repair and normally enhances the removal of UV radiation products and some other types of damage from actively transcribed genes. However, there is no experimental evidence supporting a direct coupling deficiency in Cockayne syndrome and, hence, this view is highly disputed (see chapter 9).

So far, Cockayne syndrome patients have been assigned to five different complementation groups. Two complementation groups (CS-A and CS-B) define the classical form of the disease with no defect in overall repair of genomic DNA but a severe deficiency in transcription-coupled repair.[89] The other Cockayne syndrome complementation groups are intimately associated with xeroderma pigmentosum and are characterized by the simultaneous manifestations of both syndromes, Cockayne syndrome and xeroderma pigmentosum.[90,91] Patients within this category were assigned to the genetic complementation groups B, D and G, though not all xeroderma pigmentosum individuals from these complementation groups manifest Cockayne syndrome. At the cellular level, these individuals differ from canonical Cockayne syndrome patients in having reduced unscheduled DNA synthesis.

The gene correcting complementation group B of Cockayne syndrome (*CSB*) was originally isolated by correction of the UV sensitivity of a Chinese hamster ovary cell line designated UV61.[92] Subsequent analysis of the *CSB* gene revealed that it encodes a polypeptide that is homologous to Snf2, a component of the yeast Swi/Snf transcriptional activation complex.[93] The predicted amino acid sequence of the *CSA* gene product, on the other hand, indicates that it encodes a WD protein.[94] Members of this class of proteins are made up of highly conserved repeat units usually ending with Trp-Asp (WD) and are found in all eukaryotes, where they regulate multiple and diverse aspects of cellular metabolism, including cell division cycle, transcription, RNA processing, signal transduction or vesicular trafficking.[95] In vitro translated CSA protein was found to interact with CSB protein, and with at least one subunit of the RNA polymerase II basal transcription factor TFIIH.[94] In addition, extracts from Cockayne syndrome cell lines are reportedly deficient in RNA polymerase II-dependent transcription.[94] Collectively, these observations led to the hypothesis that Cockayne, like trichothiodystrophy, is a transcription syndrome in which molecular defects of the *CSA*-, *CSB*-, *XPB*-, *XPD* or *XPG*-encoded proteins disturb transcription by RNA polymerase II. Impaired transcription may affect expression of essential genes in neuronal tissues (leading to dysmyelination in both trichothiodystrophy and Cockayne syndrome) as well as in skin (leading to the characteristic signs of trichothiodystrophy). A primary defect in transcription may also accommodate the sensitivity of Cockayne cells to UV light and other DNA damaging agents because, as discussed in greater detail in chapter 9, RNA polymerase II activity is strictly required for the preferential repair of actively expressed genes.

FANCONI'S ANEMIA

Fanconi's anemia is an inherited autosomal recessive disease that results in progressive pancytopenia (depression of all blood cells), that is eventually followed by leukemia and death at a young age.[96,97] The disease is rare with a prevalence of approximately 1:200,000, but the incidence of acute leukemia in Fanconi's anemia patients is about 15,000-fold higher than that observed in children of the general population.[98,99] The disease also causes developmental abnormalities, especially growth retardation

involving microcephaly as well as aplasia of the thumbs and radius.

Cells derived from Fanconi's anemia patients display hypersensitivity to DNA crosslinking agents such as nitrogen mustard, mitomycin or photoactivated psoralens, and show an increased level of spontaneous chromosomal aberrations.[100,101] It has been proposed that the molecular defect in Fanconi's anemia may involve a cellular pathway that deals with the repair of DNA interstrand crosslinks. Biochemical evidence for crosslink removal in normal but not in Fanconi's anemia cells has been obtained in studies with mitomycin C,[102-104] leading to the hypothesis that this inherited disorder is caused by dysfunction of a recombination repair pathway essential for the removal of DNA interstrand crosslinks. Subsequently, this hypothesis was further extended with the suggestion that the recombination repair deficiency in Fanconi's anemia may provoke a compensatory chromosomal rearrangement mechanism, perhaps involving illegitimate V(D)J recombination, that is eventually responsible for the observed genomic instability.[105]

There are five complementation groups, designated FA-A through FA-E.[106] The *FAC* gene has been cloned and shown to confer wild type resistance to DNA crosslinking agents upon transfection into Fanconi's anemia complementation group C cells.[107] Unfortunately, the predicted amino acid sequence of the FAC protein is not informative with respect to its potential biochemical function. No homology is found to known proteins, and no obvious functional domain is apparent. However, immunoprecipitation and immunofluorescence studies with an antibody to the FAC protein indicate that the gene product is localized in the cytoplasm of lymphoblast cells.[108,109] This observation challenges the prediction that FAC protein operates in nuclear DNA crosslink repair. In a recent review, Digweed and Sperling proposed the alternative hypothesis that Fanconi's anemia patients may be deficient in a regulatory feedback loop that is active during S phase of the cell cycle.[110]

Regardless of the precise molecular pathology of Fanconi's anemia, the amenability of bone marrow cells to genetic manipulation may soon offer therapeutic approaches to effectively treat patients suffering from this disorder.

BLOOM'S SYNDROME

Bloom's is a rare autosomal recessive syndrome that displays the following clinical manifestations: low birth weight, stunted growth, sun-induced facial erythema that extends to form a typical "butterfly lesion", immunodeficiency and predisposition to cancer.[111] The incidence of Bloom's syndrome is very low in the general population, but occurs in about 1 in 60,000 Ashkenazi Jews.[112] Homozygosity for mutations in the Bloom's syndrome locus constitutes a mutator genotype, indicating that the gene plays an important role in the maintenance of genomic stability.[113] In fact, cells from individuals with Bloom's syndrome accumulate mutations, excessive chromosome gaps, breaks and translocations, and exhibit increased levels of sister chromatid exchanges. Some, though not all, cultured Bloom's syndrome cells display increased sensitivity to DNA damaging agents.[114-117]

In vitro studies on Bloom's syndrome cells have revealed abnormally slow replication fork progression, suggesting that the primary defect affects DNA replication.[118] It had also been hypothesized that a DNA ligase I activity may be affected in Bloom's syndrome individuals.[119,120] However, no mutations could be found in the DNA ligase I gene of several Bloom's syndrome patients examined.[121] When the Bloom's syndrome gene (*BLM*) was cloned and sequenced, it became clear that it encodes a protein with the characteristic signatures of a DNA helicase, i.e., an enzyme that uses the energy from ATP hydrolysis to separate duplex DNA.[122] Based on this finding, it is now thought that the *BLM* gene product is required to recognize or resolve specific DNA structures arising during DNA replication. In the absence of functional *BLM*, chromosomes may not become properly separated

before their segregation into daughter cells, and this defect may result in abnormal chromosome fragmentation and genetic instability.[122]

PATIENT 46BR: A DNA LIGASE DEFECT

Patient 46BR is at present the sole representative of a human DNA ligase disorder. At the clinical and cellular level, patient 46BR shares many features with Bloom's syndrome, including immunodeficiency, retarded growth, cancer predisposition, and hypersensitivity to DNA damaging agents. The patient died at the age of 19 with lymphoma. The enzymatic function of the affected protein is clearly different from that of the *BLM* gene product: DNA ligation.[123] In 46BR, both DNA ligase I gene copies were found to be mutated. In one allele, a highly conserved glutamic acid in the active site of the enzyme is replaced by lysine. This change causes complete inhibition of enzyme activity in vitro.[124] The mutation in the second allele affects a conserved arginine that is changed to a tryptophane. This alteration affects enzyme function only partially, indicating that DNA ligase I is an essential enzyme, such that only leaky mutations may be found in patients.[123,124] This view is supported by the observation that DNA ligase I is essential for embryonic development in mice (see Table 1.1).

HEREDITARY NONPOLYPOSIS COLORECTAL CARCINOMA

Natural mutations in genes controlling DNA metabolizing processes are restricted to those that permit development and survival of the affected individual long enough for characterization of the disorder. As a consequence, such mutations are extremely rare. One exception to this rule are human disorders that disrupt the mismatch correction system. Recent discoveries showed that a frequently occurring form of inherited colon cancer, hereditary nonpolyposis colorectal carcinoma (HNPCC), is caused by defects in a DNA metabolizing pathway that corrects mismatches. Mismatch repair is concerned with the enzymatic correction of nucleotide sequence errors that arise during replication.

In the prokaryote *Escherichia coli*, the mechanism of replication error correction is rather complex, as it depends on at least 10 different proteins.[125-127] The prokaryotic pathway is initiated by MutS, which binds to all possible base-base mispairs or to loops formed by unpaired (extrahelical) nucleotides.[128,129] Binding of MutS to the heteroduplex is followed by the addition of MutL.[130] Assembly of this protein-DNA complex leads to activation of an endonuclease associated with the MutH protein, which incises the newly replicated DNA strand.[131] Completion of this pathway requires several other enzymes such as an ATPase/DNA helicase (MutU, also called UvrD), 5'-3' exonucleases (ExoVII or RecJ), a 3'-5' exonuclease (ExoI), DNA polymerase III and DNA ligase.[132] Interestingly, this system is not only involved in the correction of replication errors, but is also required to control and suppress illegitimate crossovers between evolutionary diverged sequences.[133]

Several human homologs of these bacterial mismatch correction genes were found to be associated with HNPCC, a syndrome characterized by familial clustering of tumors of the proximal colon at early age. Cancers of other parts of the gastrointestinal tract and of the genitourinary tract are also found in HNPCC patients.[134] The majority of HNPCC cases are attributable to mutations in mismatch correction genes, including *hMSH2* (a human homolog of MutS),[135,136] and a family of MutL homologs (*hMLH1*, *hPMS1* and *hPMS2*).[137] Cells affected by either one of these mutations loose the ability to correct replication errors such as base pair mismatches or insertion/deletion mismatches, and accumulate high levels of mutations.[138,139] The term RER+ has been introduced to indicate that HNPCC is associated with an abnormal susceptibility to replication errors. HNPCC cells display a remarkable instability of microsatellite sequences, i.e., short (di- or trinucleotide) repeats that are interspersed throughout the genome.[140-143] Apparently, replicative DNA

polymerases are very slippery on templates containing such repeats and produce frameshift mutations that must be corrected by the mismatch repair system to avoid genetic instability.[144]

HNPCC is inherited in an autosomal dominant fashion. Normal cells from affected individuals contain one functional and one defective copy of the gene in question.[136,137,145,146] Tumor cells from these individuals, on the other hand, are defective in both copies of the gene in question,[136,137] although the precise mechanism by which the second gene copy becomes inactivated or lost is not fully understood.[127] In contrast to xeroderma pigmentosum or Cockayne syndrome, with relative frequencies in the population ranging around 1/100,000, it is estimated that one person in every two hundred is heterozygous for a locus involved in HNPCC and therefore considered at risk to develop this form of cancer. Diagnostic kits may soon become available to test predisposition to this form of cancer, and those at risk may be advised to take possible measures to prevent colon cancer progression, including a low fat/high fiber diet and improved medical monitoring.[127]

HUMAN CELL CYCLE CHECKPOINT DISORDERS

Just as important as having mechanisms for DNA repair is the ability to coordinate repair with the cell cycle. The eukaryotic cell cycle is composed of four phases: the gap before DNA replication (G_1), the DNA synthesis phase (S), the gap after DNA replication (G_2) and the mitotic phase (M) which leads to cell division (Fig. 1.3). In most organisms, these cell cycle phases are ordered into highly dependent pathways in which the initiation of late events (for example DNA replication in S phase) is dependent on the completion of earlier events (for example repair of DNA strand breaks during G_1).[20,147]

Eukaryotes possess DNA damage-inducible checkpoints at various phases of the cell cycle. A primary function of these regulatory mechanisms is to delay cell cycle progression when structural defects are detected in the genome. Such cell cycle checkpoints require at least three distinct activities:

(1) a damage recognition system that senses the presence of structural defects,
(2) a signal pathway that transmits and amplifies the information and
(3) an effector mechanism that interacts with the cell cycle machinery.

The resulting block of cell cycle is thought to permit repair of damaged DNA prior to replication or chromosome segregation, thereby preventing the propagation of genetically damaged cells. Thus, the ultimate function of checkpoint controls is to ensure that, when a cell divides, both daughter cells receive complete duplicate sets of the genetic information. A poor execution of cell cycle control pathways is associated with genetic instability and increased frequencies of tumorigenesis.[148-150]

An important cellular response to DNA damage is the arrest of cell cycle progression before DNA replication (at the G_1/S transition) and before chromosome segregation (at the G_2/M transition). In the presence of DNA damage, these checkpoints delay entry into S or M phase, respectively (Fig. 1.3). A major factor implicated in the G_1/S checkpoint in mammalian cells is p53.[151,152] After exposure to genotoxic stress (UV or ionizing radiation, bleomycin, alkylating agents), p53 accumulates and functions as a transcription factor to activate the expression of several downstream effector genes (*p21WAF1/CIP1*, *GADD45*) that block cell cycle at the G_1/S boundary.[153-159] The level of p53 in cells is normally low because the protein is unstable with a half-life of usually less than 30 min.[160] Exposure to DNA damaging agents such as UV light or ioniz-

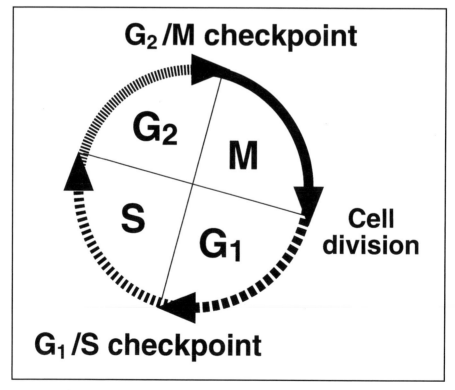

Fig. 1.3. Eukaryotic cell cycle. G_1/S and G_2/M checkpoints delay cell cycle progression in response to structural defects of DNA. A damage-induced checkpoint during S phase has also been reported (reviewed in ref. 253).

ing radiation causes a transient increase in p53 stability which, in turn, increases its cellular concentration.[161-163] This stabilization of p53 is thought to occur mainly by posttranslational mechanisms. As a consequence of elevated p53 levels, cell cycle progression from G_1 to S is inhibited for approximately 75 h.[163]

DNA DAMAGE RECOGNITION BY P53

Several observations suggest that p53 may serve as a cellular sensor of DNA damage. Studies on human p53 have identified at least three distinct protein domains. The N-terminal acidic region is able to activate transcription.[164] The large central portion of p53 binds DNA in a sequence-specific manner.[165-168] The C-terminal basic domain is responsible for the assembly of p53 in stable oligomers[166,169,170] and binds DNA in a damage-specific fashion.[162,171-173]

The C-terminal domain of p53 displays an increased binding affinity for DNA damaged by restriction digestion, DNase I treatment or ionizing radiation.[162,171] p53 protein also binds single-stranded DNA and insertion/deletion mismatches in double-stranded DNA, again by means of its C-terminal domain.[174-177] It has been postulated that association of p53 with damaged DNA may trigger its accumulation in response to genotoxic insults. This model predicts that tight binding to damaged DNA would lead to reduced degradation through changes in p53 conformation, posttranslational modifications such as phosphorylation,[178] association with other factors or compartmentalization within the nucleus. Subsequent DNA repair could release p53 from damaged DNA and restore normal degradation processes.[171] A recent report shows that the p53 core domain also displays an intrinsic 3' to 5' exonuclease activity, suggesting that p53 may be directly involved in the processing of DNA damage.[179]

P53: A TUMOR SUPPRESSOR GENE

As their name suggests, wild type *p53* or other tumor suppressor genes keep cell growth in check and act as a safeguard against cancer. Early cytogenetic studies showed that many sporadic (nonhereditary) human tumors display deletions of a small region in chromosome 17p. These observations led to the conclusion that chromosomal regions which are frequently lost in cancer cells may harbor tumor suppressor genes.[180,181] Subsequently, it became clear that chromosome 17p includes *p53*, and that *p53* is the most commonly altered gene yet identified in human or rodent cancers. Mutations and allele loss of *p53* have been associated with tumors from a wide variety of human organs, including lung, breast, colon, oesophagus, liver, bladder, ovary, brain and the hematopoietic system.[182-184] Conceivably, *p53* mutations offer a selective advantage to cancer cells because the growth-suppressive effect of wild type p53 normally limits the rate of cellular proliferation. The tumor suppressor function of *p53* is supported by the finding that the wild type form of the gene is able to inhibit growth of human or rodent cancer cells, indicating suppression of the malignant phenotype.[185-189]

Inactivation of the G_1/S checkpoint by *p53* mutations strongly correlates with genetic instability. In fact, there are several reports demonstrating that mammalian cells lacking functional *p53* show increased frequencies of chromosome losses, chromosome aberrations as well as intrachromosomal genetic recombination and gene amplification events.[190-192] The *p53* gene product is found in the cell as an oligomeric structure.[166,169,170,193,194] This observation implies that mutant p53 may exert a dominant-negative effect by inactivating any wild type protein that is complexed with it.[195-197] A possible dominant-negative mechanism operating by protein oligomerization is illustrated in Figure 1.4. The resulting genetic instability favors a subsequent deletion of chromosome 17 (loss of heterozygosity for *p53*), thereby eliminating the residual wild type p53 protein and allowing complete escape from p53-dependent checkpoint control.[197,198] In most cases, inactivation of p53 in tumor cells involves a point mutation in one allele accompanied by complete loss of the second allele.[199-202]

LI-FRAUMENI SYNDROME

In addition to the many sporadic cancers associated with *p53* mutations, there are examples of germ-line *p53* mutations that have been detected in families predisposed to multiple types of tumors. In 1969, Li and Fraumeni discovered a new cancer susceptibility syndrome by reviewing medical records of 648 childhood rhabdomyosarcoma patients. This work led to the identification of four families in which siblings or cousins had a childhood sarcoma.[203,204] About half of individuals in Li-Fraumeni families develop cancer by the age of 30, in contrast to a 1% incidence in the general population. The spectrum of cancers observed in these families includes breast carcinomas, soft tissue sarcomas, brain tumors, osteosarcomas, leukemia, adrenocortical carcinoma and other neoplasms. Li-Fraumeni is rare as fewer than 100 families affected by this syndrome have been identified worldwide.[182,183,205,206] The molecular pathology underlying Li-Fraumeni involves point mutations in evolutionary highly conserved regions of the *p53* tumor suppressor gene.[205,206] The germ cells from affected individuals carry one wild type and one mutated *p53* gene. In the tumors of these patients, p53 function is eliminated by a second (somatic) mutation or by complete loss of the remaining wild type gene copy.

ATAXIA TELANGIECTASIA

Ataxia telangiectasia is an autosomal recessive disorder characterized by progressive mental retardation, uneven gait (ataxia) resulting from cerebellar degeneration, and permanently dilated blood vessels (telangiectasia) in the eyes, ears and areas of the face.[207] In addition, ataxia telangiectasia patients suffer from immune deficiencies, premature aging and an approximately 100-fold increase in cancer susceptibility, primarily with tumors of the lymphoreticular system.[208,209] Although this recessive disease is relatively rare (1 case per 40,000-100,000 life births),[210] heterozygous carriers (about 1% of the population) also have an increased cancer risk. In particular, the ataxia telangiectasia locus may account for a significant fraction of breast cancers.[211]

At the cellular level, the phenotype of ataxia telangiectasia is highly pleiotropic. Cells are abnormally sensitive to killing by ionizing radiation and by X-ray-mimetic chemicals such as bleomycin or nigromycin.[212,213] The exquisite radiation hypersensitivity was discovered by the frequent appearance of lethal adverse reactions when ataxia telangiectasia patients affected by tumors were treated with standard doses of radiotherapy. Chromosomal instability is very characteristic of ataxia telangiectasia cells, and involves elevated frequencies of spontaneous or radiation-induced chromosome breaks accompanied by aberrant genomic rearrangements.[214-216] Interestingly, heterozygote cells display an intermediate radiation sensitivity.[212,217] This phenotype may be critical considering the increased susceptibility of ataxia telangiectasia heterozygote women to breast cancer. In fact, their tumor risk may be increased by X-ray mammography, i.e., the procedure that is routinely used to detect breast cancer.[210,211]

One peculiar aspect of the ataxia telangiectasia phenotype had been termed radioresistant DNA synthesis to describe the abnormal response of these cells following exposure to ionizing radiation or X-ray-mimetic drugs. In normal cells, ionizing radiation damage causes a marked inhibition of DNA synthesis, but ataxia telangiectasia cells continue to initiate new replication origins despite the presence of DNA damage.[218,219] This observation is indicative of a severely diminished G_1/S checkpoint function, allowing entry into S phase even under conditions of high DNA damage. In addition, X-ray irradiation also fails to arrest ataxia telangiectasia cells at the G_2/M transition, suggesting that these cells are deficient in multiple checkpoint responses to damaged DNA.[220,221]

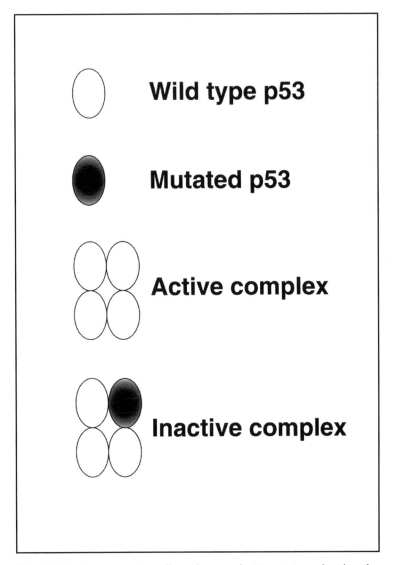

Fig. 1.4. Dominant-negative effect of mutated p53 protein molecules after their assembly into oligomeric complexes.

Another aspect of the pleiotropic ataxia telangiectasia phenotype is evidenced by histological analysis of tissues taken from homozygotes at autopsy, which revealed an abnormal loss of Purkinje and granule cells in the cerebellum as well as depletion of other neurons in the central nervous system. Other organs reported to be atrophic or hypoplastic in ataxia telangiectasia patients are the thymus, gonads, thyroid, and adrenals.[207,222-225] Careful inspection of the cerebellum revealed a high frequency of cells that exhibit morphological features of physiological cell death. Following up on these early observations, Meyn and coworkers recently reported that cells obtained from ataxia telangiectasia patients undergo apoptotic death in culture after exposure to

Table 1.4. Human hereditary disorders related to deficiencies in DNA damage recognition and processing pathways (see text for references)

Disorder	Clinical syndrome	Cellular phenotype	Mutated gene(s)	Pathway affected
Ataxia telangiectasia	Cerebellar ataxia, telangiectasia, immune deficiencies, cancer predisposition	Hypersensitivity to ionizing radiation, genetic instability	ATM	Damage-specific checkpoint control
Bloom's syndrome	Small size, facial erythema, immunodeficiency, cancer predisposition	Hypermutability, hyperrecombinability	BLM	Repair of aberrant DNA replication products?
"46BR"	Retarded growth, immunodeficiency immunodeficiency, cancer predisposition	Hypersensitivity to DNA damaging agents	LIG 1	DNA ligation
Cockayne syndrome	Cachectic dwarfism, mental retardation, premature ageing	Hypersensitivity to UV radiation	XPB, XPD, XPG, CSA, CSB	Nucleotide excision repair, transcription-repair coupling
Fanconi's anemia	Pancytopenia, leukemia, developmental abnormalities, predisposition to leukemia	Hypersensitivity to DNA crosslinking agents, increased chromosomal aberrations	FAC	Unknown
Hereditary non-polyposis colorectal cancer	Predisposition to tumors in the colon and other locations	Instability of microsatellite repeats	hMSH2, hMLH1, hPMS1, hPMS2	Mismatch correction
Li-Fraumeni syndrome	Predisposition to a broad spectrum of tumors	Various forms of genetic instability	p53	Damage-specific checkpoint control
Trichothiodystrophy with photosensitivity	Sensitivity to sunlight, ichthyosis, brittle hair, mental retardation, growth arrest	Hypersensitivity to UV radiation, hypermutable	XPD, XPB, TTDA	Nucleotide excision repair, transcription
Xeroderma pigmentosum	Sensitivity to sunlight, skin cancer, occasional neurological abnormalities	Hypersensitivity to UV radiation and bulky chemicals, hypermutable	XPA, XPB, XPC, XPD, XPE, XPF, XPG	Nucleotide excision repair

low radiation or streptonigrin doses that do not normally induce significant apoptosis in control cells.[225,226] These results indicate that inappropriate apoptosis in response to DNA damage may play a primary role in the pathology of ataxia telangiectasia. According to this hypothesis, DNA damage occurring spontaneously or induced by endogenous agents may cause ongoing loss of cells by apoptosis and eventually lead to cerebellar ataxia, thymic atrophy and lymphocytopenia. Similarly, triggering of apoptosis by otherwise nonlethal DNA damage may be responsible for the exceptional radiation sensitivity of homozygous patients.

The gene mutated in ataxia telangiectasia patients was cloned and termed *ATM* (for *a*taxia *t*elangiectasia *m*utated).[227] *ATM* falls into a class of genes that are related to phosphatidylinositol 3-kinases (PI 3-kinases), but many members of this gene family also exhibit serine/threonine protein kinase activity.[228] A common feature of these enzymes is their involvement in a variety of signal transduction pathways. In fact, some representatives of this gene family related to PI 3-kinases have already been shown to be implicated in checkpoint responses to DNA damage.[229-232] The *Saccharomyces cerevisiae* gene *MEC1* (for *m*itosis *e*ntry *c*heckpoint) illustrates a possible function of this novel family of kinases. Cells with mutations in *MEC1* cannot block entry into mitosis when DNA replication is incomplete or when they are exposed to genotoxic agents.[231,233] More recent experimentation indicates that *MEC1* is also involved in the G_1/S checkpoint that regulates entry into S phase.[234] Thus, the observed sequence homology of the *ATM* gene product to a family of protein kinases is suggestive of a role in monitoring DNA damage or signaling its presence to the cell cycle control machinery. This putative role of the ATM protein is very similar to the postulated function of DNA-PK, a DNA-dependent protein kinase that binds to DNA strand breaks and, presumably, initiates a signal transduction pathway by phosphorylating serine/threonine residues on various acceptor proteins.

Several phenotypic similarities suggest that the ATM protein functions in the same G_1/S checkpoint pathway as p53.[153,235-237] It has also been reported that certain ataxia telangiectasia cell lines lack the p53 response after exposure to ionizing radiation, indicating that the *ATM* gene product may operate upstream of p53 in the G_1/S checkpoint.[153,236,238] However, these findings apparently do not apply to all ataxia telangiectasia cell lines.[235,239,240]

CONCLUSIONS

It has become clear that cells must undergo numerous genetic changes to generate a malignant tumor.[241] In fact, cancer is considered the endpoint of a multistep process during which cells acquire malignancy by cumulative genetic or epigenetic alterations that confer proliferative, invasive and metastatic potentials. In colorectal cancer, for example, at least six independent mutations in oncogenes or tumor suppressor genes are required for a cell clone to develop into a malignant tumor.[242] The variable order of appearance of these genetic changes indicates that their cumulative number, rather than the precise order of occurrence, is important for tumor formation.

The basal mutation rate observed in normal mammalian cells ($\sim 10^{-10}$ per nucleotide and cell generation)[243] appears insufficient to sustain these multiple genetic changes involved during tumorigenesis. Also, chromosomal aberrations (translocations, deletions, amplifications) that are commonly associated with late stages of tumor progression are extremely rare in normal diploid cells,[244,245] indicating that such genetic rearrangements are usually suppressed by the DNA damage response network shown in Figure 1.1. These observations led to the conclusion that mutations in genes participating in either DNA repair, replication fidelity (including DNA polymerase proofreading and mismatch correction) or cell cycle control are necessary to trigger genetic instability in cancer cells.[243]

The significance of defects in biochemical functions that maintain genetic stability is exemplified by a set of genetic disorders characterized by deficiencies in various pathways of DNA repair and cell cycle control (summarized in Table 1.4). Inherited human cancer-prone syndromes such as xeroderma pigmentosum, HNPCC, Fanconi's anemia, Bloom's, Li-Fraumeni, and ataxia telangiectasia demonstrate that cells with enhanced genomic instability will traverse the many genetic changes required for tumor formation more rapidly than cells that maintain normal genomic stability. Of course, a DNA repair- or cell cycle control-deficient cell is more likely than normal cells to acquire genetic defects at multiple, unrelated loci. Thus, a considerable number of human diseases have unequivocally established the causative link between DNA damage and cancer development. In many cases, these genetic diseases affect the capacity to recognize DNA damage in the genome. Examples of such damage recognition disorders include xeroderma pigmentosum complementation groups A, B, D and perhaps C and E (with a defect in the recognition of bulky DNA adducts), Fanconi's anemia (with a possible defect in the recognition of DNA crosslinks), Bloom's syndrome (with a putative defect in the recognition of structural aberrations resulting from DNA replication), and Li-Fraumeni syndrome or ataxia telangiectasia (with a defect in the checkpoint circuit that arrests cell cycle progression in response to DNA damage).

REFERENCES

1. Friedberg EC, Walker GC, Siede W. DNA Repair and Mutagenesis. Washington, D.C.: American Society for Microbiology, 1995.
2. Grisham JW, Greenberg DS, Kaufman DG et al. Cycle-related toxicity and transformation in 10T1/2 cells treated with N-methyl-N'-nitro-N-nitroso-guanidine. Proc Natl Acad Sci USA 1980; 77:4813-4817.
3. Chu G. Cellular responses to cisplatin. J Biol Chem 1994; 269:787-790.
4. Williams GT, Smith CA. Molecular regulation of apoptosis: genetic controls on cell death. Cell 1993; 74:777-779.
5. Strauss BS. The "A rule" of mutagen specificity: a consequence of DNA polymerase bypass of non-instructional lesions? BioEssays 1991; 13:79-84.
6. McBride TJ, Preston BD, Loeb LA. Mutagenic spectrum resulting from DNA damage by oxygen radicals. Biochemistry 1991; 30:207-213.
7. Sancar A. Mechanisms of DNA excision repair. Science 1994; 266:1954-1956.
8. Hoeijmakers JHJ. Human nucleotide excision repair syndromes: molecular clues to unexpected intricacies. Europ J Cancer 1994; 30A:1912-1921.
9. Kelner A. Effect of visible light on the recovery of *Streptomyces griseus conidea* from ultra-violet irradiation injury. Proc Natl Acad Sci USA 1949; 35:73-79.
10. Rupert CS, Goodgal SH, Herriott RM. Photoreactivation in vitro of ultraviolet-inactivated transforming factor in *Haemophilus influenzae*. J Gen Physiol 1958; 41:451-471.
11. Hanawalt PC. Evolution and concepts in DNA repair. Environ Mol Mutagen 1994; 23 (Suppl 24):78-85.
12. Li YF, Kim ST, Sancar A. Evidence for lack of DNA photoreactivating enzyme in humans. Proc Natl Acad Sci USA 1993; 90:4389-4393.
13. Sutherland BM, Bennett PV. Human white blood cells contain cyclobutyl pyrimidine dimer photolyase. Proc Natl Acad Sci USA 1995; 92:9732-9736.
14. Pegg AE, Byers TL. Repair of DNA containing O^6-alkylguanine. FASEB J 1992; 6:2302-2310.
15. Setlow RB, Carrier WL. The disappearance of thymine dimers from DNA: an error-correcting mechanism. Proc Natl Acad Sci USA 1963; 51:226-231.
16. Boyce RP, Howard-Flanders P. Release of ultraviolet light-induced thymine dimers from DNA in *E. coli* K-12. Proc Natl Acad Sci USA 1964; 51:293-300.
17. Yajima H, Takao M, Yasuhira S et al. A eukaryotic gene encoding an endonuclease that specifically repairs DNA damaged by ultraviolet light. EMBO J 1995; 14:2393-2399.
18. Bowman KK, Sidik K, Smith CA et al. A new ATP-independent DNA endonuclease from *Schizosaccharomyces pombe* that recognizes cyclobutane pyrimidine

dimers and 6-4 photoproducts. Nucleic Acids Res 1994; 22:3026-3032.

19. Dunderdale HJ, West SC. Recombination genes and proteins. Curr Opin Genet Dev 1994; 4:221-228.

20. Hartwell LH, Weinert TA. Checkpoints: controls that ensure the order of cell cycle events. Science 1989; 246:629-634.

21. Naegeli H. Roadblocks and detours during DNA replication: mechanisms of mutagenesis in mammalian cells. BioEssays 1994; 16:557-564.

22. Lindahl T. Instability and decay of the primary structure of DNA. Nature 1993; 362:709-715.

23. Sancar A, Franklin KA, Sancar GB. *Escherichia coli* DNA photolyase stimulates UvrABC excision nuclease in vitro. Proc Natl Acad Sci USA 1984; 81:7397-7401.

24. Duckett DR, Drummond JT, Murchie AIH et al. Human MutSα recognizes damaged DNA base pairs containing O^6-methylguanine, O^4-methylthymine, or the cisplatin-d(GpG) adduct. Proc Natl Acad Sci USA 1996; 93:6443-6447.

25. Turchi JJ, Henkels K. Human Ku autoantigen binds cisplatin-damaged DNA but fails to stimulate human DNA-activated protein kinase. J Biol Chem 1996; 271:13861-13867.

26. Schiestl RH, Prakash S. RAD10, an excision repair gene of *Saccharomyces cerevisiae*, is involved in the RAD1 pathway of mitotic recombination. Mol Cell Biol 1990; 10:2485-2491.

27. Adamczewski JP, Rossignol M, Tassan J-P et al. MAT1, cdk7 and cyclin H form a kinase complex which is UV-light sensitive upon association with TFIIH. EMBO J 1996; 15:1877-1884.

28. Wang XW, Yeh H, Schaeffer L et al. p53 modulation of TFIIH-associated nucleotide excision repair activity. Nature Genetics 1995; 10:188-195.

29. Schaeffer L, Roy R, Humbert S et al. DNA repair helicase: a component of BTF2 (TFIIH) basic transcription factor. Science 1993; 260:58-63.

30. Coverley D, Kenny MK, Munn M et al. Requirement for the replication protein SSB in human DNA excision repair. Nature 1991; 349:538-541.

31. Stürzbecher H-W, Donzelmann B,

Henning W et al. p53 is linked directly to homologous recombination processes via RAD51/RecA protein interaction. EMBO J 1996; 15:1992-2002.

32. Goldmacher VS, Cuzick RA, Thilly WG. Isolation and partial characterization of human cell mutants differing in sensitivity to killing and mutation by methyl-nitrosourea and *N*-methyl-*N'*-nitro-*N*-nitrosoguanidine. J Biol Chem 1986; 261:12462-12471.

33. Hawn MT, Umar A, Carethers JM et al. Evidence for a connection between the mismatch repair system and G2 cell cycle checkpoint. Cancer Res 1995; 55:3721-3725.

34. Mellon I, Rajpal D, Koi M et al. Transcription-coupled repair deficiency and mutations in human mismatch repair genes. Science 1996; 272:557-560.

35. Arlett CF. Mutagenesis in repair-deficient human cell strains. In: Alacevic M, ed. Progress in Environmental Mutagenesis. Amsterdam: Elsevier, 1980:161-174.

36. Cleaver JE. Defective repair replication of DNA in xeroderma pigmentosum. Nature 1968; 218:652-656.

37. Kraemer KH, Levy DD, Parris CN et al. Xeroderma pigmentosum and related disorders: examining the linkage between defective DNA repair and cancer. J Invest Dermatol 1994; 103:96-101.

38. Kraemer KH. Xeroderma pigmentosum. In: Demis DJ, Dobson RL, McGuire J, eds. Clinical Dermatology. Hagerstown: Harper and Row, 1980:1-33.

39. Cleaver JE, Kraemer KH. Xeroderma pigmentosum. In: Criver CR, Beaudet AL, Sly WS, Valle D, eds. The Metabolic Basis of Inherited Human Disease. 6th ed. New York: McGraw Hill, 1989:2949-2971.

40. Kraemer KH, Lee M-M, Andrews AD et al. The role of sunlight and DNA repair in melanoma and nonmelanoma skin cancer. Arch Dermatol 1994; 130:1018-1021.

41. English JS, Swerdlow AJ. The risk of malignant melanoma, internal malignancy and mortality in xeroderma pigmentosum patients. Br J Dermatol 1987; 117:457-461.

42. Scotto J, Fears TR, Fraumeni JF. Incidence of non-melanoma skin cancer in the United States. Bethesda, Md: US Dept of Health and Human Services

1982; NIH publication No. 82-2433.

43. Kraemer KH, Lee MM, Scotto J. Xeroderma pigmentosum. Cutaneous, ocular and neurological abnormalities in 830 published cases. Arch Dermatol 1987; 123:241-250.

44. Stich HF, San RHC, Kawazoe Y. Increased sensitivity of xeroderma pigmentosum cells to some chemical carcinogens and mutagens. Mutat Res 1973; 17:127-137.

45. Maher VM, Birch N, Otto JR et al. Cytotoxicity of carcinogenic aromatic amides in normal and xeroderma pigmentosum fibroblasts with different DNA repair capabilities. J Natl Cancer Inst 1975; 54:1287-1294.

46. Maher VM, McCormick JJ. Effect of DNA repair on the cytotoxicity and mutagenicity of UV irradiation and of chemical carcinogens in normal and xeroderma pigmentosum cells. In: Yuhas JM, Tennant RW, Regan JD, eds. Biology of Radiation Carcinogens. New York: Raven Press, 1976:129-145.

47. Maher VM, Ouellette LM, Curren RD et al. Frequency of ultraviolet light-induced mutations is higher in xeroderma pigmentosum variant cells than in normal human cells. Nature 1976; 261:593-595.

48. Dumaz N, Drougard C, Sarasin A et al. Specific UV-induced mutation spectrum in the p53 gene of skin tumors from DNA-repair-deficient xeroderma pigmentosum patients. Proc Natl Acad Sci USA 1993; 90:10529-10533.

49. Brash DE, Rudolph JA, Simon JA et al. A role for sunlight in skin cancer: UV-induced p53 mutations in squamous cell carcinoma. Proc Natl Acad Sci USA 1991; 88:10124-10128.

50. Lehmann AR, Kirk-Bell S, Arlett CF et al. Xeroderma pigmentosum cells with normal levels of excision repair have a defect in DNA synthesis after UV-irradiation. Proc Natl Acad Sci USA 1975; 72:219-223.

51. Stefanini M, Lagomarsini P, Arlett CF et al. Xeroderma pigmentosum (complementation group D) mutation is present in patients affected by trichothiodystrophy with photosensitivity. Human Genet 1986; 74:107-112.

52. Stefanini M, Lagomarsini P, Giliani S et al. Genetic heterogeneity of the excision repair defect associated with trichothiodystrophy. Carcinogenesis 1993; 14:1101-1105.

53. Stefanini M, Vermeulen W, Weeda G et al. A new nucleotide-excision-repair gene associated with the disorder trichothiodystrophy. Am J Hum Genet 53:817-821.

54. Johnson RT, Squires S. The XPD complementation group. Insights into xeroderma pigmentosum, Cockayne syndrome and trichothiodystrophy. Mutat Res 1992; 273:97-118.

55. Broughton BC, Steingrimsdottir H, Weber CA et al. Mutations in xeroderma pigmentosum group D DNA repair/transcription gene in patients with trichothiodystrophy. Nature Genet 1994; 7:189-194.

56. Takayama K, Salazar EP, Broughton B et al. Defects in the DNA repair and transcription gene ERCC2(XPD) in trichothiodystrophy. Am J Hum Genet 1996; 58:263-270.

57. Itin PH, Pittelkow MR. Trichothiodystrophy: review of sulfur-deficient brittle hair syndromes and association with the ectodermal dysplasias. J Am Acad Dermatol 1990; 22:705-717.

58. Kleijer WJ, Beemer FA, Boom BW. Intermittent hair loss in a child with PIBI(D)S syndrome and trichothiodystrophy with defective DNA repair-xeroderma pigmentosum group D. Am J Med Genet 1994; 52:227-230.

59. Van Neste D, Caulier B, Thomas P et al. PIBIDS: Tay's syndrome and xeroderma pigmentosum. J Am Acad Dermatol 1985; 12:372-373.

60. Sarasin A, Blanchet-Bardot C, Renault G et al. Prenatal diagnosis in a subset of trichothiodystrophy patients defective in DNA repair. Br J Dermatol 1992; 127:485-491.

61. Mazdak C, Armier J, Stary A et al. UV-induced mutations in a shuttle vector replicated in repair deficient trichothiodystrophy cells differ with those in genetically-related cancer prone xeroderma pigmentosum. Carcinogenesis 1993; 14:1255-1260.

62. Marionnet C, Benoit A, Benhamou S. Characteristics of UV-induced mutation

spectra in human *XP-D/ERCC2* gene-mutated xeroderma pigmentosum and trichothiodystrophy cells. J Mol Biol 1995; 252:550-562.

63. Lehmann AR. Cockayne's syndrome and trichothiodystrophy: defective repair without cancer. Cancer Rev 1987; 7:82-103.

64. Vuillaume M, Daya-Grosjean L, Vincens P et al. Striking differences in cellular catalase activity between two DNA repair-deficient diseases: xeroderma pigmentosum and trichothiodystrophy. Carcinogenesis 1992; 13:321-328.

65. Norris PG, Limb GA, Hamblin AS et al. Impairment of natural-killer-cell activity in xeroderma pigmentosum. N Engl J Med 1988; 319:1668-1669.

66. Morison WL, Bucana C, Hashem N et al. Impaired immune function in patients with xeroderma pigmentosum. Cancer Res 1985; 45:3929-3931.

67. Wysenbeek AJ, Weiss H, Duczyminer-Kahana M et al. Immunologic alterations in xeroderma pigmentosum patients. Cancer 1986; 58:219-221.

68. Kripke ML, Cox PA, Alas LG et al. Pyrimidine dimers in DNA initiate systemic immunosuppression in UV-irradiated mice. Proc Natl Acad Sci USA 1992; 89:7516-7520.

69. Nishigori C, Yarosh DB, Ullrich S et al. Evidence that DNA damage triggers interleukin 10 cytokine production in UV-irradiated murine keratinocytes. Proc Natl Acad Sci USA 1996; 93:10354-10359.

70. Vermeulen W, van Vuuren AJ, Chipoulet L et al. Three unusual repair deficiencies associated with transcription factor BTF2 (TFIIH). Evidence for the existence of a transcription syndrome. Cold Spring Harbor Symp Quant Biol 1994; 59:317-329.

71. Guzder SN, Sung P, Prakash S. Lethality in yeast of trichothiodystrophy (TTD) mutations in the human xeroderma pigmentosum group D gene. J Biol Chem 1995; 270:17660-17663.

72. Friedberg EC, Bardwell AJ, Bardwell L et al. Transcription and nucleotide excision repair-reflections, considerations and recent biochemical insights. Mutat Res 1994; 307:5-14.

73. Mullenders LHF, Sakker RJ, van Hoffen A et al. Genomic heterogeneity of UV-induced repair: relationship to chromatin structure and transcriptional activity. In: Bohr VA, Wasserman K, Kraemer KH, eds. DNA Repair Mechanisms. Copenhagen: Munksgaard Press, 1992:247-254.

74. van Hoffen A, Natarajan AT, Mayne LV et al. Deficient repair of the transcribed strand of active genes in Cockayne's syndrome. Nucleic Acids Res 1993; 21:5890-5895.

75. Venema J, Mullenders LH, Natarajan AT et al. The genetic defect in Cockayne syndrome is associated with a defect in repair of UV-induced DNA damage in transcriptionally active DNA. Proc Natl Acad Sci USA 1990; 87:4707-4711.

76. Cockayne EA. Dwarfism with retinal atrophy and deafness. Arch Dis Child 1936; 11:1-8.

77. Cockayne EA. Dwarfism with retinal atrophy and deafness. Arch Dis Child 1946; 21:52-54.

78. Cantani A, Bamonte G, Bellioni P et al. Rare syndromes. I. Cockayne syndrome: a review of the 129 cases so far reported in the literature. Riv Eur Sci Med Pharmacol 1987; 9:9-17.

79. Nance MA, Berry SA. Cockayne syndrome: review of 140 cases. Am J Med Genet 1992; 42:68-84.

80. Schmickel RD, Chu EHY, Trosko JE et al. Cockayne syndrome: a cellular sensitivity to ultraviolet light. Pediatrics 1977; 60:135-139.

81. Andrews AD, Barrett SF, Yoder FW et al. Cockayne's syndrome fibroblasts have increased sensitivity to ultraviolet light but normal rates of unscheduled DNA synthesis. J Invest Dermatol 1978; 70:237-239.

82. Hoar DI, Waghorne C. DNA repair in Cockayne syndrome. Am J Hum Genet 1978; 30:590-601.

83. Wade MH, Chu EHY. Effects of DNA damaging agents on cultured fibroblasts derived from patients with Cockayne syndrome. Mutat Res 1979; 59:49-60.

84. Marshall RR, Arlett CF, Harcourt SA. Increased sensitivity of cell strains from Cockayne's syndrome to sister chromatid exchange induction and cell killing by UV light. Mutat Res 1980; 69:107-112.

85. Ahmed FE, Setlow RB. Excision repair in

ataxia telangiectasia, Fanconi's anemia, Cockayne syndrome, and Bloom's syndrome after treatment with ultraviolet radiation and N-acetoxy-2-acetylaminofluorene. Biochim Biophys Acta 1978; 521:805-817.

86. Mayne LV, Lehmann AR. Failure of RNA synthesis to recover after UV irradiation: an early defect in cells from individuals with Cockayne's syndrome and xeroderma pigmentosum. Cancer Res 1982; 42:1473-1478.

87. Lehmann AR, Kirk-Bell S, Mayne L. Abnormal kinetics of DNA synthesis in ultraviolet light-irradiated cells from patients with Cockayne's syndrome. Cancer Res 1979; 39:4237-4241.

88. Mayne LV, Lehmann AR, Waters R. Excision repair in Cockayne syndrome. Mutat Res 1982; 106:179-189.

89. Tanaka K, Kawai KY, Kumahara Y et al. Genetic complementation groups in Cockayne syndrome. Som Cell Genet 1981; 7:445-456.

90. Vermeulen W, Jaeken J, Jaspers NG et al. Xeroderma pigmentosum complementation group G associated with Cockayne syndrome. Am J Human Genet 1993; 53:185-192.

91. Itoh T, Cleaver JE, Yamaizumi M. Cockayne syndrome complementation group B associated with xeroderma pigmentosum phenotype. Hum Genet 1996; 97:176-179.

92. Troelstra C, Odijk H, de Wit J et al. Molecular cloning of the human DNA excision repair gene *ERCC-6*. Mol Cell Biol 1990; 10:5806-5813.

93. Troelstra C, Van Gool A, De Wit J et al. *ERCC6*, a member of a subfamily of putative helicases, is involved in Cockayne's syndrome and preferential repair of active genes. Cell 1992; 71:1-15.

94. Henning KA, Li L, Iyer N et al. The Cockayne syndrome group A gene encodes a WD repeat protein that interacts with CSB protein and a subunit of RNA polymerase II TFIIH. Cell 1995; 82:555-564.

95. Neer EJ, Schmidt CJ, Nambudripad R et al. The ancient regulatory-protein family of WD-repeat proteins. Nature 1994; 371:297-300.

96. Fanconi G. Familial constitution panmyelocytopathy, Fanconi's anemia (F.A.).I. Clinical aspects. Semin Hematol 1967; 4:233-240.

97. Schroeder TM, Tigen D, Kruger J et al. Formal genetics of Fanconi's anemia. Hum Genet 1976; 32:257-288.

98. Auerbach AD. Fanconi anemia and leukemia: tracking the genes. Leukemia 1992; 6 (Suppl 1):1-4.

99. Auerbach AD. Leukemia and preleukemia in Fanconi anemia patients: a review of the literature and report of the international Fanconi Anemia Registry. Cancer Genet Cytogenet 1991; 51:1-12.

100. Schroeder TM, Anschütz F, Knopp A. Spontane Chromosomenaberrationen bei familiärer Panmyelopathie. Humangenetik 1964; 1:194-196.

101. Sasaki MS, Tonomura A. A high susceptibility of Fanconi's anemia to chromosome breakage by DNA cross linking agents. Cancer Res 1973; 33:1829-1836.

102. Fujiwara Y, Tatsumi M, Sasaki MS. Crosslink repair in human cells with its possible defects in Fanconi's anemia cells. J Mol Biol 1977; 113:635-649.

103. Fujiwara Y. Defective repair of mitomycin C crosslinks in Fanconi's anemia and loss in confluent normal human and xeroderma pigmentosum cells. Biochim Biophys Acts 1982; 699:217-225.

104. Fujiwara Y, Kano Y, Yamamoto Y. DNA interstrand cross-linking, repair and SCE mechanism in human cells in special reference to Fanconi anemia. In: Tice R, Hollaender A, eds. Sister Chromatid Exchanges. New York, Plenum Publishing Corp, 1984:787-800.

105. Laquerbe A, Moustacchi E, Fuscoe JC et al. The molecular mechanism underlying formation of deletions in Fanconi anemia cells may involve a site-specific recombination. Proc Natl Acad Sci USA 1995; 92:831-835.

106. Strathdee CA, Duncan AMV, Buchwald M. Evidence for at least four Fanconi's anemia genes including *FACC* on chromosome 9. Nat Genet 1992; 1:196-198

107. Strathdee CA, Gavish H, Shannon WR et al. Cloning of cDNAs for Fanconi's anemia by functional complementation. Nature 1992; 356:763-767.

108. Yamashita T, Barber DL, Zhu Y et al. The Fanconi anemia polypeptide FACC is localized to the cytoplasm. Proc Natl Acad Sci USA 1994; 91:6712-6716.

109. Youssoufian H. Localization of Fanconi anemia C protein to the cytoplasm of mammalian cells. Proc Natl Acad Sci USA 1994; 91:7975-7979.

110. Digweed M, Sperling K. Molecular analysis of Fanconi anaemia. BioEssay 1996; 18:579-585.

111. Bloom D. Congenital telangiectatic erythema resembling lupus erythematosis in dwarfs. Am J Dis Child 1954; 88:754-758.

112. German J, Bloom D, Passarge E et al. Bloom's syndrome. VI. the disorder in Israel, and an estimate of the gene frequency in the Ashkenazim. Am J Hum Genet 1977; 29:553-562.

113. Kuhn EM, Therman E. Cytogenetics of Bloom's syndrome. Cancer Genet Cytogenet 1986; 22:1-18.

114. Krepinsky AB, Heddle JA, German J. Sensitivity of Bloom's syndrome lymphocytes to ethyl methanesulfonate. Hum Genet 1979; 50:151-156.

115. Heddle JA, Krepinsky AB, Marshall RR. Cellular sensitivity to mutagens and carcinogens in the chromosome-breakage and other cancer-prone syndromes. In: German J, ed. Chromosome Mutation and Neoplasia. New York: Alan R. Liss, 1983:203-234.

116. Kurihara T, Inoue M, Tatsumi K. Hypersensitivity of Bloom's syndrome fibroblasts to N-ethyl-N-nitrosourea. Mutat Res 1987; 184:147-151.

117. Langlois RG, Bigbee WL, Jensen RH et al. Evidence for increased in vivo mutation and somatic recombination in Bloom's syndrome. Proc Natl Acad Sci USA 1989; 86:670-674.

118. Lonn U, Lonn S, Nylen U et al. An abnormal profile of DNA replication intermediates in Bloom's syndrome. Cancer Res 1990; 50:3141-3145.

119. Chan JY, Becker FF, German J et al. Altered DNA ligase I activity in Bloom's syndrome cells. Nature 1987; 325:357-359.

120. Willis AE, Lindahl T. DNA ligase I deficiency in Bloom's syndrome. Nature 1987; 325:355-357.

121. Petrini JHJ, Huwiler KG, Weaver DT. A wild-type DNA ligase I gene is expressed in Bloom's syndrome cells. Proc Natl Acad Sci USA 1991; 88:7615-7619.

122. Ellis NA, Groden J, Ye T-Z et al. The Bloom's syndrome gene product is homologous to RecQ helicases. Cell 1995; 83:655-666.

123. Barnes DE, Tomkinson AE, Lehmann AR et al. Mutations in the DNA ligase I gene of an individual with immunodeficiencies and cellular hypersensitivity to DNA damaging agents. Cell 1992; 69:495-503.

124. Prigent C, Satoh MS, Daly G et al. Aberrant DNA repair and DNA replication due to an inherited enzymatic defect in human DNA ligase I. Mol Cell Biol 1994; 14:310-317.

125. Modrich P. Mismatch repair, genetic stability, and cancer. Science 1994; 266:1959-1960.

126. Modrich P. Mechanisms and biological effects of mismatch repair. Annu Rev Genet 1991; 25:229-253.

127. Jiricny J. Colon cancer and DNA repair: have mismatches met their match? Trends Genet 1994; 10:164-168.

128. Su S-S, Modrich P. Escherichia coli mutS-encoded protein binds to mismatched DNA base pairs. Proc Natl Acad Sci USA 1986; 83:5057-5061.

129. Parker BO, Marinus MG. Repair of heteroduplexes containing small heterologous sequences in Escherichia coli. Proc Natl Acad Sci USA 1992; 89:1730-1734.

130. Grilley M, Welsh KM, Su S-S et al. Isolation and characterization of the Escherichia coli mutL gene product. J Biol Chem 1989; 264:1000-1004.

131. Au KG, Welsh K, Modrich P. Initiation of methyl-directed mismatch repair. J Biol Chem 1992; 267:12142-12148.

132. Grilley M, Griffith J, Modrich P. Bidirectional excision in methyl-directed mismatch repair. J Biol Chem 1993; 268:11830-11837.

133. Rayssiguier C, Thaler DS, Radman M. The barrier to recombination between Escherichia coli and Salmonella typhimurium is disrupted in mismatch-repair mutants. Nature 1989; 342:396-401.

134. Lynch HT, Smyrk TC, Watson P et al.

Genetics, natural history, tumor spectrum, and pathology of hereditary nonpolyposis colorectal cancer: an updated review. Gastroenterology 1993; 104:1535-1549.

135. Fishel R, Lescoe MK, Rao MRS et al. The human mutator gene homolog *MSH2* and its association with hereditary nonpolyposis colon cancer. Cell 1993; 75:1027-1038.

136. Leach FS, Nicolaides NC, Papadopoulos N et al. Mutation of a *mutS* homolog in hereditary nonpolyposis colorectal cancer. Cell 1993; 75:1215-1225.

137. Papadopoulos N, Nicolaides NC, Wei Y-F et al. Mutation of a *mutL* homolog in hereditary colon cancer. Science 1994; 263:1625-1629.

138. Parsons R, Li GM, Longley MJ et al. Hypermutability and mismatch repair deficiency in RER+ tumor cells. Cell 1993; 75:1227-1236.

139. Bhattacharyya NP, Skandalis A, Ganesh A et al. Mutator phenotype in human colorectal carcinoma cell lines. Proc Natl Acad Sci USA 1994; 91:6319-6323.

140. Peinado MA, Malkhosyan S, Velazquez A et al. Isolation and characterization of allelic losses and gains in colorectal tumors by arbitrarily primed polymerase chain reaction. Proc Natl Acad Sci USA 1992; 89:10065-10069.

141. Thibodeau SN, Gren G, Schaid D. Microsatellite instability in cancer of the proximal colon. Science 1993; 260:816-819.

142. Ionov Y, Peinado MA, Malkhosyan S et al. Ubiquitous somatic mutations in simple repeated sequences revel a new mechanism for colonic carcinogenesis. Nature 1993; 363:558-560.

143. Lindblom A, Tannergard P, Werelius B et al. Genetic mapping of a second locus predisposing to hereditary non-polyposis colon cancer. Nature Genet 1993; 5:279-282.

144. Kunkel TA. Nucleotide repeats. Slippery DNA and diseases. Nature 1993; 365:207-208.

145. Bronner CE, Baker PT, Morrison G et al. Mutation in the DNA mismatch repair gene homologue *hMLH1* is associated with hereditary non-polyposis colon cancer. Nature 1994; 368:258-261.

146. Nicolaides VC, Papadopoulos N, Liu B et al. Mutations of two *MMS* homologues in hereditary nonpolyposis colon cancer. Nature 1994; 371:75-80.

147. Murray AW. Creative blocks: cell-cycle checkpoints and feedback controls. Nature 1992; 359:599-604.

148. Hartwell LH, Kastan MB. Cell cycle control and cancer. Science 1994; 266:1821-1828.

149. Hunter T. Braking the cycle. Cell 1993; 75:839-841.

150. Hartwell LH. Defects in a cell cycle checkpoint may be responsible for the genomic instability of cancer cells. Cell 1992; 71:543-546.

151. Kuerbitz SJ, Plunkett BS, Walsh WV et al. Wild-type p53 is a cell cycle checkpoint determinant following irradiation. Proc Natl Acad Sci USA 1992; 89:7491-7495.

152. Kastan MB, Onykwere O, Sidransky D et al. Participation of p53 protein in the cellular responses to DNA damage. Cancer Res 1991; 51:6304-6311.

153. Kastan MB, Zhan Q, El-Deiry WS et al. A mammalian cell cycle checkpoint pathway utilizing p53 and GADD45 is defective in ataxia-telangiectasia. Cell 1992; 71:587-597.

154. Harper JW, Adami GR, Wei N et al. The p21 Cdk-interacting protein Cip1 is a potent inhibitor of G1 cyclin-dependent kinases. Cell 1993; 75:805-816.

155. Dulic V, Kaufmann WK, Wilson SJ et al. p53-dependent inhibition of cyclin-dependent kinase activities in human fibroblasts during radiation-induced G_1 arrest. Cell 1994; 76:1013-1023.

156. El-Deiry WS, Harper JW, O'Connor PM et al. *WAF1/CIP1* is induced in *p53*-mediated G_1 arrest and apoptosis. Cancer Res 1994; 54:1169-1174.

157. Waga S, Hannon GJ, Beach D et al. The p21 inhibitor of cyclin-dependent kinases controls DNA replication by interaction with PCNA. Nature 1994; 369:574-578.

158. Zhan Q, Lord KA, Alamo I et al. The *gadd* and *MyD* genes define a novel set of mammalian genes encoding acidic proteins that synergistically suppress cell growth. Mol Cell Biol 1994; 14:2361-2371.

159. El-Deiry WS, Tokino T, Velculescu VE et al. WAF1, a potential mediator of p53

tumor suppression. Cell 1993; 75:817-825.

160. Gronostajski RM, Goldberg AJ, Pardee AB. Energy requirement for degradation of tumor-associated protein p53. Mol Cell Biol 1984; 4:442-448.

161. Maltzman W, Czyzyk L. UV irradiation stimulates levels of p53 cellular tumor antigen in nontransformed mouse cells. Mol Cell Biol 1984; 4:1689-1694.

162. Nelson WG, Kastan MB. DNA strand breaks: the DNA template alterations that trigger p53-dependent DNA damage response pathways. Mol Cell Biol 1994; 14:1815-1823.

163. Zambetti GP, Levine AJ. A comparison of the biological activities of wild-type and mutant p53. FASEB J 1993; 7:855-865.

164. Fields S, Jang SK. Presence of a potent transcription activating sequence in the p53 protein. Science 1990; 249:1046-1049.

165. Bargonetti J, Manfredi JJ, Chen X et al. A proteolytic fragment from the central region of p53 has marked sequence-specific DNA-binding activity when generated from wild-type but not from oncogenic mutant p53 protein. Genes Dev 1993; 7:2565-2574.

166. Halazonetis TD, Kandil AN. Conformational shifts propagate from the oligomerization domain of p53 to its tetrameric DNA binding domain and restore DNA binding to selected p53 mutants. EMBO J 1993; 12:5057-5064.

167. Pavletich NP, Chambers KA, Pabo CO. The DNA-binding domain of p53 contains the four conserved regions and the major mutation hot spot. Genes Dev 1993; 7:2556-2564.

168. Wang Y, Reed M, Wang P et al. p53 domains: identification and characterization of two autonomous DNA-binding regions. Genes Dev 1993; 7:2575-2586.

169. Stürzbecher HW, Brain R, Addison C et al. A C-terminal alpha helix plus basic motif is the major structural determinant of p53 tetramerization. Oncogene 1992; 7:1513-1523.

170. Wang P, Reed M, Wang Y et al. p53 domains: structure, oligomerization, and transformation. Mol Cell Biol 1994; 14:5182-5191.

171. Reed M, Woelker B, Wang P et al. The C-terminal domain of p53 recognizes DNA damaged by ionizing radiation. Proc Natl Acad Sci USA 1995; 92:9455-9459.

172. Wu L, Bayle JH, Elenbaas B et al. Alternatively spliced forms in the carboxy-terminal domain of the p53 protein regulate its ability to promote annealing of complementary single strands of nucleic acids. Mol Cell Biol 1995; 15:497-504.

173. Foord OS, Bhattacharya P, Reich Z et al. A DNA binding domain is contained in the C-terminus of wild type p53 protein. Nucleic Acids Res 1991; 19:5191-5198.

174. Huang L-C, Clarkin KC, Wahl GM. Sensitivity and selectivity of the DNA damage sensor responsible for activating p53-dependent G_1 arrest. Proc Natl Acad Sci USA 1996; 93:4827-4832.

175. Bakalkin G, Yakovleva T, Selivanova G et al. p53 binds single-stranded DNA ends and catalyzes DNA renaturation and strand transfer. Proc Natl Acad Sci USA 1994; 91:413-417.

176. Bakalkin G, Selivanova G, Yakovleva T et al. p53 binds single-stranded DNA ends through the C-terminal domain and internal DNA segments via the middle domain. Nucleic Acids Res 1995; 23:362-369.

177. Lee S, Elenbaas B, Levine A et al. p53 and its 14 kDa C-terminal domain recognize primary DNA damage in the form of insertion/deletion mismatches. Cell 1995; 81:1013-1020.

178. Knippshild U, Milne D, Campbell L et al. p53 N-terminus-targeted protein kinase activity is stimulated in response to wild type p53 and DNA damage. Oncogene 1996; 13:1387-1393.

179. Mummenbrauer T, Janus F, Müller B et al. p53 exhibits 3'-5' exonuclease activity. Cell 1996; 85:1089-1099.

180. Knudson AG. Mutation and cancer: statistical study of retinoblastoma. Proc Natl Acad Sci USA 1971; 68:820-823.

181. Stanbridge EJ. The reemergence of tumor suppression. Cancer Cells 1989; 1:31-33.

182. Vogelstein B. A deadly inheritance. Nature 1990; 348:681-682.

183. Hollstein M, Sidransky D, Vogelstein B et al. p53 mutations in human cancers. Science 1991; 253:49-53.

184. Toguchida J, Yamaguchi T, Dayton SH

et al. Prevalence and spectrum of germline mutations of the *p53* gene among patients with sarcoma. N Engl J Med 1992; 326:1301-1308.

185. Mercer WE, Shields MT, Amin M et al. Negative growth regulation in a glioblastoma tumor cell line that conditionally expresses human wild-type p53. Proc Natl Acad Sci USA 1990; 87:6166-6170.

186. Baker SJ, Markowitz S, Fearon ER et al. Suppression of human colorectal carcinoma cell growth by wild-type p53. Science 1990; 249:912-915.

187. Michalovitz D, Halevy O, Oren M. Conditional inhibition of transformation and of cell proliferation by a temperature-sensitive mutant of p53. Cell 1990; 62:671-680.

188. Eliyahu D, Michalovitz D, Eliyahu S et al. Wild-type p53 can inhibit oncogene-mediated focus formation. Proc Natl Acad Sci USA 1989; 86:8763-8767

189. Chen P-L, Chen Y, Brookstein R et al. Genetic mechanisms of tumor suppression by the human *p53* gene. Science 1990; 250:1576-1580.

190. Bischoff FZ, Yim SO, Pathak S et al. Spontaneous abnormalities in normal fibroblasts from patients with Li-Fraumeni cancer syndrome: aneuploidy and immortalization. Cancer Res 1990; 50:7979-7984.

191. Livingstone LR, White A, Sprouse J et al. Altered cell cycle arrest and gene amplification potential accompany loss of wild-type p53. Cell 1992; 70:923-935.

192. Yin Y, Tainsky MA, Bischoff FZ et al. Wild-type p53 restores cell cycle control and inhibits gene amplification in cells with mutant p53 alleles. Cell 1992; 70:937-948.

193. Stenger J, Mayr G, Mann K et al. Formation of stable p53 homotetramers and multiple of tetramers. Mol Carcinog 1992; 5:102-106.

194. Kraiss S, Quaiser A, Oren M et al. Oligomerization of oncoprotein p53. J Virol 1988; 62:4737-4744.

195. Herskowitz I. Functional inactivation of genes by dominant negative mutations. Nature 1987; 329:219-222.

196. Eliyahu D, Goldfinger N, Pinhasi-Kimhi O et al. Meth A fibrosarcoma cells express two transforming mutant p53 species. Oncogene 1988; 3:313-321.

197. Levine AJ, Momand J, Finlay CA. The *p53* tumour suppressor gene. Nature 1991; 351:453-456.

198. Lane DP, Benchimol S. *p53*: oncogene or antioncogene? Genes Dev 1990; 4:1-8.

199. Baker SJ, Fearon ER, Nigro JM et al. Chromosome 17 deletions and *p53* gene mutations in colorectal carcinomas. Science 1989; 244:217-221.

200. Oka K, Ishikawa J, Bruner JM et al. Detection of loss of heterozygosity in the *p53* gene in renal cell carcinoma and bladder cancer using the polymerase chain reaction. Mol Carcinog 1991; 4:10-13.

201. Nigro JM, Baker SJ, Preisinger AC et al. Mutations in the *p53* gene occur in diverse human tumour types. Nature 1989; 342:705-708.

202. Mulligan LM, Matlashewski GJ, Scrable HJ et al. Mechanisms of p53 loss in human sarcomas. Proc Natl Acad Sci USA 1990; 87:5863-5867.

203. Li FP, Fraumeni JF. Soft-tissue sarcomas, breast cancer, and other neoplasms: a familial syndrome? Ann Intern Med 1969; 71:747-753.

204. Li FP, Fraumeni JF, Mulvihill JJ et al. A cancer family syndrome in twenty-four kindreds. Cancer Res 1988; 48:5358-5362.

205. Malkin D, Li, FP, Strong LC et al. Germ line p53 mutations in a familial syndrome of breast cancer, sarcomas, and other neoplasms. Science 1990; 250:1233-1238.

206. Srivastava S, Zou Z, Pirollo K et al. Germ-line transmission of a mutated p53 gene in a cancer-prone family with Li-Fraumeni syndrome. Nature 1990; 348:747-749.

207. Boder E., Sedgwick RP. Ataxia-telangiectasia. A familial syndrome of progressive cerebellar ataxia, oculocutaneous telangiectasia and frequent pulmonary infection. Pediatrics 1957; 21:526-554.

208. Swift M. Genetic aspects of ataxia-telangiectasia. Immunodef Rev 1990; 2:67-81.

209. Morrell D, Cromartie E, Swift M. Mortality and cancer incidence in 263 patients with ataxia-telangiectasia. J Natl

Cancer Inst 1986; 77:89-92.

210. Swift M, Morrell D, Cromartie E et al. The incidence and gene frequency of ataxia-telangiectasia in the United States. Am J Hum Genet 1986; 39: 573-583.

211. Easton DF. Cancer risks in A-T heterozygotes. Int J Radiat Biol 1994; 66:S177-S182.

212. Taylor AMR, Harnden DG, Arlett CF et al. Ataxia-telangiectasia: a human mutation with abnormal radiation sensitivity. Nature 1975; 258:427-429.

213. Paterson MC, Smith PJ. Ataxia telangiectasia: an inherited human disorder involving hypersensitivity to ionizing radiation and related DNA-damaging chemicals. Annu Rev Genet 1979; 13:291-398.

214. Lambert C, Schultz R, Smith M et al. Functional complementation of ataxia-telangiectasia group D (AT-D) cells by microcell-mediated chromosome transfer and mapping of the AT-D locus to the region 11q22-23. Proc Natl Acad Sci USA 1991; 88:5907-5911.

215. Meyn MS. High spontaneous intrachromosomal recombination rates in ataxia-telangiectasia. Science 1993; 260:1327-1330.

216. Luo CM, Tang W, Mekeel KL et al. High frequency and error-prone DNA recombination in ataxia telangiectasia cell lines. J Biol Chem 1996; 271:4497-4503.

217. Arlett CF, Priestly A. An assessment of the radiosensitivity of ataxia-telangiectasia. In: Gatti RA, Smith M, eds. Ataxia-Telangiectasia: Genetics, Neuropathology, and Immunology of a Degenerative Disease of Childhood. New York: Alan R. Liss, 1985:1-63.

218. Houldsworth J, Lavin MF. Effect of ionizing radiation on DNA synthesis in ataxia telangiectasia cells. Nucleic Acids Res 1980; 8:3709-3720.

219. Painter RB, Young BR. Radiosensitivity in ataxia-telangiectasia: a new explanation. Proc Natl Acad Sci USA 1980; 77:7315-7317.

220. Zampetti-Bosseler F, Scott D. Cell death, chromosome damage and mitotic delay in normal human, ataxia telangiectasia and retinoblastoma fibroblasts after X-irradiation. Int J Radiat Biol 1981; 39:547-558.

221. Rudolph NS, Latt SA. Flow cytometric analysis of X-ray sensitivity in ataxia telangiectasia. Mutat Res 1989; 211:31-41.

222. Sedgwick RP, Boder E. Ataxia-telangiectasia. Handbook of Clinical Neurology 1991; 16:347-423.

223. Amromin GD, Boder E, Teplitz R. Ataxia-telangiectasia with a 32 year survival. A clinicopathological report. J Neuropathol Exp Neurol 1979; 38:621-643.

224. Agamanolis DP, Greenstein JI. Ataxia-telangiectasia: report of a case with Lewy bodies and vascular abnormalities within cerebral tissue. J Neuropathol Exp Neurol 1979; 39:475-489.

225. Meyn MS. Ataxia-telangiectasia and cellular responses to DNA damage. Cancer Res 1995; 55:5991-6001.

226. Meyn MS, Strasfeld L, Allen C. Testing the role of p53 in genetic instability and apoptosis in ataxia telangiectasia. Int J Radiat Biol 1994; 66:141-149.

227. Savitsky K, Bar-Shira A, Gilad S et al. A single ataxia telangiectasia gene with a product similar to PI-3 kinase. Science 1995; 268:1749-1753.

228. Stack JH, Emr SD. Vps34p required for yeast vacuolar protein sorting is a multiple specificity kinase that exhibits both protein kinase and phosphatidylinositol-specific PI 3-kinase activities. J Biol Chem 1994; 269:31552-31562.

229. Al-Khodairy F, Carr AM. DNA repair mutants defining G2 checkpoint pathways in *Schizosaccharomyces pombe*. EMBO J 1992; 11:1343-1350.

230. Rowley R, Subramani S, Young PG. Checkpoint controls in *Schizosaccharomyces pombe*: *rad1*. EMBO J 1992; 11:1335-1342.

231. Kato R, Ogawa H. An essential gene, *ESR1*, is required for mitotic cell growth, DNA repair and meiotic recombination in *Saccharomyces cerevisiae*. Nucleic Acids Res 1994; 22:3104-3112.

232. Hari KL, Santerre A, Sekelsky JJ et al. The *mei-41* gene of D. melanogaster is a structural and functional homolog of the human ataxia telangiectasia gene. Cell 1995; 82:815-821.

233. Weinert TA, Kiser GL, Hartwell LH. Mitotic checkpoint genes in budding yeast and the dependence of mitosis on

DNA replication and repair. Genes Dev 1994; 8:652-665.

234. Paulovich AG, Hartwell LH. A checkpoint regulates the rate of progression through S phase in *S. cerevisiae* in response to DNA damage. Cell 1995; 82:841-847.

235. Lu X, Lane DP. Differential induction of transcriptionally active p53 following UV or ionizing radiation: defects in chromosome instability syndromes. Cell 1993; 75:765-778.

236. Khanna KK, Lavin MF. Ionizing radiation and UV induction of p53 protein by different pathways in ataxia-telangiectasia cells. Oncogene 1993; 8:3307-3312.

237. Canman CE, Wolff AC, Chen CY et al. The p53-dependent G1 cell cycle checkpoint pathway and ataxia-telangiectasia. Cancer Res 1994; 54:5054-5058.

238. Artuso M, Esteve A, Bresil H et al. The role of the ataxia telangiectasia gene in the p53, WAF1/CIP1 (p21)- and GADD45-mediated response to DNA damage produced by ionising radiation. Oncogene 1995; 11:1427-1435.

239. Downes C, Wilkins AS. Cell cycle checkpoints, DNA repair and DNA replication strategies. BioEssays 1994; 16:75-79.

240. Blattner C, Knebel A, Radler-Pohl C et al. DNA damaging agents and growth factors induce changes in the program of expressed gene products through common routes. Environ Mol Mutagen 1994; 24:3-10.

241. Armitage P, Doll R. The age distribution of cancer and a multi-stage theory of carcinogenesis. Br J Cancer 1954; 8:1-12.

242. Vogelstein B, Kinzler K. The multistep nature of cancer. Trends Biochem Sci 1993; 9:138-141.

243. Loeb LA. Microsatellite instability: marker of a mutator phenotype in cancer. Cancer Res 1994; 54:5059-5063.

244. Tlsty TD. Normal diploid human and rodent cells lack a detectable frequency of gene amplification. Proc Natl Acad Sci USA 1990; 87:3132-3136

245. Wright JA, Smith HS, Watt FM et al. DNA amplification is rare in normal human cells. Proc Natl Acad Sci USA 1990; 87:791-1795.

246. Cavazzana-Calvo M, Le Deist F, De Saint Basile G et al. Increased radiosensitivity of granulocyte macrophage colony-forming units and skin fibroblasts in human autosomal recessive severe combined immunodeficiency. J Clin Invest 1993; 91:1214-1218.

247. Bentley DJ, Selfridge J, Millar JK et al. DNA ligase I is required for fetal liver erythropoiesis but is not essential for mammalian cell viability. Nature Genetics 1996; 13:489-491.

248. Umar A, Buermeyer AB, Simon JA et al. Requirement for PCNA in DNA mismatch repair at a step preceding DNA resynthesis. Cell 1996; 87:65-73.

249. Park MS, Ludwig DL, Stigger E et al. Physical interaction between human RAD52 and RPA is required for homologous recombination in mammalian cells. J Biol Chem 1996; 271:18996-19000.

250. Longhese MP, Neecke H, Paciotti V. The 70 kDa subunit of replication protein A is required for the G1/S and intra-S damage checkpoints in budding yeast. Nucleic Acids Res 1996; 24:3533-3537.

251. Yang YY, Johnson AL, Johnston LH et al. A mutation in a *Saccharomyces cerevisiae* gene (*RAD3*) required for nucleotide excision repair and transcription increases the efficiency of mismatch correction. Genetics 1996; 144:459-466.

252. Leadon SA, Cooper PK. Preferential repair of ionizing radiation-induced damage in the transcribed strand of an active human gene is defective in Cockayne syndrome. Proc Natl Acad Sci USA 1993; 90:10499-10503.

253. Kaufmann WK, Paules RS. DNA damage and cell cycle checkpoints. FASEB J 1996; 10:238-247.

CHAPTER 2

ANIMAL MODELS OF DNA DAMAGE RECOGNITION AND PROCESSING DISORDERS

TRANSGENIC MODELS

Gene targeting by homologous recombination in embryonic stem cells provides a powerful tool to insert exogenous segments of DNA into specific mammalian genes, thereby producing mutant mice in which the function of the targeted gene is changed or inactivated.[1,2] This methodology has already been exploited to generate germ line mutations in genes that participate in various aspects of DNA damage recognition and metabolism. At the time of writing, transgenic mice models have been reported for null (or loss of function) mutations in several nucleotide excision repair genes (*XPA, XPC, ERCC1*), in mismatch correction genes (*PMS2, MSH2*), in the gene coding for O^6-methylguanine-DNA methyltransferase, and in the cell cycle checkpoint regulatory genes *p53* and *ATM*. A spontaneously mutated mouse strain (SCID) carries a genetic defect in *DNA-PK$_{CS}$*, a gene required for DNA double strand break repair. A knock-out of the regulatory Ku86 subunit of DNA-PK has also been established. Finally, an example of *Drosophila melanogaster* mutant deficient in the transcription/nucleotide excision repair factor XPB will be mentioned because of its striking phenotype.

Many targeted gene disruptions affecting specific pathways of DNA damage metabolism produced offspring that developed normally and were viable. In other cases, however, disruption of genes implicated in DNA damage metabolism was not compatible with normal embryonic development and led to a lethal phenotype. Inactivation of different base excision repair enzymes (DNA polymerase β, AP endonuclease) or knock-out of the mouse homolog of the yeast recombination gene *RAD51* provides prototypical examples of embryonic lethality caused by the loss of a DNA damage processing mechanism. Embryonic lethality was also observed after inactivation of DNA ligase I.

Compared to lower eukaryotes, gene targeting in mammalian cells is complicated by the propensity to randomly integrate foreign DNA by a cut and join reaction, referred to as illegitimate recombination, that requires little or no sequence homology.[3] Since homologous recombination is extremely rare in mammalian cells, it is necessary to exploit effective molecular and genetic strategies to select for the desired targeted recombination event against

Mechanisms of DNA Damage Recognition in Mammalian Cells, by Hanspeter Naegeli.
© 1997 R.G. Landes Company.

a high background of nontargeted integrations. A successful targeting method and the corresponding positive/negative selection technique is illustrated below by describing in detail the approach used to inactivate *XPA* in mice.

XPA-DEFICIENT MICE

The human inherited disorder xeroderma pigmentosum outlines the phenotypic consequences of a deficiency in the normal cellular response to UV radiation. The disease comprises 8 complementation groups designated XP-A through XP-G, and XP-V. Complementation groups A, B, C, D, E, F and G reflect mutations in different nucleotide excision repair genes and are deficient in excision repair of UV radiation damage (see chapter 1). Several mutant mice models have been generated to study the pathophysiology of this human nucleotide excision repair disorder.

Mice lacking functional XPA protein were established by disrupting the *XPA* gene in embryonic stem cells with appropriate vectors.[4,5] In one study,[4] exon 4 of the mouse *XPA* gene was altered by insertion of a sequence that confers neomycin resistance (*neo*). This was achieved by transfection of embryonic stem cells with a targeting vector that contains a 5.7-kilobase long genomic fragment of the mouse *XPA* gene, comprising exons 3 and 4. The targeting vector is schematically shown in Figure 2.1. Exon 4 in this vector was interrupted by a 1.1-kilobase long *neo* cassette that serves two functions: it is used to physically disrupt the target gene and allows for positive selection of transfectants that have stably integrated vector sequences, resulting in neomycin resistance. The herpes simplex virus thymidine kinase gene (HSV-*tk*) was inserted on the 3' side of the *XPA* sequence and used for enrichment of specific integration events (Fig. 2.1). HSV-*tk* allows direct selection against transfectants that contain random integrations generated by illegitimate recombination. Most cells containing such random integrations express viral thymidine kinase and, as a consequence, become hy-

persensitive to the negative selection marker gancyclovir. The second paper reporting on a transgenic XPA knock-out[5] is based on the use of an almost identical targeting vector containing positive and negative selection cassettes. In this second study, however, homologous recombination resulted in the complete loss of exons 3 and 4 from the endogenous *XPA* gene.

The transfected embryonic stem cells were cultured in the presence of G418 (to select for neomycin resistance) and gancyclovir (to select against random integrations). Genomic DNA of neomycin- and gancyclovir-resistant colonies was screened by polymerase chain reaction and positive clones were verified by southern blotting. Selected embryonic stem cell clones were injected into host blastocysts, and the modified preimplantation embryos were transferred into the uteri of pseudopregnant foster mothers. Breeding of chimeric males showed germline transmission of the mutant *XPA* allele. Heterozygous *XPA* (+/-) mice (F_1 animals) were then obtained and intercrossed to generate homozygous (-/-) individuals. In the first study,[4] the frequencies of weaned pups with (+/+), (+/-) and (-/-) *XPA* genotypes were 24.5%, 49.8% and 25.7%, respectively. This Mendelian ratio of 1:2:1 indicates that XPA-deficient animals developed normally. In addition, the (-/-) mice showed no obvious clinical or pathological abnormalities and were fertile when they were not exposed to DNA damaging agents. The absence of XPA protein in the (-/-) animals was confirmed by western blotting.[4]

In the second study involving deletion of exons 3 and 4 of *XPA*, (-/-) offspring were not obtained in the expected Mendelian ratio.[5] Of 732 pups born, only 81 (11%) were of the *XPA* (-/-) genotype. Growth retardation in the (-/-) embryos was observed after day 13 post coitum, and blood smears of these embryos and their histopathological examination indicated anemia and disturbed liver hematopoiesis. Apparently, about 50% of the *XPA* (-/-) embryos died of severe anemia at mid gestation. However,

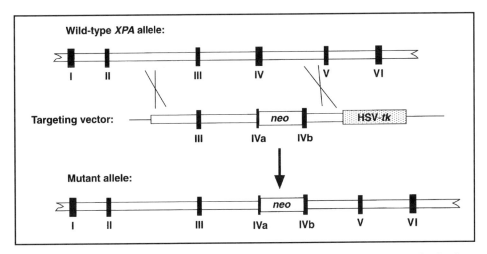

Fig. 2.1. Molecular strategy used by Nakane et al[4] to disrupt the mouse *XPA* gene. The thin line represents the targeting vector backbone; filled boxes labeled I through VI represent exons of the *XPA* gene.

those *XPA* (-/-) animals that did come to term developed normally and were fertile. In the surviving offspring, no histological abnormalities were found in liver, blood, kidney, heart, skin, spleen, lung or in the brain. The authors confirmed the absence of *XPA* transcripts in these animals by northern blotting.[5]

The homozygous mutant *XPA* mice were highly sensitive to UV-B light (290-320 nm wavelength). A single UV-B irradiation induced more severe inflammatory edema in the ears of *XPA* (-/-) mice than in those of (+/-) or (+/+) mice.[4] To test their susceptibility to skin carcinogenesis, *XPA* (-/-) mice and their (+/-) or (+/+) littermates were challenged by UV-B radiation or treatment with the chemical carcinogen 9,10-dimethyl-1,2-benz[*a*]anthracene.[4,5] The shaved back skins of these mice were irradiated three times a week with UV-B at a dose of 2,000 J/m². All XPA-deficient animals responded to this treatment with skin ulcers within the first two weeks. After 20 weeks the first tumors were detected in the *XPA* (-/-) mice, exclusively squamous cell carcinomas, and by 34 weeks of UV irradiation all (-/-) mice developed such tumors. Abnormalities of the eyes such as corneal opacity were also observed. The *XPA* (+/+) and

(+/-) mice, on the other hand, did not develop any significant skin abnormalities during these UV exposure experiments. As expected, the skin of XPA-deficient mice was also abnormally prone to tumorigenesis by 9,10-dimethyl-1,2-benz[*a*]anthracene.

To monitor the cellular phenotype after exposure to DNA damaging agents, fibroblasts were derived from newborn *XPA* (-/-) mice and exposed to UV-C radiation (254 nm wavelength) or, alternatively, to the 9,10-dimethyl-1,2-benz[*a*]anthracene.[4,5] The observed responses are fully consistent with the prediction that *XPA* (-/-) mice are deficient in nucleotide excision repair. No differences could be observed between heterozygous (+/-) and wild type (+/+) cells. In particular, cells obtained from *XPA* (-/-) animals were:

(1) hypersensitive to killing by UV radiation,

(2) hypersensitive to killing by 9,10-dimethyl-1,2-benz[*a*]anthracene,

(3) deficient in unscheduled DNA synthesis following exposure to UV radiation,

(4) deficient in the removal of cyclobutane pyrimidine dimers and pyrimidine(6-4)pyrimidone photoproducts

after exposure to UV light. Table 2.1 summarizes UV sensitivity, UV-specific DNA repair synthesis and the kinetics of UV damage excision measured in the different cell lines.

In summary, XPA-deficient mice provide a good in vivo model to study skin and ocular abnormalities as well as enhanced skin carcinogenesis associated with xeroderma pigmentosum. In contrast to human patients of complementation group A, however, XPA-deficient mice lack overt neurological manifestations. Also, pathological analysis of their neural tissues failed to reveal any neurodegeneration up to 18 months of age. The cause of this important difference between human patients and the mouse model remains to be determined.

XPC-DEFICIENT MICE

Similar to the *XPA* disruption strategy, mice deficient for XPC were generated using a homologous replacement vector that deletes approximately 4.5 kilobases of the mouse *XPC* locus.[6] The hypoxanthine-guanine phosphoribosyl transferase (*HPRT*) and HSV-*tk* genes were used for positive and negative selection, respectively. The resulting deletion in the mutated allele eliminates a coding portion of mouse *XPC* of about 500 nucleotides in length, and results in a message containing a frameshift and premature stop codons. This deletion is located earlier in the coding region than any of the mutations described so far in cell lines derived from xeroderma pigmentosum complementation group C patients, indicating that the mutated allele is likely to be functionally inactive.

Blastocyst microinjection of embryonic stem cell clones containing the predicted replacement led to germ line chimeras, which were used to transmit the mutant allele to F_1 animals. Of 150 offspring from *XPC* (+/-) intercrosses, 25% were wild type (+/+), 46% were heterozygous (+/-) and 29% were homozygous for the null (-/-) allele. These numbers correspond approximately to the expected Mendelian ratios. Thus, mice homozygous for the *XPC* knockout developed normally and were viable. Also, the *XPC* (-/-) animals did not exhibit any increased susceptibility to spontaneous tumor generation or any other overt abnormality.

To measure susceptibility to UV-induced tumor formation, littermates with *XPC* (-/-), (+/-) and (+/+) genotypes were exposed daily to UV-B light. After 20 weeks of treatment, all *XPC* (-/-) mice developed at least one type of skin tumor, and some mice developed multiple tumors, predominantly squamous cell carcinomas. In contrast, mice heterozygous for the mutant allele (+/-) and wild type controls (+/+) did not develop any tumors by the end of the 20-weeks treatment period. XPC-deficient mice also displayed pathological eye changes including keratitis and corneal ulceration that are reminiscent of ocular abnormalities found in xeroderma pigmentosum patients.[6]

Table 2.1. UV-sensitive phenotype of XPA-deficient mouse cells (refs. 4 and 5). The differential repair rates of (6-4) photoproducts and cyclobutane pyrimidine dimers are discussed in chapter 6.

XPA Genotype	UV dose for 37% survival (D_{37})	Unscheduled DNA synthesis (grains/cell)	$t_{1/2}$ of (6-4) photoproducts	$t_{1/2}$ of cyclobutane dimers
(+/+)	~1.5 J/m²	58 ± 14	~4 h	~38 h
(+/-)	~1.5 J/m²	61 ± 14	~4 h	~38 h
(-/-)	~0.3 J/m²	3 ± 2	No detectable repair	No detectable repair

Embryonic fibroblasts obtained from XPC-deficient (-/-) animals were hypersensitive to treatment with UV light, whereas heterozygous (+/-) cells responded normally. Surprisingly, the XPC deficiency did not alter significantly cell survival in response to 4-nitroquinoline 1-oxide, a known mutagen that induces bulky base adducts. This finding differs from previous reports indicating increased 4-nitroquinoline 1-oxide sensitivity of human xeroderma pigmentosum cells of various complementation groups.[7,8] This discrepancy may have profound impacts on genotoxicity testing procedures in rodents, but biochemical studies on the mechanism of mouse and human nucleotide excision repair are necessary to resolve this issue.

ERCC1-Deficient Mice

The very first DNA repair-deficient transgenic mouse was reported in 1993.[9,10] ERCC1 (for *excision repair cross complementing 1*) is a mammalian nucleotide excision repair factor involved in DNA incision (see chapter 6). The *ERCC1* gene was knocked out by insertion of a 2.4-kilobase long *neo* cassette into exon 5 of *ERCC1*. The expected homologous recombination event resulted in a transcript in which part of exon 5 and the remaining exons 6-10 of *ERCC1* were missing. Chimeras were produced by blastocyst injection of targeted embryonic stem cell clones, and these chimeric animals were used to generate F_1 offspring with the expected insertion in the germline.[10] Heterozygous carriers of the targeted *ERCC1* allele were interbred, and *ERCC1* (-/-) pups were obtained in the expected Mendelian ratio. However, these (-/-) mice were runted at birth and died before weaning. Primary embryonic fibroblasts were isolated from *ERCC1* (-/-) pups and used to confirm the expected hypersensitivity to killing by UV radiation and treatment with the cross-linking agent mitomycin C.

Interestingly, histological analysis of perinatal and 3-week old mice showed enhanced variations in liver cell nuclear size in *ERCC1* (-/-) animals when compared to the uniform, regular size of nuclei from their (+/+) or (+/-) littermates. This abnormality was observed in the absence of any carcinogen treatment, and indicated an extremely variable degree of polyploidy in the perinatal liver, progressing to aneuploidy by 3 weeks of age. It was concluded that this cytogenetic heterogeneity may suggest enhanced genomic instability in the liver. Serum samples from 2- to 3-week old pups showed 15-fold increased alanine aminotransferase activity in *ERCC1* (-/-) relative to (+/+) mice, supporting the hypothesis of liver cell injury. There was also a significant increase in the level of alkaline phosphatase and bilirubin. Finally, immunohistochemical studies using an antibody raised against murine p53 yielded positive nuclei in liver, kidney or brain of *ERCC1* (-/-) mice but not in control sections from (+/+) controls.[10] The increased level of p53 strengthened the hypothesis that *ERCC1* (-/-) mice are abnormally susceptible to genome destabilizing reactions.

In summary, the complex phenotype of transgenic *ERCC1* null mice implies that nucleotide excision repair may protect liver, kidney and brain from endogenous DNA damage caused, perhaps, by oxygen radicals that arise during normal cellular metabolism. This conclusion is prompted by the fact that these organs are metabolically very active; in addition, liver and kidney also contain cytochrome P450 biotransformation enzymes which can generate oxygen radicals (see chapter 3). Alternatively, the ERCC1 protein may, like other repair factors, have a second separate function which is particularly important for normal liver, kidney or brain physiology. In any case, the deleterious phenotypic consequences observed in mice may explain the lack of a human equivalent for ERCC1 deficiency. None of the xeroderma pigmentosum complementation groups with impaired nucleotide excision repair activity carry a mutation in the *ERCC1* gene. Thus, ERCC1-deficient mice provide a particularly useful model system to study the biological function of this nucleotide excision repair factor.

DNA Repair Knock-Outs
with Lethal Outcome

DNA polymerase β functions in the base excision repair pathway of mammalian cells.[11] Deletion of the promoter and the first exon of the DNA polymerase β gene in the mouse germ line results in a lethal phenotype.[12] Homozygous DNA polymerase β (-/-) embryos do not survive beyond day 10.5 of gestation. Embryonic mice that are homozygous for a null mutation in another base excision repair enzyme, AP endonuclease (AP-1), fail to develop beyond the blastocyst stage.[13] Thus it is tempting to speculate that the whole base excision repair system is required for normal embryonic development in mammals. Lethality after targeted disruption of specific genes involved in DNA damage processing has also been observed in other occasions. For example, mice with a null mutation in the *Rad51* recombination gene die during early embryonic development and rarely reach the morula stage.[14] Additionally, embryonic stem cell lines in which both copies of the *Rad51* gene were knocked out fail to grow in culture,[14] indicating that this DNA recombination gene is essential for normal development and cell proliferation.

In the case of the DNA polymerase β knock-out, (-/-) fibroblast could be obtained at 10 days of gestation and DNA polymerase β–deficient cell lines were immortalized by transformation with an expression vector for the simian virus 40 large-T antigen.[15] The resulting DNA polymerase β-deficient cells are normal in viability and growth characteristics, but exhibit increased sensitivity to the monofunctional DNA alkylating agents methyl methanesulphonate, ethyl methanesulphonate, *N*-nitroso-*N*-methylurea or *N*-methyl-*N*'-nitro-*N*-nitrosoguanidine. These results are consistent with the concept that alkylation damage induced by these electrophilic agents is mainly processed by the base excision repair pathway. On the other hand, polymerase β-deficient cells were not hypersensitive to treatment with UV, ionizing radiation, oxygen radical-generating agents, cisplatin, or nitrogen mustard.[15]

Mismatch Correction-Deficient
Mice

The mismatch repair system is absolutely required for maintenance of genomic stability because it corrects replication errors and prevents recombination between nonhomologous sequences, thereby safeguarding the genome from "promiscuous" rearrangements.[16-20] The biological significance of mismatch correction was confirmed in mouse models by knocking out both copies of either the *PMS2*[21] or the *MSH2*[22] gene.

Mice with a null mutation in the *PMS2* gene were obtained by disrupting exon 2 with an appropriate targeting vector.[21] The resulting *PMS2* (-/-) homozygotes developed normally and were viable. However, these animals were defective in base mismatch correction. The phenotypic consequences of this loss of function include genomic instability at microsatellite loci (see chapter 1) and an elevated frequency of tumors, mainly sarcomas and lymphomas, relative to wild type littermates. Another prominent aspect of the PMS2-deficient phenotype was male infertility, presumably resulting from abnormalities in chromosome synapsis during prophase of meiosis I.

Like the *PMS2* knock-outs, mice deficient for the *MSH2* gene have lost their mismatch correction capability, have acquired genomic instability at microsatellite loci, and show increased rates of tumor incidence, predominantly lymphomas.[22] Previously, it had been shown that the presence of subtle (less than 1%) nucleotide difference between DNA segments strongly reduces homologous recombination in mammalian cells.[23] Interestingly, *MSH2* (-/-) cells have completely lost this sequence heterology-dependent suppression of recombination, indicating that the mismatch correction system recognizes heteroduplexes between not perfectly complementary strands during genetic recombination processes. These results are comparable to those obtained previously with mismatch correction-deficient yeast[24,25] or bacterial cells.[26] Thus, both mouse models for mismatch correction deficiency confirm the important role of this system in preserving genomic stability. In particular, the rodent models support the idea that a defect in mismatch correction strongly accelerates malignant transformation.

INACTIVATION OF METHYLGUANINE-DNA METHYLTRANSFERASE

The enzyme O^6-methylguanine-DNA methyltranserase (MGMT) is responsible for the removal of highly mutagenic and cytotoxic alkyl groups from O^6-guanine and O^4-thymine.[27,28] The gene coding for this enzyme was disrupted by replacing part of exon 2 and the adjacent intron with a *neo* cassette.[29] Homozygous MGMT (-/-) mice were obtained in the expected Mendelian ratio, and their tissues contained essentially no methyltransferase activity. These MGMT (-/-) mice appeared normal, although there was some retardation of growth; body weight of homologous mutant animals was about 85% of that of wild type littermates.[29]

A single intraperitoneal administration of methylnitrosourea (50 mg/kg body weight) to MGMT (-/-) mice led to death within 18 days, while wild type animals treated in the same manner showed no negative effects. The MGMT (-/-) genotype led to a dramatic decrease in the number of leukocytes and platelets after methylnitrosourea treatment. The size of spleen and thymus was considerably reduced. Systematic histological analysis revealed suppression of hematopoietic cells in the bone marrow and dysplastic changes in the intestinal mucosa. It was concluded that MGMT plays an important role in protecting the lymphoreticular system and the intestinal mucosa from the cytotoxic effects of alkylating agents.[29]

P53-DEFICIENT MICE

To investigate the function of p53 in mammalian development and tumorigenesis, a null mutation was introduced into the mouse p53 gene by homologous recombination.[30] The targeting vector contained a 3.7-kilobase portion of the genomic p53 gene interrupted in exon 5 by a *neo* expression cassette. A clone of embryonic stem cells containing the correct insertion was used to generate germ-line chimeras. Intercrossed heterozygote animals yielded homozygous mutant mice at the expected Mendelian proportion. These p53 (-/-) mice appeared healthy and lacked any observable defect both morphologically and by histopathological examination. Thus, intact p53 protein is dispensable for normal mouse development. The absence of functional p53 transcripts in these offspring was confirmed by northern blotting, RNA polymerase chain reaction assays, and protein immunoblotting.[30]

Although p53 (-/-) mice developed normally, they were clearly susceptible to spontaneous formation of a variety of tumors as early as 6 months of age. Some tumors appeared very early (before 10 weeks of age), and tumor frequency increased dramatically between 15 and 25 weeks of age. Frequently, p53 (-/-) animals developed multiple primary neoplasms of different cell types of origin, but the tumors most frequently observed were malignant lymphomas.[30]

The heterozygous mice that contained one wild type and one null p53 allele were also prone to developing tumors but at a reduced frequency and at a slower rate.[31] In p53 (+/-) mice, osteosarcomas and soft tissue sarcomas were the most frequent tumors. Southern blot analysis showed that a large fraction (55%) of the tumors in p53 (+/-) mice exhibited loss of heterozygosity, as was previously found in many human cancer cases either sporadic or associated with familial Li-Fraumeni syndrome (chapter 1). These observations with p53 (+/-) mutant mice confirmed the dominant-negative effect of a p53 mutation by which inactivation of one p53 allele is sufficient to predispose an organism to spontaneous neoplasms (see Fig. 1.3).

Heterozygous p53 (+/-) mice were also hypersensitive to carcinogen treatment. Low dose levels of dimethylnitrosamine, a potent inducer of liver hemangiosarcomas, were added to the drinking water of wild type (+/+), heterozygous (+/-) or p53-deficient (-/-) animals for a period of 12 months. In this experiment, *p53* (-/-) homozygous developed spontaneous tumors too rapidly to monitor the effects of the carcinogen. On the other hand, *p53* heterozygous animals were significantly more susceptible to the carcinogen than wild type controls. The mean survival time of (+/-) heterozygous mice was 29 weeks, whereas the mean survival time of wild type animals was 42 weeks.[31] In the future, the elevated mutagen sensitivity of these p53 (+/-) heterozygous mice may be employed to facilitate in vivo carcinogenesis assays.

ATM-Deficient Mice

Ataxia telangiectasia is a highly pleiotropic human syndrome that predisposes to lymphoreticular cancer[32] and is caused by mutations in *ATM* (for *a*taxia *t*elangiectasia *m*utated; chapter 1). Mouse *ATM* (a gene of approximately 150 kilobases in length and spread over 66 exons) was disrupted in embryonic stem cells by placing a *neo* expression cassette at nucleotide position 5790.[33] The modified *ATM* transcript contains a frameshift and results in a truncated product. The site of protein truncation corresponds to that found in several human patients. All three *ATM* genotypes were detected in litters of F_1 intercrosses. *ATM* (+/+), (+/-) and (-/-) offspring were obtained in the expected Mendelian ratio, indicating that the *ATM* knock-out is compatible with embryonic development and postnatal viability.

Mice homozygous for the disrupted allele displayed growth retardation, neurologic dysfunction, infertility, immunodeficiency (mainly associated with reduced T cell maturation) and extreme hypersensitivity to ionizing radiation.[33] No evidence of neuronal degeneration was observed in histological sections, but neurologic deficits were nevertheless assessed by several tests of motor function. The majority of animals developed lymphomas of the thymus be-

tween two and four months of age. The death of *ATM* (-/-) mice following exposure to γ-irradiation resulted mainly from acute radiation toxicity to the gastrointestinal tract with severe epithelial degeneration.

In analogy to cells obtained from human ataxia telangiectasia patients, fibroblasts from ATM-deficient mice exhibited radioresistant DNA synthesis indicative of a deficient G_1/S checkpoint function.[33] This phenotypic trait is consistent with the idea that *ATM* may be involved in an important cellular response to DNA double strand breaks arising physiologically (during V(D)J recombination of immunoglobulin and T cell receptor genes) or by exposure to ionizing radiation and radiomimetic chemicals. Thus, ATM-deficient mice provide an excellent mammalian model in which to study the complex pathophysiology of ataxia telangiectasia. In addition, *ATM* (+/-) mice will help understand the possible correlation between ataxia telangiectasia heterozygosity and breast cancer in women.[34] As mentioned in chapter 1, it has been postulated that the tumor risk of ataxia telangiectasia in heterozygous women may be increased by routine X-ray mammography.

SPONTANEOUS ANIMAL MODELS

With exception of the human disorders described in chapter 1, spontaneous mutations affecting a DNA damage processing pathway are extremely rare in mammalian species. At present, the only true examples of spontaneous animal models involve the repair of DNA strand breaks.

Severe Combined Immunodeficiency in Mice

In 1983, Bosma et al[35] identified a spontaneous mutation in CB-17 mice which yielded the characteristic phenotype of severe combined immunodeficiency (SCID). The mouse SCID defect is inherited in an autosomal recessive manner, and the affected mice lack lymphocytes of both the B and T cell lineages because their V(D)J recombination mechanism is severely impaired.[36-39]

V(D)J recombination is a genetic rearrangement process that results in the production of complete immunoglobulins by B lymphocytes and the expression of T-cell receptors by T lymphocytes. In rodents and humans, the juxtaposition of the variable (V), diversity (D) and joining (J) gene segments of immunoglobulins and T cell receptors is required to generate the wide molecular diversity (10^{12}) associated with these essential mediators of specific immune responses.[40] The V(D)J reorganization mechanism is initiated by cleavage of DNA at specific recombination signal sequences that flank the V, D and J coding segments. In a subsequent step, these double strand DNA cuts are religated. SCID mice can initiate V(D)J recombination by forming site-specific double strand breaks, but are impaired in the subsequent religation step.

The factor that is deficient in SCID mice not only impairs V(D)J rearrangement but also affects the ability to repair double strand breaks induced by ionizing radiation or X-ray-mimetic chemicals such as bleomycin.[41-44] Murine SCID cells are also hypersensitive to chemicals that form DNA interstrand crosslinks, suggesting that crosslink repair involves the formation of double strand breaks as reaction intermediates.[44] A prominent feature of these studies is that phenotypic characterization of the SCID mice established a link between V(D)J recombination, a process that is specific for lymphoid cells, and the ubiquitous mechanism by which genotoxin-induced DNA breaks are repaired in essentially all mammalian tissues. It appears that the mutation that causes murine SCID affects a factor that is required for the final DNA ligation step in lymphoid-specific processes as well as for the general pathway of DNA double strand break repair. Several reports indicate that the molecular defect in SCID mice resides in the catalytic subunit of DNA-PK, a multiprotein complex with protein kinase activity that binds to DNA ends or hairpin loops and is stimulated by these structural discontinuities (see for example refs. 45-47).

In addition to the catalytic subunit mutated in SCID, DNA-PK contains a regulatory DNA binding component called Ku, which itself is a heterodimer of 70 and 86 kDa polypeptides.[48] Transgenic *Ku86* knock-outs have been recently obtained.[49,50] These *Ku86* null mutations are mainly characterized by severe deficiencies in V(D)J recombination, causing arrest of B and T lymphocyte development at early progenitor stages. *Ku86* (-/-) mice were viable and fertile, but were 40-60% of the size of littermate controls. With the exception of lymphoid organs, histological sections from *Ku86* (-/-) animals appeared normal, and their organ weights were proportional to the reduced body weight. However, spleen, lymph nodes and thymus were considerably smaller than in control animals and were devoid of lymphocytes. It was noted that the DNA-PK defect in SCID mice is not associated with growth retardation. Also, the immunodeficiency in SCID is less severe than in *Ku86* (-/-) mice. As a likely explanation for this phenotypic differences, the spontaneous mutation in SCID affects DNA-PK function only partially whereas the transgenic knock-out completely abolished DNA-PK activity.

SEVERE COMBINED IMMUNODEFICIENCY IN ARABIAN FOALS

Severe combined immunodeficiency in Arabian foals was initially reported in 1973.[51] Like its murine counterpart, equine SCID is an autosomal recessive mutation which results in immunodeficiency.[52] The equine SCID phenotype is almost completely identical to that found in the spontaneous mouse SCID model. Foals affected by this disease have severely depressed B and T lymphocytes, whereas natural killer cell activity is normal.[53,54] SCID foals have essentially no immunoglobulin production, and display severe hypoplasia of the thymus and peripheral lymphoid tissues. The affected foals normally die before reaching the age of 3 months.

Cells obtained from Arabian SCID foals are defective in V(D)J recombination and are hypersensitive to ionizing radiation.[55] The analogy between murine and equine SCID was extended to the biochemical level.

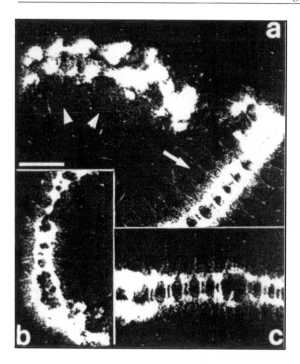

Fig. 2.2. Immunofluorescence microscopy visualizing nerve cord defects in *Drosophila haywire* mutants. Panel (a) shows a comparison between the wild type (arrow) and a severely affected nerve cord (arrowheads); (b) and (c) show additional examples of defective nerve cords. Reprinted with permission from: Mounkes LC, Jones RS, Liang B-C et al. A *Drosophila* model for xeroderma pigmentosum and Cockayne's syndrome: *haywire* encodes the fly homolog of *ERCC3*, a human excision repair gene. Cell 1992; 71:925-937. © 1992 Cell Press.

Like their murine counterparts, equine SCID cells are deficient in DNA-PK, i.e., they lack DNA end binding activity and fail to display detectable DNA-PK protein kinase activity. Thus, severe combined immunodeficiency of Arabian foals defines a second spontaneous animal model for defective V(D)J recombination and DNA strand break repair.

A *DROSOPHILA* DNA REPAIR/ TRANSCRIPTION MUTANT

A single example from a lower eukaryotic species (the *haywire* mutation in *Drosophila melanogaster*) is included here because of its striking implications with regard to the multifunctionality of nucleotide excision repair proteins. Another class of *Drosophila* strains that have been identified on the basis of their hypersensitivity to various genotoxic agents (*mus* mutants) is presented in chapter 10.

Unexpected insights into the function of the human xeroderma pigmentosum complementation group B gene emerged a few years ago when the fruitfly *haywire* allele was studied. This gene encodes a protein with 66% identity to the product of the human *XPB* gene, which is associated with xeroderma pigmentosum, Cockayne syndrome and trichothiodystrophy (see chapter 1).[56] Mutant alleles of this gene were originally isolated by their inability to complement mutations in *B2t*, a gene which encodes a testis-specific isoform of β2-tubulin in *Drosophila*.[57] In a subsequent study, a large number of different *haywire* mutants were analyzed and most of them were found to be lethal.[56] However, 3 out of 27 *haywire* mutations were viable as hemizygotes, and these alleles were found to mimic typical phenotypic traits of the human syndromes xeroderma pigmentosum, Cockayne and trichothiodystrophy. The analogies between the *haywire* phenotype and these human syndromes include UV hypersensitivity, central nervous system (nerve cord) abnormalities, motor defects suggestive of ataxia, impaired sexual development and reduced life span. Figure 2.2 shows a typical immunofluorescence analysis of the developing nervous system in wild type and mutant *Drosophila* embryos.

Table 2.2. Transgenic deficiencies in DNA damage processing pathways

Mutant mice	Viability	Cancer proneness	References
XPA, XPC	Yes	Carcinogen-induced	4,5,6
ERCC1	Limited	NA	9,10
Pol β, AP-1	No	NA	12, 13,15
Rad51	No	NA	14
PMS2, MSH2	Yes	Spontaneous	21,22
DNA ligase I	No	NA	64
MGMT	Yes	NR	29
p53	Yes	Spontaneous and carcinogen-induced	30,31,65,66
ATM	Yes	Spontaneous	33,67
Ku86	Yes	NR	49,50

NA, not applicable; NR, not reported.

β2-tubulin is required for microtubule-based processes, including flagellar elongation, nuclear shaping, and meiotic spindle formation. Spermatogenesis in *Drosophila* is particularly sensitive to the level of β2 tubulin.[58] The observation that a mutation in the *Drosophila XPB* homolog affected tubulin expression and, hence, caused male sterility, provided a first indication that this gene may be involved in transcription. Similarly, defective transcription may also lead to the observed neurological signs. In the meantime, it is well known that XPB and several other nucleotide excision repair factors are components of the multisubunit transcription factor TFIIH required for RNA polymerase II-dependent transcription.[59-61]

CONCLUSIONS

The availability of powerful methods to generate mutant mice has rapidly expanded the number of animal models for the loss of function in DNA damage recognition or processing pathways (summarized in Table 2.2). Reports on additional knockouts of genes implicated in the recognition

of DNA lesions or their metabolism are in progress, involving for example the mouse homologs of *CSA* and *CSB* which, in humans, are implicated in Cockayne syndrome. It is likely that mice models for Fanconi's anemia and Bloom's syndrome will also be presented soon.

In the future, introduction of more subtle deletions or even single base mutations will be performed using more sophisticated techniques. Methods are also available to construct transgenic animals in which a particular gene can be inactivated during specific stages of development or in specific tissues. These animal models are important to understand the function of a particular gene product within a complex organism, but also provide a valuable source of new cell lines that may be used to investigate the cellular and biochemical implications of a defect or deficiency in this gene product. In addition, animal models can mimic human diseases and provide systems to study the pathogenic mechanisms underlying a particular syndrome. For example, XPA-deficient mice are being used to test the hypothesis that systemic immunosuppression contributes to the high frequency of skin cancer in human patients suffering from reduced nucleotide excision repair capacity.[62] Animal models also provide attractive systems to develop new therapeutic strategies to treat or prevent such human diseases.

Additionally, some transgenic animal models may open completely new perspectives in toxicological risk assessment. Current methods used to evaluate carcinogenicity in rodents are likely to generate a large number of false positive results. In surveying a wide range of synthetic and natural chemicals, Ames and coworkers[63] reported that about half of all compounds tested could act as carcinogens in rodents and they suggested that exposure to chemicals at high concentrations (close to the maximum tolerated doses) may be carcinogenic due to chronic tissue injury and consequent cell replacement, thereby generating additional risk factors for cancer such as inflammation and cell proliferation. To improve current risk assessment procedures, we need a picture of how a chemical behaves in the body at doses several order of magnitude smaller than those normally administered to laboratory animals. The dramatically increased tumor susceptibility of rodents defective in a particular DNA damage processing pathway (Table 2.2) makes them potentially valuable test animals for screening suspected carcinogens at considerable lower doses, thereby avoiding the induction of inflammatory or proliferative reactions.

REFERENCES

1. Koller BH, Smithies O. Altering genes in animals by gene targeting. Annu Rev Immunol 1992; 10:705-730.
2. Joyner AL. Gene targeting: a practical Approach. Oxford: IRL Press, 1993.
3. Roth D, Wilson J. Illegitimate recombination in mammalian cells. In: Kucherlapati R, Smith GR, eds. Genetic Recombination. Washington DC: Am Soc Microbiol, 1988:621-653.
4. Nakane H, Takeuchi S, Yuba S et al. High incidence of ultraviolet-B- or chemical-carcinogen-induced skin tumours in mice lacking the xeroderma pigmentosum group A gene. Nature 1995; 377:165-168.
5. de Vries A, van Oostrom CTM, Hofhuis FMA et al. Increased susceptibility to ultraviolet-B and carcinogens of mice lacking the DNA excision repair gene XPA. Nature 1995; 377:169-173.
6. Sands AT, Abuin A, Sanchez A et al. High susceptibility to ultraviolet-induced carcinogenesis in mice lacking *XPC*. Nature 1995; 377:162-165.
7. Cleaver JE. DNA repair with purines and pyrimidines in radiation- and carcinogen-damaged normal and xeroderma pigmentosum human cells. Cancer Res 1973; 33:362-369.
8. Takebe H, Furuyama JI, Miki Y et al. High sensitivity of xeroderma pigmentosum cells to the carcinogen 4-nitroquinoline-1-oxide. Mutat Res 1972; 15:98-100.
9. Selfridge J, Pow AM, McWhir J et al. Gene targeting using a mouse *HPRT* minigene/*HPRT*-deficient embryonic stem cell system: inactivation of the mouse *ERCC-1* gene. Som Cell Molec Genet 1992; 18:325-336.
10. McWhir J, Selfridge J, Harrison DJ et al. Mice with DNA repair gene (*ERCC1*) deficiency have elevated levels of p53, liver nuclear abnormalities and die before weaning. Nature Genet 1993; 4:217-223.
11. Dianov G, Price A, Lindahl T. Generation of single-nucleotide repair patches following excision of uracil residues from DNA. Mol Cell Biol 1992; 12:1605-1612.
12. Gu H, Marth JD, Orban PC et al. Deletion of a DNA polymerase β gene segment in T cells using cell type-specific gene targeting. Science 1994; 265:103-106.
13. Xanthoudakis S, Smeyne RJ, Wallace JD et al. The redox/DNA repair protein, Ref-1, is essential for early embryonic development in mice. Proc Natl Acad Sci USA 1996; 93:8919-8923.
14. Tsuzuki T, Fujii Y, Sakumi K et al. Targeted disruption of the *Rad51* gene leads to lethality in embryonic mice. Proc Natl Acad Sci USA 1996; 93:6236-6240.
15. Sobol RB, Horton JK, Kühn R et al. Requirement of mammalian DNA polymerase-β in base-excision repair. Nature 1996; 379:183-186.
16. Modrich P. Mismatch repair, genetic sta-

bility, and cancer. SAnimal Models of DNA Damage Recognition and Processing DisordersAnimal Models of DNA Damage Recognition and Processing Disorderscience 1994; 266:1959-1960.

17. Modrich P. Mechanisms and biological effects of mismatch repair. Annu Rev Genet 1991; 25:229-253.

18. Jiricny J. Colon cancer and DNA repair: have mismatches met their match? Trends Genet 1994; 10:164-168.

19. Lynch HT, Smyrk TC, Watson P et al. Genetics, natural history, tumor spectrum, and pathology of hereditary nonpolyposis colorectal cancer: an updated review. Gastroenterology 1993; 104:1535-1549.

20. Fishel R, Lescoe MK, Rao MRS et al. The human mutator gene homolog *MSH2* and its association with hereditary non-polyposis colon cancer. Cell 1993; 75:1027-1038.

21. Baker SM, Bronner CE, Zhang L et al. Male mice defective in the DNA mismatch repair gene *PMS2* exhibit abnormal chromosome synapsis in meiosis. Cell 1995; 82:309-319.

22. de Wind N, Dekker M, Berns A et al. Inactivation of the mouse *Msh2* gene results in mismatch repair deficiency, methylation tolerance, hyperrecombination, and predisposition to cancer. Cell 1995; 82:321-330.

23. te Riele H., Robanus Maandag E, Berns A. Highly efficient gene targeting in embryonic stem cells through homologous recombination with isogenic DNA constructs. Proc Natl Acad Sci USA 1992; 89:5128-5132.

24. Alani E, Reenan RAG, Kolodner R. Interaction between mismatch repair and genetic recombination in *Saccharomyces cerevisiae*. Genetics 1994; 137:19-39.

25. Selva E, New L, Crouse G et al. Mismatch correction acts as a barrier to homeologous recombination in *Saccharomyces cerevisiae*. Genetics 1995; 139:1175-1188.

26. Rayssiguier C, Thaler DS, Radman M. The barrier to recombination between *Escherichia coli* and *Salmonella typhimurium* is disrupted in mismatch-repair mutants. Nature 1989; 342:396-401.

27. Pegg AE, Dolan ME, Moschel RC. Structure, function, and inhibition of O^6-alkylguanine-DNA alkyltransferase. Progr Nucleic Acid Res Mol Biol 1995; 51:167-223.

28. Preuss I, Thust R, Kaina B. Protective effect of O^6-methylguanine-DNA methyltransferase (MGMT) on the cytotoxic and recombinogenic activity of different antineoplastic drugs. Int J Cancer 1996; 65:506-512.

29. Tsuzuki T, Sakumi K, Shiraishi A. Targeted disruption of the DNA repair methyltransferase gene renders mice hypersensitive to alkylating agents. Carcinogenesis 1996; 17:1215-1220.

30. Donehower LA, Harvey M, Slagle BL et al. Mice deficient in p53 are developmentally normal but susceptible to spontaneous tumours. Nature 1992; 356:215-221.

31. Harvey M, McArthur MJ, Montgomery CA et al. Spontaneous and carcinogen-induced tumorigenesis in p53-deficient mice. Nature Genet 1993; 5:225-229.

32. Paterson MC, Smith PJ. Ataxia telangiectasia: an inherited human disorder involving hypersensitivity to ionizing radiation and related DNA-damaging chemicals. Annu Rev Genet 1979; 13:291-398.

33. Barlow C, Hirotsune S, Paylor R et al. *Atm*-deficient mice: a paradigm of ataxia telangiectasia. Cell 1996; 86:159-171.

34. Morrell D, Cromartie E, Swift M. Mortality and cancer incidence in 263 patients with ataxia-telangiectasia. J Natl Cancer Inst 1986; 77:89-92.

35. Bosma GC, Custer RP, Bosma MJ. A severe combined immune deficiency mutation in the mouse. Nature 1983; 301:527-530.

36. Lieber MR, Hesse JE, Lewis et al. The defect in murine severe combined immunodeficiency: joining of signal sequences but not coding segments in V(D)J recombination. Cell 1988; 55:7-16.

37. Schuler W, Weiler IJ, Schuler A et al. Rearrangement of antigen receptor genes is defective in mice with severe combined immunodeficiency. Cell 1986; 46:963-972.

38. Malynn BA, Blackwell TK, Fulop GM et al. The *scid* defect affects the final step of the immunoglobulin VDJ recombinase mechanism. Cell 1988; 54:453-460.

39. Bosma MJ, Carroll AM. The SCID mouse mutant: definition, characterization, and potential uses. Annu Rev Immunol 1991; 9:323-350.

40. Lieber MR. Site-specific recombination in immune system. FASEB J 1991; 5:2934-2944.

41. Fulop GM, Phillips RA. The *scid* mutation in mice causes a general defect in DNA repair. Nature 1990; 347:479-482.

42. Biedermann KA, Sun J, Giaccia AJ et al. *scid* mutation in mice confers hypersensitivity to ionizing radiation and a deficiency in DNA double-strand break repair. Proc Natl Acad Sci USA 1991; 88:1394-1397.

43. Hendrickson EA, Qin X-Q, Bump ET et al. A link between double-strand break-related repair and V(D)J recombination: the scid mutation. Proc Natl Acad Sci USA 1991; 88:4061-4065.

44. Tanaka T, Yamagami T, Oka Y et al. The scid mutation in mice causes defects in the repair system for both double-strand DNA breaks and DNA cross-links. Mutat Res 1993; 288:277-280.

45. Blunt T, Finnie NJ, Taccioli GE et al. Defective DNA-dependent protein kinase activity is linked to V(D)J recombination and DNA repair defects associated with the murine scid mutation. Cell 1995; 80:813-823.

46. Kirchgessner CU, Patil CK, Evans JW et al. DNA-dependent kinase (p350) as a candidate gene for the murine SCID defect. Science 1995; 267:1178-1183.

47. Peterson SR, Kurimasa A, Oshimura M et al. Loss of the catalytic subunit of the DNA-dependent protein kinase in DNA double-strand-break-repair mutant mammalian cells. Proc Natl Acad Sci USA 1995; 92:3171-3174.

48. Gottlieb TM, Jackson SP. The DNA-dependent protein kinase: requirements for DNA ends and association with Ku antigen. Cell 1993; 72:131-142.

49. Nussenzweig A, Chen C, da Costa Soares V et al. Requirement for Ku80 in growth and immunoglobulin V(D)J recombination. Nature 1996; 382:551-555.

50. Zhu C, Bogue MA, Lim D-S et al. Ku86-deficient mice exhibit severe combined immunodeficiency and defective processing of V(D)J recombination intermediates. Cell 1996; 86:379-389.

51. McGuire TC, Poppie MJ. Hypogammaglobulinemia and thymic hypoplasia in horses: a primary combined immunodeficiency disorder. Infect Immun 1973; 8:272-277.

52. Poppie MJ, McGuire TC. Combined immunodeficiency in foals in Arabian breeding: evaluation of mode of inheritance and estimation of prevalence of affected foals and carrier mares and stallion. J Am Vet Med Assoc 1977; 170:31-33.

53. Felsburg PJ, Somberg RL, Perryman LE. Domestic animal models of severe combined immunodeficiency: canine X-linked severe combined immunodeficiency and severe combined immunodeficiency in Immunodefic Rev 1992; 3:277-303.

54. Lunn DP, McClure JT, Schobert CS et al. Abnormal patterns of equine leukocyte differentiation antigen expression in severe combined immunodeficiency foals suggests the phenotype of normal equine natural killer cells. Immunology 1995; 84:495-499.

55. Wiler R, Leber R, Moore BB et al. Equine severe combined immunodeficiency: a defect in V(D)J recombination and DNA-dependent protein kinase activity. Proc Natl Acad Sci USA 1995; 92:11485-11489.

56. Mounkes LC, Jones RS, Liang B-C et al. A Drosophila model for xeroderma pigmentosum and Cockayne's syndrome: *haywire* encodes the fly homolog of ERCC3, a human excision repair gene. Cell 1992; 71:925-937.

57. Regan CL, Fuller MT. Interacting genes that affect microtubule function: the *nc2* allele of the *haywire* locus fails to complement mutations in the testis-specific β-tubulin gene of *Drosophila*. Genes Dev 1988; 2:82-92.

58. Kemphues KJ, Kaufman TC, Raff RA et al. The testis-specific beta-tubulin subunit in *Drosophila melanogaster* has multiple functions in spermatogenesis. Cell 1982; 31:655-670.

59. Schaeffer L, Roy R, Humbert S et al. DNA repair helicase: a component of BTF2 (TFIIH) basic transcription factor. Science 1993; 260:58-63.

60. Drapkin R, Reardon JT, Ansari A et al.

Dual role of TFIIH in DNA excision repair and in transcription by RNA polymerase II. Nature 1994; 368:769-772.

61. Feaver WJ, Svejstrup JQ, Bardwell L et al. Dual roles of a multiprotein complex from *Saccharomyces cerevisiae* in transcription and DNA repair. Cell 1993; 75:1379-1387.

62. Miyauchi-Hashimoto H, Tanaka K, Horio T. Enhanced inflammation and immunosuppression by ultraviolet radiation in xeroderma pigmentosum group A (XPA) model mice. J Invest Dermatol 1996; 107:343-348.

63. Gold LS, Slone TH, Stern BR et al. Rodent carcinogens: setting priorities. Science 1992; 258:261-265.

64. Bentley DJ, Selfridge J, Millar JK et al. DNA ligase I is required for fetal liver erythropoiesis but is not essential for mammalian cell viability. Nature Genetics 1996; 13:489-491.

65. Lowe SW, Schmitt EM, Smith SW et al. p53 is required for radiation-induced apoptosis in mouse thymocytes. Nature 1993; 362:847-849.

66. Clarke AR, Purdie CA, Harrison DJ et al. Thymocyte apoptosis induced by p53-dependent and independent pathways. Nature 1993; 362:849-852.

67. Xu Y, Ashley T, Brainerd EE et al. Targeted disruption of *ATM* leads to growth retardation, chromosomal fragmentation during meiosis, immune defects, and thymic lymphoma. Genes Dev 1996; 10:2411-2422.

CHAPTER 3

DNA STRUCTURE: INHERENT INSTABILITY AND GENOTOXIC REACTIONS

The principal goal of this chapter is to provide examples of biologically significant lesions in DNA that are (1) chemically and structurally well characterized and (2) serve as experimental models to understand how DNA damage is recognized in mammalian chromosomes.

Native DNA consists of long threads of deoxyribonucleotides, each composed of a sugar (deoxyribose), a phosphate and a nitrogenous base.[1] The backbone of DNA, which is invariant throughout the molecule, is formed by deoxyriboses linked together by phosphate groups. In this repetitive sugar-phosphate backbone, phosphodiester bridges join the 3'-hydroxyl group of the sugar residue of one deoxyribonucleotide to the 5'-hydroxyl of the adjacent sugar (Fig. 3.1). The nitrogenous bases constitute the variant part of DNA, and are attached to the 1'-position of the deoxyribose residues, forming N-glycosylic linkages (Fig. 3.1). The genetic information is carried by the linear sequence of purine (adenine, guanine) and pyrimidine bases (cytosine, thymine) along the DNA molecule. All these constituents of DNA, i.e., phosphate, deoxyribose and nitrogenous bases, are potential targets of spontaneous decay or genotoxic reagents.

MACROMOLECULAR INTERACTIONS IN NATIVE DNA

At the three-dimensional level, native DNA is a helical macromolecule consisting of two strands running in opposite polarity.[1-3] The DNA double helix forms two types of grooves differing in their size: the major groove is 1.2 nm wide, while the minor groove is only 0.6 nm wide. The planar bases are located on the inside of the DNA double helix, while the sugar-phosphate backbone is on the outside.

COMPLEMENTARY BASE PAIRING

The two strands of the DNA double helix are held together by hydrogen bonds between pairs of bases, whereby adenine is normally paired with thymine and cytosine is normally paired with guanine. In fact, continuous base pairing throughout the DNA molecule is of crucial significance for most nucleic acid functions. Since each strand contains a sequence that is complementary to that of the other strand, both strands actually carry the same genetic information. This complementary nature of the double helix provides the molecular basis by which genetic information is maintained with high fidelity. For example, during semiconservative DNA replication the two strands of the duplex are unwound and separated,

Mechanisms of DNA Damage Recognition in Mammalian Cells, by Hanspeter Naegeli.
© 1997 R.G. Landes Company.

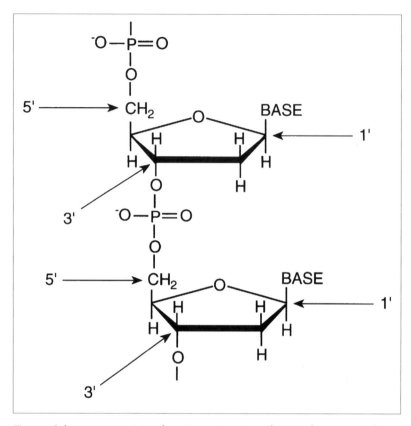

Fig. 3.1. Scheme summarizing the primary structure of DNA. The arrows indicate the 1', 3' and 5' positions of deoxyribose moieties.

and each parent strand acts as a template for the synthesis of a new complementary strand by DNA polymerases.[1,3] DNA excision repair mechanisms also rely on the redundant information in the double helix to excise a damaged base, or an oligonucleotide segment containing a damaged base, and replace it by DNA synthesis using the complementary strand as a template.[4]

Disruption of complementary base pairing, on the other hand, constitutes a strong signal in DNA damage recognition for example by attracting the mismatch correction system to unpaired bases in duplex DNA. Also, improper base pairing is thought to facilitate the binding of DNA photolyase or certain DNA glycosylases to their specific substrates (see for example the case of uracil-DNA glycosylase in chapter 4). Additionally, disruption of complementary base pairing is exploited in mammalian cells to preferentially target the nucleotide excision repair complex to those lesions that display the highest cytotoxic and mutagenic potentials (see chapter 7). Only in the rare event that both DNA strands are damaged simultaneously and in close proximity is the cell left without any template for DNA replication or repair, but recombinational mechanisms have evolved to handle this critical situation.[5]

DNA-PROTEIN INTERACTIONS

The DNA double helix displays three types of noncovalent bonds for interactions with sequence- or damage-specific recognition proteins.[1,6]

(1) The major and the minor grooves are lined with oxygen and nitrogen atoms that serve as hydrogen bond donor and acceptor sites for potential interactions with proteins.

(2) The negatively charged phosphates in the DNA backbone can form ion bonds with positively charged amino acid side chains such as arginine or lysine.

(3) Hydrophobic or van der Waal's forces are available for stacking interactions between the nitrogenous bases of DNA and the aromatic side chains of phenylalanine, tryptophan or tyrosine. Additionally, hydrophobic or van der Waal's attractions may also occur between specific functional groups of bases, such as the methyl group of thymine or 5-methylcytosine, and nonpolar amino acid side chains.

INHERENT INSTABILITY OF DNA

The primary structure of DNA has only limited chemical stability.[7] In fact, spontaneous defects resulting from the intrinsic chemistry of DNA are usually many times more frequent than genetic injuries arising from environmental sources. DNA is mainly subject to spontaneous depurination and deamination.

DEPURINATION

The *N*-glycosylic bond that joins the bases to the sugar-phosphate backbone is particularly labile under physiological conditions. As a consequence, purine bases dissociate spontaneously to generate abasic sites. As many as 10,000 purine bases are lost per day from the DNA of each mammalian cell, with guanine being released slightly more rapidly than adenine.[7,8] This rate has not been actually measured in cells, but was estimated from the spontaneous depurination of purified DNA molecules in solution and under physiological conditions (pH 7.4, 37°C). Cytosine and thymine are lost at approximately 5% the rate of the purines.

Apurinic/apyrimidinic sites are intrinsically mutagenic because of their failure to provide hydrogen bonding information during DNA synthesis. Moreover, at sites of base loss the sugar-phosphate backbone becomes unstable and undergoes spontaneous endonucleolytic cleavage with a half-life of 190 hours.[9] Within cells, extensive DNA fragmentation at these sites is prevented by an excision repair process that is rapidly initiated by apurinic/apyrimidinic (AP) endonucleases.[10-12] This ubiquitous class of enzymes is highly abundant in mammalian tissues and is able to handle large loads of DNA depurination.

DEAMINATION

In addition to the lability of *N*-glycosylic bonds, DNA bases are intrinsically susceptible to hydrolytic deamination. Three of the four bases normally present in DNA (cytosine, adenine and guanine) contain exocyclic amino groups. The loss of these amino groups and their replacement with an oxygen atom (deamination) results in the conversion of the affected bases into uracil (from cytosine), hypoxanthine (from adenine) and xanthine (from guanine); deamination of 5-methylcytosine, on the other hand, produces thymine (Fig. 3.2). Cytosine and its methylated derivative, 5-methylcytosine, are particularly prone to this reaction.[13-15]

These deamination events are mutagenic: deamination of cytosine generates G·U mismatches, deamination of 5-methylcytosine G·T mismatches. Uracil bases are eliminated by an excision repair pathway initiated by the enzyme uracil-DNA glycosylase;[16] G·T base pairs are processed by mismatch correction.[17] If inappropriate uracil and thymine residues are not removed, the next round of replication would produce G·C→A·T transitions at these sites. The finding that G·C→A·T transitions at CpG dinucleotides account for 35% of single-base pair mutations in several inherited human diseases (including hemophilia, familial Alzheimer, colon cancer, retinoblastoma or Gerstmann-Sträussler syndrome) supports

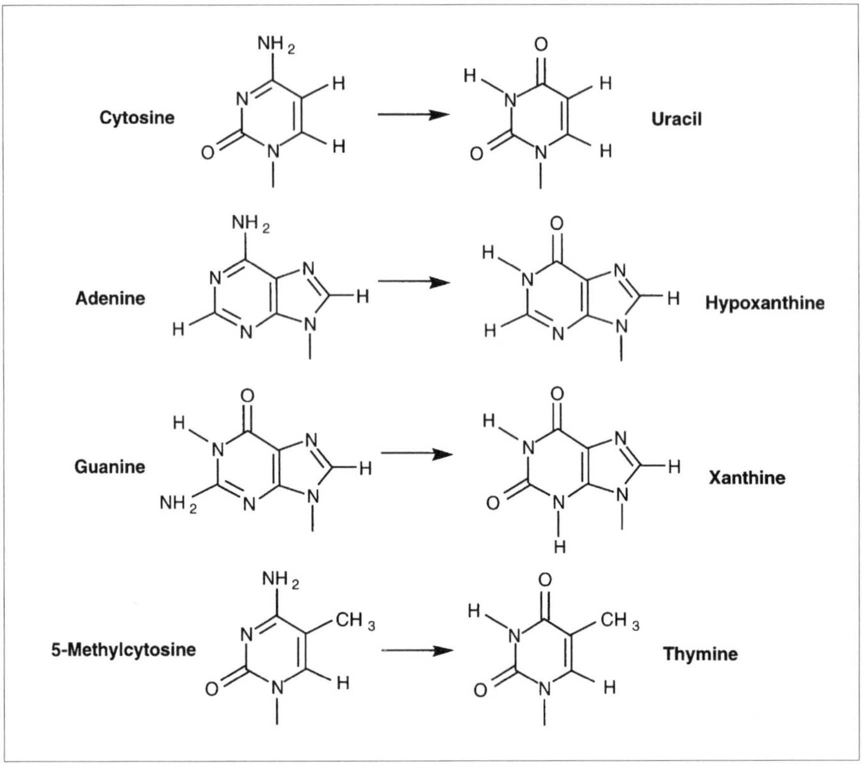

Fig. 3.2. Hydrolytic deamination of cytosine, adenine, guanine and 5-methylcytosine.

the idea that cytosine and 5-methylcytosine deamination events are significant mutagenic reactions.[18]

The rates of cytosine deamination in both single-stranded and double-stranded DNA were determined at physiologically relevant conditions.[19] According to these measurements, approximately 500 spontaneous cytosine deamination events per day are expected in a double-stranded mammalian genome, but deamination is 4,000 times faster in single-stranded DNA, suggesting that it may occur preferentially in actively transcribed genes and replication forks.[19,20] Also, cytosine deamination is greatly accelerated by several mutagens such bisulfite ions or nitrous acid,[21] and occurs with increased frequency in DNA exposed to UV radiation.[22]

Spontaneous deamination of 5-methylcytosine is thought to be particularly deleterious: it occurs three to four times more rapidly than cytosine deamination, and the mutagenic potential of this reaction may be further accentuated by the slower repair of the resulting G·T mismatches.[7,13-15] Uracil bases in DNA also arise by misincorporation of dUMP during DNA synthesis. Over 10,000 deoxyuridine incorporation events are believed to occur per replicative cycle of a mammalian genome.[20] These uracil residues are in A·U base pairs and, hence, are not mutagenic. However, uracil in place of thymine may nevertheless display negative effects for example by preventing specific DNA-protein interactions in regulatory sequences.[23]

As stated before, deamination also involves purine bases: adenine is deaminated to produce hypoxanthine, and guanine is deaminated to xanthine (Fig. 3.2). Deamination of adenine or guanine is a quantitatively minor reaction that occurs at 2-3% of the rate of cytosine deamination,[7,24] but purine deamination is enhanced by nitrous acid[25] or nitric oxide (see section on nitric oxide genotoxicity).[26,27] The formation of hypoxanthine from adenine is expected to be mutagenic because hypoxanthine preferentially pairs with cytosine rather than with thymine, and may therefore generate A·T→G·C transitions.[14] Like uracil, removal of hypoxanthine is initiated by a DNA glycosylase, while no specific repair enzymes for xanthine residues have been identified.[28]

DNA DAMAGE INDUCED BY ENDOGENOUS AGENTS

In addition to spontaneous decay, DNA is constantly faced with intracellular agents that arise during normal metabolism or by pathological reactions such as those occurring during chronic inflammatory processes. The main sources of endogenous DNA damage include oxygen radicals generated as by-products of aerobic metabolism, nitric oxide and alkylating agents,[29,30] but many other endogenous compounds may react with DNA and are potentially genotoxic. Even certain amino acids or reducing sugars such as glucose-6-phosphate and glucose have been implicated in DNA damaging reactions (see ref. 31 and references therein).

OXYGEN

The use of oxygen by living organisms is always accompanied by the formation of highly reactive by-products, with hydroxyl radical (·OH) being the ultimate agent for the majority of oxidative damage to DNA.[7,32-34] Because of their unpaired electron, ·OH and other free radicals are extremely reactive and exert deleterious effects on a broad range of cellular macromolecules. Estimates of the number of oxidative DNA lesions formed on a daily basis are in the range of 10^5 per mammalian cell.[29,30] Nearly 100 different free radical damages have been identified in DNA,[33-36] too many to be summarized in the context of this review. Oxygen-induced DNA modifications may be divided into two broad classes, i.e., base and deoxyribose lesions. Oxidative base damage includes 8-hydroxyguanine, fragmented purines (formamidopyrimidines), and various ring-saturated pyrimidine de-

rivatives (for example thymine hydrates and glycols). Oxidative damage to the sugar-phosphate backbone generates single- or double-stranded breaks.

OXIDATIVE DAMAGE TO DNA BASES

8-hydroxyguanine (Fig. 3.3) is considered the major mutagenic lesion generated by active oxygen species.[32,37] In fact, 8-hydroxyguanine pairs preferentially with adenine rather than cytosine and thus generates G·C→T·A transversions during DNA replication.[38] Oxidation of guanine residues to form 8-hydroxyguanine is thought to occur at similar rates (approximately 500 events per mammalian cell per day) as cytosine deamination.[7] Formamidopyrimidines are derived from both adenine and guanine, and consist of ruptured imidazole rings joined to an intact pyrimidine ring (Fig. 3.3).[32,39,40] These lesions are produced by oxygen radicals in similar amounts as 8-hydroxyguanine,[7,36,40] but formamidopyrimidine residues are also generated as secondary products of N^7-alkylated guanines or adenines.[30] The ring-saturated pyrimidine derivatives have lost the 5,6 double bond (Fig. 3.4) and, in vitro, are generated at rates marginally higher than 8-hydroxyguanine.[7,32,36,40,41] These pyrimidine derivatives may undergo alkali-catalyzed decomposition to yield urea and other fragmentation products linked to deoxyribose.[32,41]

Both 8-hydroxyguanine and formamidopyrimidine derivatives are processed by the same base excision repair reaction initiated by the enzyme formamidopyrimidine-DNA glycosylase.[32,39,42] Repair of ring-saturated pyrimidine derivatives, on the other hand, is initiated by a distinct DNA glycosylase designated thymine glycol-DNA glycosylase or, in *Escherichia coli*, endonuclease III.[32,43]

OXIDATIVE DAMAGE TO THE DNA BACKBONE

The sugar-phosphate backbone is highly exposed to solvent in both single-stranded and double-stranded DNA and, hence, susceptible to degradation by the action of free radicals. After free radical attack, DNA mol-

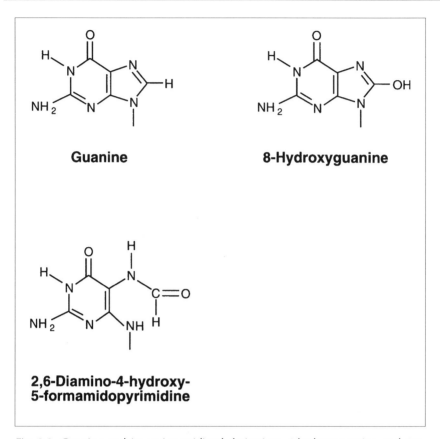

Fig. 3.3. Guanine and its major oxidized derivatives, 8-hydroxyguanine and 2,6-diamino-4-hydroxy-5-formamidopyrimidine.

ecules are subject to a series of reactions that eventually generate DNA strand breaks with or without different forms of deoxyribose fragmentation.[44-46] DNA strand breaks (mainly single strand breaks) are estimated to occur at a rate of 2,300 per hour in each mammalian cell (equivalent to over 50,000 per day).[30,46]

Major products formed by hydroxyl radical attack of the deoxyribose moiety are strand breaks with 3'-phosphoglycolate or 3'-phosphate termini. Often, various fragmented sugar derivatives are left.[33,44,45] Such DNA ends are unable to serve as substrates for DNA polymerases or DNA ligases. Thus, if present, these unusual or damaged termini are removed by specialized exonucleases that also possess apurinic-apyrimidinic (AP) endonuclease and 3'-phosphatase activ-ity.[32,45] The formation of double strand breaks presents particular problems for DNA repair because the two strands must be realigned to restore the correct nucleotide sequence. In principle, this task can be solved by recruiting the homologous recombination machinery to the break, provided that the cell contains another identical copy of the broken chromosomal region.[47]

SOURCES OF OXYGEN FREE RADICALS

Three endogenous sources appear to account for most oxidative DNA damage in mammalian cells, i.e., aerobic metabolism, inflammation and microsomal P450 systems.[29]

In all aerobic organisms, cellular metabolism is accompanied by the formation of reactive oxygen species as inevitable by-

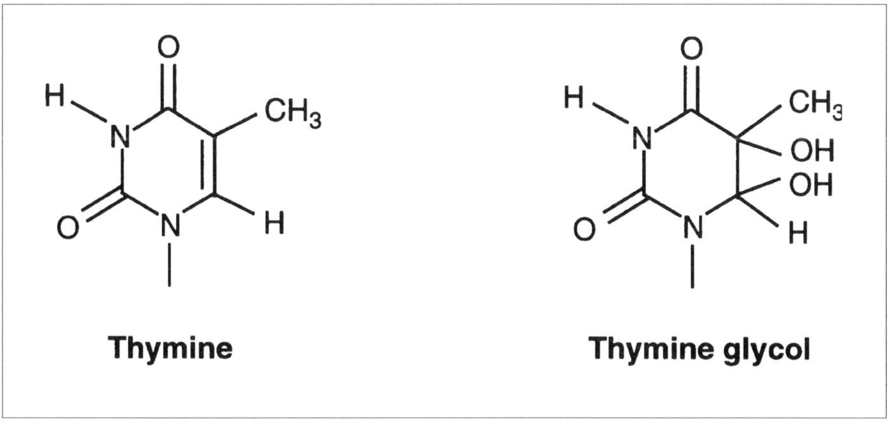

Fig. 3.4. Example of oxidative damage to pyrimidine bases: thymine glycol is generated by ring saturation of thymine.

products of mitochondrial electron carriers, which reduce O_2 to form H_2O.[48] Studies on human diploid fibroblasts indicate that the level of oxidative DNA damage arising from aerobic metabolism increases with aging.[49] For example, senescent human cells maintained in culture excise from nuclear DNA four times more 8-hydroxyguanine per day than early-passage young cells, and the steady state level of 8-hydroxyguanine in DNA is approximately 35% higher in senescent cells than in the corresponding early-passage cells. Defective mitochondria in senescent cells generate higher amounts of toxic by-products and, possibly, cause increased levels of oxidative DNA damage upon aging.[49]

Activated macrophages and cytotoxic lymphocytes kill bacteria, parasites and virus-infected cells with oxidative bursts that are likely to contribute significantly to the level of oxidative DNA damage, particularly during chronic inflammatory processes.[50] Therefore, chronic infections by viruses (for example hepatitis B and C), parasites (schistosomiasis, liver flukes) or bacteria (*Helicobacter pylori*) constitute major risk factors for cancer in humans.[29,51]

Biotransformation by P450 enzymes is a third intracellular source of oxidative DNA damage. This inducible system is a primary defense system against toxic chemicals, but also results in oxidant by-products that damage DNA.[29] In addition to the load of

reactive oxygen radicals generated endogenously, several exogenous oxidant sources have been recognized, of which cigarette smoke plays a primary role in lung cancer induction.[52-55] Also, the mechanism of toxicity of many natural or man-made poisons involves the formation of oxygen radicals. A classical example of such compounds is the redox-cycling agent paraquat, which diverts electrons from NAD(P)H to O_2 and generates intracellular superoxide.[56] Finally, the clinical efficacy of several therapeutic agents depends on the ability to produce DNA-reactive oxygen species or other free radicals. The antiparasitic/antimicrobial drug metronidazole[57] and the antitumor agent bleomycin[58] are two representative compounds that exploit this therapeutic principle.

NITRIC OXIDE GENOTOXICITY

Nitric oxide (NO·) is a potent free radical that has received much attention during recent years as a cigarette smoke component and air pollutant,[29,52,54,55] but particularly as a physiological messenger and cytotoxic agent.[59,60] Under aerobic conditions, nitric oxide can form reactive nitrosyl donors such as peroxynitrite (ONOO⁻)[61] which result in the deamination of purine and pyrimidine bases at physiological pH.[26,27] In the presence of nitric oxide, guanine is deaminated more rapidly than cytosine or adenine.[26,27] This hierarchy of reactivity contrasts with spontaneous hydrolytic deamination in which the reaction kinetics decrease in the order cytosine > adenine > guanine.

The hypothesis that nitric oxide-induced deamination could, in principle, cause genetic alterations in living cells was confirmed by exposing *Salmonella typhimurium* strain TA1535 to various nitric oxide-releasing compounds, including the therapeutic agent nitroglycerin.[26] Virtually all the mutants obtained were G•C→A•T transitions. Wink and coworkers[26] proposed that these transition mutations may be generated mainly by nitric oxide-induced deamination of cytosine (to uracil) or 5-methylcytosine (to thymine). When human lymphoblastoid cells were exposed to nitric oxide, mutations in two different genes (hypoxanthine-guanine phosphoribosyl transferase and thymidine kinase) increased up to 18-fold above background levels.[27] In parallel, nitric oxide induced DNA strand breaks in a dose-dependent manner.[27] The relatively high concentration of nitric oxide required for these effects (> 5 mM) may be related to the inefficient method of delivery, as a large fraction of nitric oxide is likely to escape into the gas phase without undergoing any chemical reaction. Nguyen et al[27] proposed another mechanism for nitric oxide-induced mutagenesis involving mainly deamination of guanine (to xanthine). This inappropriate base is unstable, leading to depurination and formation of abasic sites that are prone to mutagenesis. Moreover, depurination favors subsequent strand breakage.[9] It should be stated, however, that the observed susceptibility of the guanine deamination product (xanthine) to spontaneous depurination, and the fact that no specific repair enzyme exists for xanthine, have been interpreted in very different ways. Nguyen et al[27] concluded that rapid depurination of xanthine, promoted by nitric oxide, may constitute a major mechanism of genetic instability. Conversely, the relatively poor cellular responses to deaminated guanine have been used as an argument against the idea that nitric oxide may represent an important endogenous mutagen in mammals.[7]

ENDOGENOUS DNA METHYLATION

All heteroatoms in DNA (*N, O, P*) are potentially susceptible to the covalent addition of alkyl substituents such as methyl groups. Therefore, alkylating agents generate a broad spectrum of mutagenic and cytotoxic lesions in the genome.[62]

The quantitatively major products of base methylation are *N*-methylpurines, primarily N^3-methyladenine and N^7-methylguanine (Fig. 3.5). N^3-methyladenine has been partly implicated in the cytotoxicity exerted by methylating agents because this base lesion is able to inhibit DNA synthesis.[63] N^7-methylguanine is considered to be an innocuous lesion,[30,64] but it can generate secondary products either by spontaneous cleavage of the *N*-glycosylic bond to form an apurinic site, or by the opening of the imidazole ring to yield 2,6-diamino-4-hydroxy-5-*N*-alkylformamidopyrimidine.[65,66] Both apurinic and formamidopyrimidine residues are potentially genotoxic and mutagenic. However, the most mutagenic lesions induced by methylating agents are the quantitatively minor products O^6-methylguanine and O^4-methylthymine (Fig. 3.5). O^6-methylguanine preferentially pairs with thymine during replication and, as a consequence, causes G•C→A•T transition mutations.[67-69] O^4-methylthymine, on the other hand, induces T•A→C•G transitions upon replication.[70,71] O^6-methylguanines and, presumably, a few other alkylation products are thought to enhance the cytotoxic effects of alkylating agents by an unexpected mechanism. Because of their abnormal base pairing properties, they are able to stimulate the postreplicative mismatch correction system, thereby inducing repeated cycles of futile DNA incision and degradation processes that culminate in cell death.[72,73]

The ubiquitous presence of DNA glycosylases capable of excising *N*- and *O*-methylated bases in virtually all cell types suggests

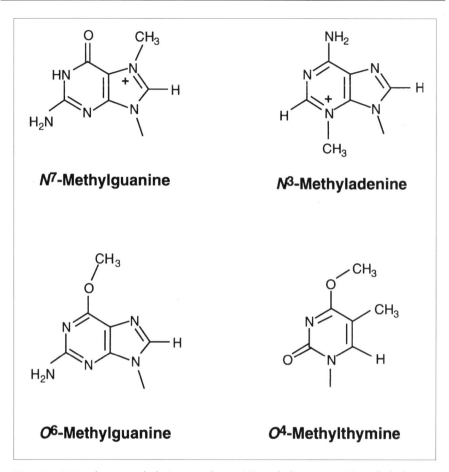

Fig. 3.5. Major base methylation products: N^7-methylguanine, N^3-methyladenine, O^6-methylguanine and O^4-methylthymine.

that DNA alkylation is a common event.[7] Candidates for endogenous DNA methylating agents include quaternary amines such as betaine or cholin, and the trialkylsulfonium agent S-adenosylmethionine.[74,75] S-adenosylmethionine is a methyl group donor that is used as cofactor in many cellular transmethylation reactions. Incubation of DNA with this compound in vitro produced methylation products at N^7, N^3 and O^6 of guanine and N^3 of adenine.[74,75] From these experiments, it was estimated that about 600 3-methyladenine derivatives per day may be generated by S-adenosylmethionine in a single mammalian cell.[74] The hypothesis

that cellular DNA may suffer from a considerable load of endogenous alkylating agents is supported by the fact that methylated DNA bases have been detected in biological samples without prior exposure to exogenous alkylating agents.[76-78]

Methylated bases may be eliminated from DNA by three distinct mechanisms. First, the presence of a methyl group on N^7 of guanine or N^3 of adenine labilizes the glycosylic bond, and these methylated bases undergo spontaneous depurination.[79] Second, O^6-methylguanines and, to a minor extent, other oxygen alkylation products, are a substrate for direct DNA damage reversal by the suicide enzyme O^6-methylguanine-DNA methyltransferase.[80] Third, N-methyl-purines are substrates for base excision repair initiated by DNA glycosylases, and may also be processed by nucleotide excision repair.[81,82] The repair of O^4-alkylthymines appears to be slow in mammalian tissues, suggesting that alkylated thymine derivatives may constitute a serious threat to genetic integrity.[83]

DNA DAMAGE INDUCED BY PHYSICAL ENVIRONMENTAL AGENTS

Most forms of life are permanently exposed to ultraviolet (UV) and visible radiation emanating from the sun. Presumably, UV photoproducts are the most frequent damage inflicted on DNA by the environment and the main cause of skin cancer in humans.[84,85] UV radiation is also one of the most common genotoxic agents used to investigate damage response pathways in the laboratory.

UV Spectrum

Already in 1877, Downes and Blunt reported that sunlight is capable of killing microorganisms (summarized in ref. 84). This cytotoxic property is associated with UV photons of wavelengths below 400 nm and the capacity for damaging cells becomes increasingly pronounced at shorter wavelengths, or higher energies.

UV light is divided into three categories depending on wavelength: 190-290 nm is classified as UV-C; 290-320 nm as UV-B; and 320-400 nm as UV-A. Sunlight consists of 0.3% UV-B, 5.1% UV-A, 62.7% visible light, and 31.9% infrared.[84,86] The most hazardous component of sunlight (wavelengths shorter than 290 nm) is completely absorbed by the ozone (O_3) layer in the upper atmosphere. However, with the gradual depletion of stratospheric ozone a higher flux of UV with short wavelengths is expected to reach the Earth's surface and, hence, more UV lesions in DNA will be formed. Thus, continued ozone layer depletion is likely to eventually result in an increased incidence of skin cancer in human populations.[87,88]

Direct UV Radiation Damage

The absorption spectrum of DNA falls into the UV-C and UV-B range. In most studies, UV-C radiation was used because of the availability of ordinary germicidal lamps and the corresponding equipment to measure their intensities. For that reason, the photodamage produced when DNA is irradiated with light in the UV-C range has been studied in greatest detail.

The two major types of photoproducts generated in DNA by short-wavelength UV radiation are covalent linkages between adjacent pyrimidines: cyclobutane pyrimidine dimers and pyrimidine(6-4)pyrimidone photoproducts (Fig. 3.6).[84,85,89] To a lesser extent, pyrimidine hydrates,[90] mixed purine-pyrimidine[91,92] and dipurine adducts are also generated.[93] The relative frequency of the two major products depends on the wavelength and dose of the incident light. Typically, the yield of pyrimidine(6-4)photoproducts in DNA is about 0.1-0.5 times that of cyclobutane pyrimidine dimers.[85,94]

Cyclobutane pyrimidine dimers are characterized by the formation of four-membered ring structures, or cyclobutyl rings, by which adjacent pyrimidine bases become covalently linked (Fig. 3.6). Pyrimidine(6-4)pyrimidone dimers, on the other hand, are produced by a covalent linkage between the C^6 position of one pyrimidine base and the C^4 position of an adjacent pyrimidine (Fig. 3.6). Both types of pyrimidine dimers are inhibitory to DNA replication are cytotoxic,[95] and are mutagenic.[96,97] The most frequent mutations induced by UV-C are C·G→T·A and T·A→C·G transitions. Tandem CC→TT transitions are often considered to be a characteristic feature of UV mutagenesis,[98] although such tandem mutations may also be caused by oxygen free radicals[99] and certain chemicals.[21] The pyrimidine(6-4) photoproduct, while being chemically stable, undergoes photoisomerization to Dewar valence isomers (Fig. 3.6),[84,94] a reversible process with UV-B

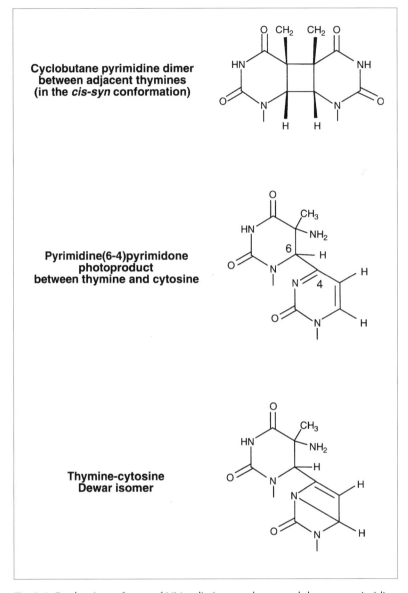

Cyclobutane pyrimidine dimer between adjacent thymines (in the *cis-syn* conformation)

Pyrimidine(6-4)pyrimidone photoproduct between thymine and cytosine

Thymine-cytosine Dewar isomer

Fig. 3.6. Predominant forms of UV radiation products: cyclobutane pyrimidine dimer, pyrimidine(6-4)pyrimidone photoproduct and Dewar isomer.

wavelengths favoring formation of the Dewar structure and UV-C favoring the conventional pyrimidine(6-4)photoproduct. DNA exposed to natural sunlight is expected to contain a mixture of pyrimidine(6-4)pyrimidone photoproducts and their Dewar valence isomers, raising considerable uncertainty regarding the actual species responsible for the major cytotoxic and mutagenic effects of natural UV radiation.

As outlined in chapter 1, multiple strategies are employed to repair UV radiation damage. Many microorganisms possess a DNA photolyase, or photoreactivating enzyme, that catalyzes direct reversal of cyclobutane pyrimidine dimers to regenerate monomeric pyrimidines in the presence of visible light.[100] A photolyase that repairs pyrimidine(6-4)pyrimidone photoproducts has also been identified in some species.[101] In *Micrococcus luteus* and bacteriophage T4-infected *Escherichia coli*, cyclobutane pyrimidine dimers are processed by an excision repair pathway initiated by a specific DNA glycosylase.[102] In the yeast *Schizosaccharomyces pombe*, in the fungus *Neurospora crassa* and perhaps other organisms, pyrimidine dimers are processed by a specialized UV endonuclease.[103] In contrast to these microorganisms, mammalian cells are highly dependent on the complex multisubunit nucleotide excision repair system to process the major forms of UV radiation damage.[104]

Indirect UV Radiation Damage

Photons with wavelengths in the UV-A region (> 320 nm) comprise a larger proportion of the total energy in sunlight than UV-C and UV-B.[84] In addition, UV-A is not absorbed by ozone and is much more efficient in penetrating the actively dividing basal layer of the epidermis.[105,106] UV-A causes DNA damage through the action of photosensitizing chemicals. Such reactions can be mediated either by endogenous (porphyrines, flavines) or exogenous agents (psoralens, tetracycline, promazine, methylene blue),[107] and may damage DNA directly (type I photosensitized reactions) or indirectly through the formation of reactive oxygen species (type II photosensitized reactions).[108] For example, absorption of UV light by the amino acid tryptophan has been shown to generate peroxide radicals that damage DNA.[109]

UV-A is a complete carcinogen in albino hairless mice[110] and is mutagenic in mammalian cells maintained in culture, where irradiation in the UV-A range has been shown to induce a large proportion of T·A→G·C transversions in the adenine phosphoribosyltransferase gene.[111] These findings indicate that UV-A may contribute significantly to the dramatically increased incidence of skin cancer in western countries. The role of UV-A in the patho-

genesis of human skin cancer is presumably enhanced by the widespread use of UV-B-specific sunscreens over the last decades[112] and, additionally, by exposure to artificial light sources during recreational sunning.

Ionizing Radiation

Ionizing radiation produces DNA lesions mainly through production of free radicals from radiolysis of H_2O, i.e., scission of an O-H bond of water to generate the highly reactive hydroxyl radical ($^{\cdot}OH$).[113] Formation of DNA-reactive free radicals is potentiated by the presence of O_2.[32,33] Because DNA damage results primarily from hydroxyl radicals, ionizing radiation produces many of the same lesions that are also formed by endogenous oxygen by-products generated during metabolic processes (see previous sections on oxidative damage). These lesions consist of single and double strand breaks, and a plethora of base modifications including 8-hydroxyguanine, formamidopyrimidines and ring-saturated pyrimidines. A distinct signature of ionizing radiation is the frequent induction of DNA strand breaks bearing 3' phosphates or 3'-phosphoglycolate termini.

DNA DAMAGE INDUCED BY CHEMICAL ENVIRONMENTAL AGENTS

DNA is an organic macromolecule consisting of an array of electron-rich (nucleophilic) centers that are permanently targeted by a large number of electrophilic compounds. Multiple potential reaction sites for the formation of covalent adducts have been identified in all four bases. In general, the ring nitrogens of the bases are more nucleophilic than the oxygens, with the N^7 position of guanine and the N^3 position of adenine being the most reactive.[30,62,79] The mutagenic consequences of these nitrogen and oxygen adducts are extremely variable and include base substitutions (transitions, transversions), single nucleotide insertions or deletions (frameshifts), or larger insertions or deletions.

A general awareness of the potential risk arising from environmental carcinogens has greatly stimulated studies on the mechanisms by which electrophilic chemicals react with DNA and generate cytotoxic and mutagenic products. The application of sensitive detection methods, primarily ^{32}P-postlabeling combined with thin layer chromatography,[114] has provided hard evidence for the formation of DNA adducts following exposure to environmental carcinogens. In numerous studies, detection of DNA adducts in human tissues has been reported in association with exposure to occupational, environmental, dietary, tobacco and drug-related risk factors (see for example ref. 115). As a general rule, base adducts generated by environmental carcinogens are mainly processed by DNA excision repair, particularly nucleotide excision repair.[4,82,104] A notable exception is the removal of O^6-alkylguanine and O^4-alkylthymine adducts by the suicide enzyme O^6-methylguanine methyltransferase.[80]

EXOGENOUS ALKYLATING AGENTS

Principal sources of exogenous alkylating agents are industrial processes, tobacco smoke and dietary components.[62,116,117] These chemicals mainly induce adducts at the ring nitrogens, leading to the formation of 1-alkyladenine, 3-alkyladenine, 7-alkyladenine, 3-alkylguanine, 7-alkylguanine, 3-alkylcytosine and 3-alkylthymine. In addition, there is quantitatively minor alkylation at the exocyclic oxygens of guanine (O^6-alkylguanine), thymine (O^2- and O^4-alkylthymine) and cytosine (O^2-alkylcytosine). A third fraction of alkyl substituents consists of alkylphosphates in the DNA backbone. Depending on their alkyl transfer mechanism, alkylating agents can be classified as species that react with nucleophiles by either S_N1 (for example N-methyl-N'-nitro-N-nitroso-guanidine, nitrosoureas, nitrosamines) or S_N2 mechanisms (for example dialkyl sulfates, methyl methane sulfonate).[118] One important implication of this distinction is that S_N1 agents modify the exocyclic base oxygens more frequently than S_N2 compounds. As O^6-alkylguanine and O^4-alkylthymine are the

most mutagenic lesion generated by alkylating agents, compounds that react by S_N1 mechanisms display higher mutagenic and carcinogenic potencies than those that react by S_N2 mechanisms.[62,118]

Exogenous alkylating agents can be either monofunctional or bifunctional.[116,117] The former have a single reactive group, whereas bifunctional compounds have two reactive groups and are potentially able to react with two different sites, generating intra- or interstrand crosslinks as well as DNA-protein crosslinks. One of the first alkylating agents studied, mustard gas or sulfur mustard, was a chemical warfare agent.[119] The early observation that sulfur mustard inhibits cell division provided the rational basis for the use of such alkylating agents as clinically effective antitumor drugs. Today, several categories of bifunctional alkylating agents (chlorambucil, cyclophosphamide, nitrosourea derivatives, mitomycins or platinum compounds) are widely employed against cancer.

AROMATIC AMINES

Aromatic amines have been widely used in industry (for example as dyestuff) and are common environmental pollutants.[120] Important sources of aromatic amines also include cigarette smoke, diesel exhaust, coal, and cooked food.[121,122] In particular, high temperature cooking processes produce aromatic amines from tryptophan and glutamic acid.[123] Aromatic amines require metabolic activation mainly by the microsomal P450 monooxygenase system to form ultimate carcinogens (sulfate or acetate esters) that are capable of nonenzymatic modification of nucleophilic sites on nucleic acids.[124] Once activated by metabolic processes, aromatic amines form mutagenic and carcinogenic adducts predominantly at the C^8 position of guanine.[125]

Of all DNA adducts generated by aromatic amines, those resulting from 2-aminofluorene and its acetylated N-acetyl-2-aminofluorene derivative have been studied most extensively (Fig. 3.7). These

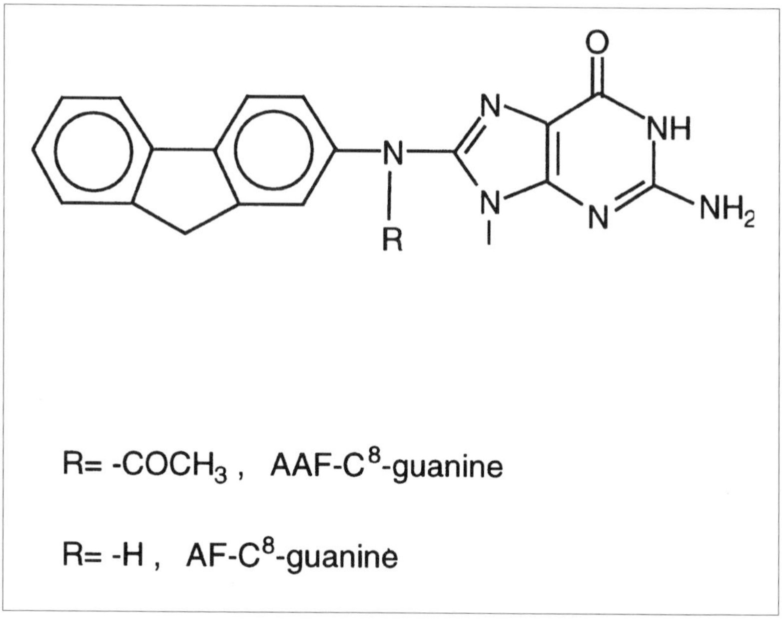

R= -COCH$_3$, AAF-C^8-guanine

R= -H , AF-C^8-guanine

Fig. 3.7. Examples of DNA adducts formed by aromatic amines: acetyl-aminofluorene-C[8]-guanine (AAF-C[8]-guanine) and aminofluorene-C[8]-guanine (AF-C[8]-guanine).

compounds were tested as potential insecticides in the 1940s and were found to induce a variety of tumors in laboratory animals.[126] During the last decades, 2-aminofluorene and *N*-acetyl-2-aminofluorene provided model chemical carcinogens to study the biological responses to DNA damaging agents. After metabolic activation, these carcinogens form C[8]-guanine adducts that are remarkably persistent in mammalian systems[127,128] and induce base substitutions or frame-shift mutations.[129-133]

The major aminofluorene-C[8]-guanine modification is also an attractive prototype lesion to exemplify the dynamic nature of carcinogen-DNA adducts. This particular C[8]-guanine lesion has been shown to adopt at least two different conformations within double-stranded DNA.[134-136] In the first (external) conformation, the fluorene moiety of aminofluorene protrudes out of the major groove with little perturbation of Watson-Crick base pairing. In the second (internal) conformation, the fluorene moiety is inserted into the DNA helix and un-

dergoes extensive stacking interactions with the neighboring bases, thereby disturbing Watson-Crick base pairing. This internal conformer of aminofluorene-C[8]-guanine is characterized by considerable displacement of the modified guanine and its partner cytosine into the minor and the major groove, respectively. Interestingly, the two structurally distinct conformers of aminofluorene-C[8]-guanine have nearly the same overall free energy. About 60% of adducts were found in the external conformation and about 40% in the fluorene-inserted conformation. Also, the two conformers interconvert readily with an estimated chemical exchange lifetime in the millisecond range at 30°C.

The existence of multiple rapidly interchangeable conformations provides a molecular basis to understand the wide range of biological responses associated with most carcinogen-DNA adducts. For example, the external conformation of aminofluorene-C[8]-guanine with its intact base pairing is consistent with the observed ability of DNA polymerases to bypass the lesion and correctly incorporate cytosines across the adduct.[137-139] The internal conformation, on the other hand, is likely to constitute a highly mutagenic substrate because its base pairing geometry is seriously disrupted. Therefore, this aminofluorene-inserted conformation may account for the mutagenic and carcinogenic effects associated with aminofluorene or its derivatives.[126,129-132] Clearly, the conformational heterogeneity of aminofluorene-C^8-guanine adducts may also have profound impacts on their recognition by excision repair processes.

POLYCYCLIC AROMATIC HYDROCARBONS

This class of carcinogens is widely distributed in the environment.[140,141] In 1915, Yamagawa and Ichikawa[142] described the first experimental induction of skin tumors in animals by the application of coal tar to their skin. In the early 1930s, several polycyclic aromatic hydrocarbons were isolated from active carcinogenic fractions of coal tar. Also the first synthetic polycyclic aromatic hydrocarbon was produced and shown to be a potent carcinogen.[143] Polycyclic aromatic hydrocarbons are activated by microsomal P450 enzymes to reactive diol epoxide intermediates that covalently modify DNA bases.

Benzo[a]pyrene is the most extensively studied member of this family of mutagenic electrophiles. Important sources of benzo[a]pyrene include cigarette smoke, automotive exhaust fumes and cooked meat or fish.[144] The metabolic conversion of benzo[a]pyrene to generate ultimate carcinogens results in many stereoisomeric diol epoxides, of which (+)-*anti*-BPDE [(+)-7β,8α-dihydroxy-9α,10α-epoxy-7,8,9,10-tetrahydrobenzo[a]pyrene] is known to be the most mutagenic and tumorigenic in mammalian cells (Fig. 3.8).[145-147] Reactive (+)- and (-)-*anti*-BPDE intermediates form base adducts primarily with the N^2-exocyclic amino group of guanine moieties via either trans or cis opening of the epoxide ring (Fig. 3.8).[148-150] As will be discussed in chapter 7, the stereochemistry of benzo[a]-pyrene diol epoxide reactions with DNA

plays a decisive role in determining the efficiency by which the resulting carcinogen adducts are recognized by mammalian nucleotide excision repair.

NATURAL TOXINS

Plants, fungi and microorganisms contain a plethora of natural compounds that are genotoxic, mutagenic and carcinogenic.[151] The aflatoxins, for example, are produced by several strains of *Aspergillus* that grow on peanuts, corn and other food crops.[152] These compounds are potential contaminants of many farm products that are stored under warm and humid conditions. Aflatoxin B_1 is the most cytotoxic and mutagenic member of the aflatoxin family and one of the most potent hepatocarcinogens in experimental systems. It is also suspected to constitute the etiologic agent in a large fraction of human hepatic cancer. Aflatoxin B_1 is activated by microsomal P450 enzymes to an epoxide intermediate that reacts with DNA, primarily at the position N^7 of guanine.[153-155] The resulting DNA adduct is unstable under physiological conditions and undergoes depurination to produce an abasic site. A fraction of aflatoxin B_1-guanine adducts also leads to opening of the imidazole ring, generating a more stable aflatoxin B_1-2,6-diamino-4-hydroxy-5-formamidopyrimidine derivative.[156]

Psoralens and other furocoumarins are normal constituents of food plants such as celery and parsley.[151] These compounds can intercalate between the base pairs of duplex DNA and, upon photoactivation by long-wavelength UV light, form covalent adducts to pyrimidine bases.[157] Certain psoralens with a totally planar molecular structure are able to react with pyrimidines at either one or both sides to form monoadducts and interstrand crosslinks, respectively. The ability of photoactivated psoralens to block DNA replication has promoted their antiproliferative use in the treatment of human psoriasis, vitiligo, cutaneous T-cell lymphoma and some viral infections.[158] Other examples of natural genotoxins that form highly mutagenic DNA adducts are safrole

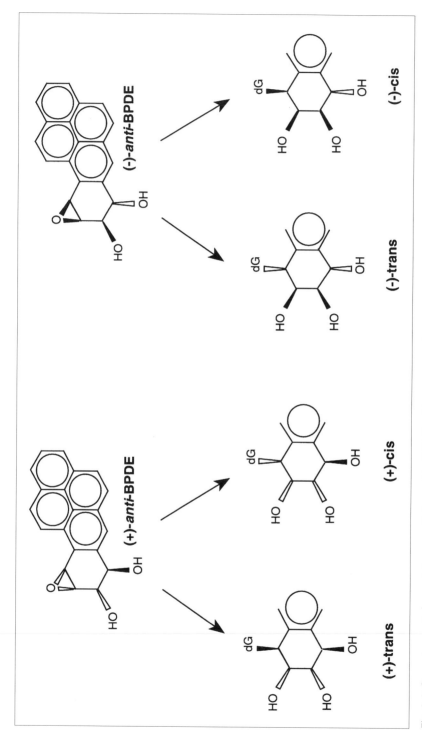

Fig. 3.8. Stereoisomeric benzo[a]pyrene diol epoxide (BPDE) adducts. Stereochemical attack of the reactive epoxides of either (+)-anti-BPDE or (-)-anti-BPDE at the position N^2 of guanine results in two pairs of enantiomeric adducts: (+)-trans-, (+)-cis-, (-)-trans- and (-)-cis-BPDE-N^2-guanine.

and estragole (produced by several plants),[159] mitomycins (produced by *Streptomyces caespitosus*),[160] anthramycin (produced by *Streptomyces refuineus*)[161] or CC-1065 (produced by *Streptomyces zelensis*).[162]

CONCLUSIONS

DNA is an extremely monotonous molecule consisting of only six different building blocks: phosphate, deoxyribose, and the bases adenine, cytosine, guanine and thymine. Despite its highly repetitive chemical structure, DNA is subject to a multitude of lesions arising from genotoxic reactions that can be of endogenous or exogenous sources. In fact, DNA is intrinsically unstable and subject to spontaneous decay. Injuries are additionally inflicted on DNA by ubiquitous genotoxic agents, predominantly oxygen, sunlight and natural carcinogens. The different lesions that can be formed result from the covalent addition of new groups (DNA adducts), the introduction of new bonds [cyclobutane pyrimidine dimers or pyrimidine(6-4)pyrimidone photoproducts] or cleavage of existing bonds (abasic sites or DNA strand breaks). The net result of these DNA damaging reactions is that there are about 10^5 lesions per mammalian cell per day or 10^7 lesions per cell per year.[30,46] If left unrepaired, these DNA lesions exert cytotoxic effects by inhibiting important nuclear functions such as transcription or replication. In addition, loss or disruption of the hydrogen bonding information is responsible for the mutagenic and carcinogenic effects of DNA damage. After briefly reviewing frequent genotoxic reactions (this chapter) and their biological endpoints (chapters 1 and 2), the next sections are focused on the mechanisms by which mammalian cells recognize damaged DNA, thereby initiating appropriate repair mechanisms or other damage processing pathways.

REFERENCES

1. Stryer L. Biochemistry. New York: W.H. Freeman and Company, 1988:71-78.
2. Watson JD, Crick FHC. Molecular structure of nucleic acid. A structure for deoxyribose acid. Nature 1953; 171:737-738.
3. Watson JD, Crick FHC. Genetic implications of the structure of deoxyribonucleic acid. Nature 1953; 171:964-967.
4. Sancar A. Mechanisms of DNA excision repair. Science 1994; 266: 1954-1956
5. Dunderdale HJ, West SC. Recombination genes and proteins. Curr Opin Genet Dev 1994; 4:221-228.
6. Van Houten B. Nucleotide excision repair in *Escherichia coli*. Microbiol Rev 1990; 54:18-51.
7. Lindahl T. Instability and decay of the primary structure of DNA. Nature 1993; 362:709-715.
8. Lindahl T, Nyberg B. Rate of depurination of native deoxyribonucleic acid. Biochemistry 1972; 11:3610-3618.
9. Lindahl T, Andersson A. Rate of chain breakage at apurinic sites in double-stranded deoxyribonucleic acid. Biochemistry 1972; 11:3618-3623.
10. Demple B, Herman T, Chen DS. Cloning and expression of *APE*, the cDNA encoding the major human apurinic endonuclease: definition of a family of DNA repair enzymes. Proc Natl Acad Sci USA 1991; 88:11450-11454.
11. Robson CN, Hickson ID. Isolation of cDNA clones encoding a human apurinic/apyrimidinic endonuclease that corrects DNA repair and mutagenesis defects in *E. coli xth* (exonuclease III) mutants. Nucleic Acids Res 1991; 19:5519-5523.
12. Doetsch PW, Cunningham RP. The enzymology of apurinic/apyrimidinic endonucleases. Mutat Res 1990; 236:173-201.
13. Lindahl T, Nyberg B. Heat-induced deamination of cytosine residues in deoxyribonucleic acid. Biochemistry 1974; 13:3405-3410.
14. Lindahl T. DNA glycosylases, endonucleases for apurinic/apyrimidinic sites, and base excision-repair. Progr Nucleic Acid Res Molec Biol 1979; 22:135-192.
15. Ehrlich M, Zhang X-Y, Inamdar NM. Spontaneous deamination of cytosine and 5-methylcytosine residues in DNA and replacement of 5-methylcytosine residues with cytosine residues. Mutat Res 1990; 238:277-286.
16. Dianov G, Price A, Lindahl T. Generation of single-nucleotide repair patches following excision of uracil residues from

DNA. Mol Cell Biol 1992; 12:1605-1612.

17. Jiricny J. Colon cancer and DNA repair: have mismatches met their match? Trends Genet 1994; 10:164-168.

18. Cooper DN, Youssoufian H. The CpG dinucleotide and human disease. Hum Genet 1988; 78:151-155.

19. Frederico LA, Kunkel TA, Shaw BR. A sensitive genetic assay for the detection of cytosine deamination: determination of rate constants and the activation energy. Biochemistry 1990; 29:2532-2537.

20. Mosbaugh DW, Bennett SE. Uracil-excision DNA repair. Progr Nucleic Acid Res Molec Biol 1994; 48:315-370.

21. Chen H, Shaw BR. Bisulfite induces tandem double CC→TT mutations in double-stranded DNA. 2. Kinetics of cytosine deamination. Biochemistry 1994; 33:4121-4129.

22. Peng W, Shaw BR. Accelerated deamination of cytosine residues in UV-induced cyclobutane pyrimidine dimers leads to CC→TT transitions. Biochemistry 1996; 35:10172-10181.

23. Verri A, Mazzarello P, Biamonti G et al. The specific binding of nuclear protein(s) to the cAMP responsive element (CRE) sequence (TGACGTCA) is reduced by the misincorporation of U and increased by the deamination of C. Nucleic Acids Res 1990; 18:5775-5780.

24. Karran P, Lindahl T. Hypoxanthine in deoxyribonucleic acid: generation by heat-induced hydrolysis of adenine residues and release in free form by a deoxyribonucleic acid glycosylase from calf thymus. Biochemistry 1980; 19:6005-6011.

25. Schuster H. The reaction of nitrous acid with deoxyribonucleic acid. Biochem Biophys Res Commun 1960; 2:320-323.

26. Wink DA, Kasprzak KS, Maragos CM, et al. DNA deamination ability and genotoxicity of nitric oxide and its progenitors. Science 1991; 254:1001-1003.

27. Nguyen T, Brunson D, Crespi CL et al. DNA damage and mutation in human cells exposed to nitric oxide in vitro. Proc Natl Acad Sci USA 1992; 89:3030-3034.

28. Dianov G, Lindahl T. Preferential recognition of I·T base-pairs in the initiation of excision-repair by hypoxanthine-DNA

glycosylase. Nucleic Acids Res 1991; 19:3829-3833.

29. Ames BN, Shinegaga MK, Hagen TM. Oxidants, antioxidants, and the degenerative diseases of aging. Proc Natl Acad Sci USA 1993; 90:7915-7922.

30. Marnett LJ, Burcham PC. Endogenous DNA adducts: potential and paradox. Chem Res Toxicol 1993; 6:771-785.

31. Lee AT, Cerami A. Elevated glucose 6-phosphate levels are associated with plasmid mutations in vivo. Proc Natl Acad Sci USA 1987; 84:8311-8314.

32. Demple B, Harrison L. Repair of oxidative damage to DNA: enzymology and biology. Annu Rev Biochem 1994; 63:915-948.

33. von Sonntag C. The Chemical Basis of Radiation Biology. London: Taylor and Francis Ltd, 1987.

34. Imlay JA, Linn S. DNA damage and oxygen radical toxicity. Science 1988; 240:1302-1309.

35. Ames BN, Gold LS. Endogenous mutagens and the causes of aging and cancer. Mutat Res 1991; 250:3-16.

36. Dizdaroglu M. Oxidative damage to DNA in mammalian chromatin. Mutat Res 1992; 275:331-342.

37. Kasai H, Nishimura S. Hydroxylation of deoxyguanosine at the C-8 position by ascorbic acid and other reducing agents. Nucleic Acids Res. 1984; 12:2137-2145.

38. Shibutani S, Takeshita M, Grollman A. Insertion of specific bases during DNA synthesis past the oxidation-damaged base 8-oxodG. Nature 1991; 349:431-434.

39. Tchou J, Kasai H, Shibutani S et al. 8-Oxoguanine (8-hydroxyguanine) DNA glycosylase and its substrate specificity. Proc Natl Acad Sci USA 1991; 88:4690-4694.

40. Aruoma OI, Halliwell B, Dizdaroglu M. Iron ion-dependent modification of bases in DNA by the superoxide radical generating system hypoxanthine/xanthine oxidase. J Biol Chem 1989; 264:13024-13028.

41. Hariharan PV, Cerutti PA. Formation of products of the 5,6-dihydroxydihydrothymine type by ultraviolet light in HeLa cells. Biochemistry 1977; 16:2791-2795.

42. Boiteux S, O'Connor TR, Laval J. Forma-

midopyrimidine-DNA glycosylase of *Escherichia coli*: cloning and sequencing of the fpg structural gene and overproduction of the protein. EMBO J 1987; 6:3177-3183.

43. Demple B, Linn S. DNA *N*-glycosylases and UV repair. Nature 1980; 287:203-208.

44. Giloni L, Takeshita M, Johnson F et al. Bleomycin-induced strand-scission of DNA. Mechanism of deoxyribose cleavage. J Biol Chem 1981, 256:8608-8615.

45. Henner WD, Grunberg SM, Haseltine WA. Enzyme action at 3' termini of ionizing radiation-induced DNA strand breaks. J Biol Chem 1983; 258:15198-15205.

46. Shapiro R. Damage to DNA caused by hydrolysis. In: Seeberg E, Kleppe K, eds. Chromosome Damage and Repair. New York: Plenum Press, 1981:3-18.

47. West SC. Enzymes and molecular mechanisms of genetic recombination. Annu Rev Biochem 1992; 61:603-640.

48. Chance B, Sies H, Boveris A. Hydroxyperoxide metabolism in mammalian organs. Physiol Rev 1979; 59:527-605.

49. Chen Q, Fischer A, Reagan JD et al. Oxidative DNA damage and senescence of human diploid fibroblast cells. Proc Natl Acad Sci USA 1995; 92:4337-4341.

50. Chanock SJ, El Benna J, Smith RM. The respiratory burst oxidase. J Biol Chem 1994; 269:24519-24522.

51. Baik SC, Youn HS, Chung MH et al. Increased oxidative damage in *Helicobacter pylori*-infected human gastric mucosa. Cancer Res 1996; 56:1279-1282.

52. Kiyosawa H, Suko M, Okudaira H et al. Cigarette smoking induces formation of 8-hydroxydeoxyguanosine, one of the oxidative DNA damages in human peripheral leukocytes. Free Radical Res Comm 1990; 11:23-27.

53. Frei B. Reactive oxygen species and antioxidant vitamins: mechanisms of action. Am J Med 1994; 97 (3A):5S-13S.

54. Reznick AZ, Cross CE, Hu M-L et al. Modification of plasma proteins by cigarette smoke as measured by protein carbonyl formation. Biochem J 1992; 286:607-611.

55. Bui MH, Sauty A, Collet F et al. Dietary vitamin C intake and concentrations in the body fluids and cells of male smokers and nonsmokers. J Nutr 1991; 122:312-316.

56. Kappus H, Sies H. Toxic drug effects associated with oxygen metabolism: redox cycling and lipid peroxidation. Experientia 1981; 37:1233-1241.

57. LaRusso NF, Tomasz M, Müller M et al. Interaction of metronidazole with nucleic acids in vitro. Mol Pharmacol 1977; 13:872-882.

58. Caspary WJ, Niziak C, Lanzo DA et al. Bleomycin A₂: a ferrous oxidase. Mol Pharmacol 1979; 16:256-260.

59. Garthwaite J, Charles SL, Chess-Williams R. Endothelium-derived relaxing factor release on activation of NMDA receptors suggests role as intercellular messenger in the brain. Nature 1988; 336:385-388.

60. Bredt DS, Hwang PM, Snyder SH. Localization of nitric oxide synthase indicating a neural role for nitric oxide. Nature 1990; 347:768-770.

61. Beckman JS, Beckman TW, Chen J et al. Apparent hydroxyl radical production by peroxynitrite: implications for endothelial injury from nitric oxide and superoxide. Proc Natl Acad Sci USA 1990; 87:1620-1624.

62. Singer B, Grunberger D. Molecular Biology of Mutagens and Carcinogens, New York: Plenum Press, 1983.

63. Larson K, Sahm J, Shenkar R et al. Methylation-induced blocks to in vitro DNA replication. Mutat Res 1985; 150:77-84.

64. Saffhill R, Margison GP, O'Connor PJ. Mechanisms of carcinogenesis induced by alkylating agents. Biochim Biophys Acta 1985; 823:111-145.

65. O'Connor TR, Boiteux S, Laval J. Ring-opened 7-methylguanine residues in DNA are a block to *in vitro* DNA synthesis. Nucleic Acids Res 1988; 16:5879-5894.

66. Tudek B, Boiteux S, Laval J. Biological properties of imidazole ring-opened N^7-methylguanine in M13mp18 phage DNA. Nucleic Acids Res 1992; 20:3079-3084.

67. Loveless A. Possible relevance of O^6-alkylation of deoxyguanosine to mutagenicity and carcinogenicity of nitrosamines and nitrosamides. Nature 1969; 233:206-207.

68. Loechler EL, Green CL, Essigmann JM. In vivo mutagenesis by O^6-methylguanine built into a unique site in a viral genome. Proc Natl Acad Sci USA 1984; 81:6271-6275.

69. Singer B, Chavez F, Goodman MF et al. Effect of 3' flanking neighbors on kinetics of pairing of dCTP or dTTP opposite O^6-methylguanine in a defined primed oligonucleotide when *Escherichia coli* DNA polymerase I is used. Proc Natl Acad Sci USA 1989; 86:8271-8274.

70. Preston BD, Singer B, Loeb LA. Mutagenic potential of O^4-methylthymine in vivo determined by an enzymatic approach to site-specific mutagenesis. Proc Natl Acad Sci USA 1986; 83:8501-8505.

71. Preston BD, Singer B, Loeb LA. Comparison of the relative mutagenicities of O-alkylthymines site-specifically incorporated into ϕX174 DNA. J Biol Chem 1987; 262:13821-13827.

72. Kat A, Thilly WG, Fang WH et al. An alkylation-tolerant, mutator human cell line is deficient in strand-specific mismatch repair. Proc Natl Acad Sci USA 1993; 90:6424-6428.

73. Branch P, Aquilina G, Bignami M et al. Defective mismatch binding and a mutator phenotype in cells tolerant to DNA damage. Nature 1993; 362:652-654.

74. Rydberg B, Lindahl T. Nonenzymatic methylation of DNA by the intracellular methyl group donor S-adenosyl-L-methionine is a potentially mutagenic reaction. EMBO J. 1982; 1:211-216.

75. Barrows LR, Magee PN. Nonenzymatic methylation of DNA by S-adenosylmethionine *in vitro*. Carcinogenesis 1982; 3:349-351.

76. Park J-W, Ames BN. 7-Methylguanine adducts in DNA are normally present at high levels and increase on aging: analysis by HPLC with electrochemical detection. Proc Natl Acad Sci USA 1988; 85:7467-7470.

77. Tan BH, Bencsath FA, Gaubatz JW. Steady-state levels of 7-methylguanine increase in nuclear DNA of postmitotic mouse tissues during aging. Mutat Res 1990; 237:229-238.

78. Prevost V, Shuker DE, Friesen MD et al. Immunoaffinity purification and gas-chromatography-mass spectrometric quantification of 3-alkylated adenines in urine: metabolism studies and basal excretion levels in man. Carcinogenesis 1993; 14:199-204.

79. Lawley PD. Methylation of DNA by carcinogens: some applications of chemical analytical methods. In: Montesano R, Bartsch H, Tomatis L, eds. Screening Tests in Chemical Carcinogenesis. Lyon: IARC Scientific Publications, 1976:181-210.

80. Zak P, Kleibl K, Laval F. Repair of O^6-methylguanine and O^4-methylthymine by the human and rat O^6-methylguanine-DNA methyltransferase. J Biol Chem 1994; 269:730-733.

81. Sakumi K, Sekiguchi M. Structures and functions of DNA glycosylases. Mutat Res 1990; 236:161-172.

82. Huang J-C, Hsu DS, Kazanzsev A et al. Substrate spectrum of human excinuclease: repair of abasic sites, methylated bases, mismatches, and bulky adducts. Proc Natl Acad Sci USA 1994; 91:12213-12217.

83. Swenberg JA, Dyroff MC, Bedell MA et al. O^4-Ethyldeoxythymidine, but not O^6-ethyldeoxyguanosine, accumulates in hepatocyte DNA of rats exposed continuously to diethylnitrosamine. Proc Natl Acad Sci USA 1984; 81:1692-1695.

84. Davies RJH. Ultraviolet radiation damage in DNA. Biochem Soc Trans 1995; 23:407-418.

85. Mitchell DL. The relative cytotoxicity of (6-4) photoproducts and cyclobutane dimers in mammalian cells. Photochem Photobiol 1988; 48:51-57.

86. Sage E. Distribution and repair of photolesions in DNA: genetic consequences and the role of sequence context. Photochem Photobiol 1993; 57:163-174.

87. Lubin D, Jensen EH. Effects of clouds and stratospheric ozone depletion on ultraviolet radiation trends. Nature 1995; 377:710-713.

88. Madronich S, de Gruijl FR. Skin cancer and UV radiation. Nature 1993; 366:23.

89. Varghese AJ. Photochemistry of nucleic acids and their constituents. Photophysiology 1972; 7:207-274.

90. Mitchell DL, Jen J, Cleaver JE. Relative induction of cyclobutane dimers and cy-

tosine photohydrates in DNA irradiated in vitro and in vivo with ultraviolet-C and ultraviolet-B light. Photochem Photobiol 1991; 54:741-746.

91. Bose SN, Davies RJ, Sethi SK et al. Formation of an adenine-thymine photoadduct in the deoxydinucleoside monophosphate d(TpA) and in DNA. Science 1983; 220:723-725.

92. Zhao X, Taylor J-S. Mutation spectra of TA*, the major photoproduct of thymidylyl-(3'-5')-deoxyadenosine, in *Escherichia coli* under SOS conditions. Nucleic Acids Res 1996; 24:1561-1565.

93. Kumar S, Joshi PC, Sharma ND et al. Adenine photodimerization in deoxyadenylate sequences: elucidation of the mechanism through structural studies of a major d(ApA) photoproduct. Nucleic Acids Res 1991; 19:2841-2847.

94. Mitchell DL, Nairn RS. The biology of the (6-4) photoproduct. Photochem Photobiol 1989; 49:805-819.

95. Chan GL, Doetsch PW, Haseltine WA. Cyclobutane pyrimidine dimers and (6-4) photoproducts block polymerization by DNA polymerase I. Biochemistry 1985; 24:5723-5728.

96. Gentil A, Le Page F, Lawrence CW et al. Mutagenicity of a unique thymine-thymine dimer or thymine-thymine pyrimidine pyrimidone (6-4) photoproduct in mammalian cells. Nucleic Acids Res 1996; 24:1837-1840.

97. Lebkowski JS, Clancy S, Miller JH et al. The *lacI* shuttle: rapid analysis of the mutagenic specificity of ultraviolet light in human cells. Proc Natl Acad Sci USA 1985; 82:8606-8610.

98. Dumaz N, Drougard C, Sarasin A et al. Specific UV-induced mutation spectrum in the *p53* gene of skin tumors from DNA-repair-deficient xeroderma pigmentosum patients. Proc Natl Acad Sci USA 1993; 90:10529-10533.

99. Reid TM, Loeb LA. Tandem double CC→TT mutations are produced by reactive oxygen species. Proc Natl Acad Sci USA 1993; 90:3904-3907.

100. Kim ST, Sancar A. Photochemistry, photophysics, and mechanism of pyrimidine dimer repair by DNA photolyase. Photochem Photobiol 1993; 57:895-904.

101. Todo T, Takemori H, Ryo H et al. A new photoreactivating enzyme that specifically repairs ultraviolet light-induced (6-4) photoproducts. Nature 1993; 361:371-374.

102. Friedberg EC, King JJ. Endonucleolytic cleavage of UV-irradiated DNA controlled by the V^+ gene in phage T4. Biochem Biophys Res Commun 1969; 37:646-651.

103. Takao M, Yonemasu R, Yamamoto K et al. Characterization of a UV endonuclease gene from the fission yeast *Schizosaccharomyces pombe* and its bacterial homolog. Nucleic Acids Res 1996; 24:1267-1271.

104. Sancar A. DNA excision repair. Annu Rev Biochem 1996; 65:43-81.

105. Bruls WA, Slaper H, van der Leun JC et al. Transmission of human epidermis and stratum corneum as a function of thickness in the ultraviolet and visible wavelengths. Photochem Photobiol 1984; 40:485-494.

106. Tyrell RM, Keyse SM. New trends in photobiology. The interaction of UVA radiation with cultured cells. J Photochem Photobiol B 1990; 4:349-361.

107. Epe B, Pflaum M, Boiteux S. DNA damage induced by photosensitizers in cellular and cell-free systems. Mutat. Res. 1993; 299:135-145.

108. Epe B. Genotoxicity of singlet oxygen. Chem Biol Interactions 1991; 80:239-260.

109. McCormick JP, Fischer JR, Pachlatko JP et al. Characterization of a cell-lethal product from the photooxidation of tryptophan: hydrogen peroxide. Science 1976; 191:468-469.

110. Sterenborg HJCM, van der Leun JC. Tumorigenesis by a long wavelength UV-A source. Photochem Photobiol 1990; 4:349-361.

111. Drobetsky EA, Turcotte J, Châteauneuf A. A role for ultraviolet A in solar mutagenesis. Proc Natl Acad Sci USA 1995; 92:2350-2354.

112. Garland CF, Garland FC, Gorham ED. Rising trends in melanoma. A hypothesis concerning sunscreen effectiveness. Ann Epidemiol 1993; 3:103-110.

113. Wallace SS, Painter RB, eds. Ionizing Radiation Damage to DNA: Molecular Aspects. Proceedings of a Radiation So-

ciety-UCLA Symposia Colloquium. New York: Wiley-Liss, 1990.

114. Randerath K, Reddy MV, Gupta RC. 32P-Labeling test for DNA damage. Proc Natl Acad Sci USA 1981; 78:6126-6129.

115. Perera FP, Hemminki K, Gryzbowska E et al. Molecular and genetic damage in humans from environmental pollution in Poland. Nature 1992; 360:256-258.

116. Lawley PD. Effects of some chemical mutagens and carcinogens on nucleic acids. Prog Nucleic Acid Res Mol Biol 1966; 5:89-131.

117. Pegg AE. DNA repair and carcinogenesis by alkylating agents. In: Cooper CS, Grover PL, eds. Chemical Carcinogenesis and Mutagenesis II. Berlin: Springer-Verlag; 1990:103-131.

118. Lawley PD. Carcinogenesis by alkylating agents. In: Searle CE, ed. Chemical Carcinogenesis, ACS Monograph 182. Washington DC: American Chemical Society, 1984:325-484.

119. Haddow A. The chemical and genetic mechanisms of carcinogenesis. In: Homburger F, Fishman WH, eds. The Physiopathology of Cancer. New York: Hoeber, 1953:475-551.

120. Basu AK, Essigmann JM. DNA damage: structural and functional consequences. In: Vos J-M H, ed. DNA Repair Mechanisms: Impact on Human Diseases and Cancer. Austin: RG Landes, 1995:1-24.

121. Patrianakos C, Hoffmann D. Chemical studies on tobacco smoke. LXIV. On the analysis of aromatic amines in cigarette smoke. J Anal Toxicol 1979; 3:150-154.

122. Rosenkranz HS, McCoy EC, Sanders DR et al. Nitropyrenes: isolation, identification, and reduction of mutagenic impurities in carbon black and toners. Science 1980; 209:1039-1043.

123. Sugimura T. Past, present and future of mutagens in cooked foods. Environ Health Perspect 1986; 67:5-10.

124. Miller JA, Miller EC. Metabolic activation of carcinogenic aromatic amines and amides via N-hydroxylation and N-hydroxy esterification and its relationship to ultimate carcinogens as electrophilic reactants. In: Bergmann ED, Pullman B, eds. The Jerusalem Symposia on Quantum Chemistry and Biochemistry, Physico-chemical Mechanism of Carcinogenesis. Jerusalem: The Israel Academy of Sciences and Humanities, 1969:237-261.

125. Humphreys WG, Kadlubar FF, Guengerich FP. Mechanisms of C^8 alkylation of guanine residues by activated arylamines: evidence for initial adduct formation at the N^7 position. Proc Natl Acad Sci USA 1992; 89:8278-8282.

126. Kriek E. Fifty years of research on N-acetyl-2-aminofluorene, one of the most versatile compounds in experimental cancer research. J Cancer Res Clin Oncol 1992; 118:481-489.

127. Gupta RC, Dighe NR. Formation and removal of DNA adducts in rat liver treated with N-hydroxy derivatives of 2-acetylaminofluorene, 4-acetylaminobiphenyl, and 2-acetylaminophenanthrene. Carcinogenesis 1984; 5:343-349.

128. Poirier MC, Ture B, Laishes BA. Formation and removal of (guan-8-yl)-DNA-2-acetylaminofluorene adducts in liver and kidney of male rats given dietary 2-acetylaminofluorene. Cancer Res 1982; 42:1317-1321.

129. Bichara M, Fuchs RP. DNA binding and mutation spectra of the carcinogen N-2-aminofluorene in Escherichia coli. A correlation between the conformation of the premutagenic lesion and the mutation specificity. J Mol Biol 1985; 183:341-351.

130. Carothers AM, Urlaub G, Mucha J et al. A mutational hot spot induced by N-hydroxy-aminofluorene in dihydrofolate reductase mutants of Chinese hamster ovary cells. Carcinogenesis 1993; 14:2181-2184.

131. Mah MC, Maher VM, Thomas H et al. Mutations induced by aminofluorene-DNA adducts during replication in human cells. Carcinogenesis 1989; 10:2321-2328.

132. Reid TM, Lee M-S, King CM. Mutagenesis by site-specific arylamine adducts in plasmid DNA: enhancing replication of the adducted strand alters mutation frequency. Biochemistry 1990; 29:6153-6161.

133. Shibutani S, Grollman AP. On the mechanism of frameshift (deletion) mutagenesis in vitro. J Biol Chem 1993; 268:11703-11710.

134. Eckel LM, Krugh TR. 2-Aminofluorene modified DNA duplex exists in two in-

terchangeable conformations. Nature Struct Biol 1994; 1:89-94.

135. Eckel LM, Krugh TR. Structural characterization of two interchangeable conformations of a 2-aminofluorene-modified DNA oligomer by NMR and energy minimization. Biochemistry 1994; 33:13611-13624.

136. Cho BP, Beland FA, Marques MM. NMR structural studies of a 15-mer DNA duplex from a ras protooncogene modified with the carcinogen 2-aminofluorene: conformational heterogeneity. Biochemistry 1994; 33:1373-1384.

137. Lutgerink JT, Retel J, Westra JG et al. Bypass of the major aminofluorene-DNA adduct during in vivo replication of single- and double-stranded φX174 DNA treated with N-hydroxy-2-aminofluorene. Carcinogenesis 1985; 6:1501-1506.

138. Michaels ML, Reid TM, King CM et al. Accurate in vitro translesion synthesis by *Escherichia coli* DNA polymerase I (large fragment) on a site-specific, aminofluorene-modified oligonucleotide. Carcinogenesis 1991; 12:1641-1646.

139. van de Poll ML, Lafleur MV, van Gog et al. N-acetylated and deacetylated 4'-fluoro-4-aminobiphenyl and 4-aminobiphenyl adducts differ in their ability to inhibit DNA replication of single-stranded M13 in vitro and of single-stranded φX174 in *Escherichia coli*. Carcinogenesis 1992; 13:751-758.

140. Harvey RG. Polycyclic Aromatic Hydrocarbons: Chemistry and Cancer. Cambridge: Cambridge University Press, 1991.

141. Min Z, Gill RD, Cortez C. Targeted A→T and G→T mutations induced by site-specific deoxyadenosine and deoxyguanosine adducts, respectively, from the (+)-*anti*-diol epoxide of dibenz[a,j]anthracene in M13mp7L2. Biochemistry 1996; 35:4128-4138.

142. Yamagawa K, Ichikawa K. Experimentelle Studie über die Pathogenese der Epithelialgewülste. Mitteilungen Med Fakultät Kaiserl Univ Tokyo 1915; 15:295-344.

143. Miller EC. Some current perspectives on chemical carcinogenesis in humans and experimental animals: Presidential address. Cancer Res 1978; 38:1479-1496.

144. Albert RE, Burns FJ. Carcinogenic atmospheric pollutants and the nature of low-level risks. In: Hiatt HH, Watson JD, Winston JA, eds. Origins of Human Cancer. Cold Spring Harbor: Cold Spring Harbor Laboratory, 1977:289-292.

145. Newbold RF, Brookes P. Exceptional mutagenicity of a benzo[a]pyrene diol epoxide in cultured mammalian cells. Nature 1976; 261:52-54.

146. Buening MK, Wislocki PG, Levin W et al. Tumorigenicity of the optical enantiomers of the diastereomeric benzo[a]pyrene 7,8-diol-9,10-epoxides in newborn mice: exceptional activity of (+)-7β,8α-dihydroxy-9α,10α-epoxy-7,8,9,10-tetrahydrobenzo[a]pyrene. Proc Natl Acad Sci USA 1978; 75:5358-5361.

147. Conney AH. Induction of microsomal enzymes by foreign chemicals and carcinogenesis by polycyclic aromatic hydrocarbons. Cancer Res 1982; 42:4875-4917.

148. Meehan T, Straub K, Calvin M. Benzo[a]pyrene diol epoxide covalently binds to deoxyguanosine and deoxyadenosine in DNA. Nature 1977; 269:725-727.

149. Mehan T, Straub K. Double-stranded DNA stereoselectively binds benzo[a]pyrene diol epoxides. Nature 1979; 277:410-412.

150. Cheng SC, Hilton BD, Roman JM et al. DNA adducts from carcinogenic and noncarcinogenic enantiomers of benzo[a]pyrene dihydrodiol epoxide. Chem. Res. Toxicol. 1989; 2:334-340.

151. Ames BN, Profet M, Gold LS. Dietary pesticides (99.99% all natural). Proc Natl Acad Sci USA 1990; 87:7777-7781.

152. Busby Jr WF, Wogan GN. Aflatoxins. In: Searle C, ed. Chemical Carcinogens. Washington DC: American Chemical Society, 1984:945-1136.

153. Swenson DH, Lin J-K, Miller EC et al. Aflatoxin B₁-2,3-oxide as a probable intermediate in the covalent binding of aflatoxin B₁ and B₂ to rat liver DNA and ribosomal RNA in vivo. Cancer Res 1977; 37:172-181.

154. Martin CN, Garner CN. Aflatoxin B₁-oxide generated by chemical or enzymic

oxidation of aflatoxin B_1 causes guanine substitution in nucleic acids. Nature 1977; 267:863-865.

155. Essigmann JM, Croy RG, Nadzan AM et al. Structural identification of the major DNA adduct formed by aflatoxin B_1 in vitro. Proc Natl Acad Sci USA 1977; 74:1870-1874.

156. Groopman JD, Croy RG, Wogan GN. In vitro reactions of aflatoxin B_1-adducted DNA. Proc Natl Acad Sci USA 1981; 78:5445-5449.

157. Hearst JE, Isaacs ST, Kanne D et al. The reaction of the psoralens with deoxyribonucleic acid. Q Rev Biophys 1984; 17:1-44.

158. Knobler RM, Hönigsmann H, and Edelson RL. Psoralen phototherapies. In:

Gasparro FP, ed. Psoralen DNA Photobiology. Boca Raton: CRC Press, 1988:117-148.

159. Miller JA, Miller EC. Carcinogens occurring naturally in foods. Federation Proc 1976; 35:1316-1321.

160. Carter SK, Crooke ST, Mitomycin C, Current Status and New Developments. New York: Academic Press, 1979.

161. Hurley LH, Petrusek R. Proposed structure of the anthramycin-DNA adduct. Nature 1979; 282:529-531.

162. Boger DL, Johnson DS. CC-1065 and the duocarmycins: unraveling the keys to a new class of naturally derived DNA alkylating agents. Proc Natl Acad Sci USA 1995; 92:3642-3649.

MOLECULAR RECOGNITION STRATEGIES I: ONE ENZYME-ONE SUBSTRATE MOTIFS

To store the genetic information and serve as the genetic link between generations, the nucleotide sequence of DNA must be faithfully maintained despite the numerous physical or chemical insults discussed in the previous chapter. To that end, all organisms from bacteria to mammals are endowed with DNA repair mechanisms, and even some viral genomes carry their own DNA repair enzymes.

DNA suffers from multiple forms of damage including inappropriate bases, base adducts, sites of base loss, DNA-DNA crosslinks, DNA-protein crosslinks, strand breaks, base pair mismatches or insertion-deletion mismatches. The cellular DNA repair pathways that have evolved to counteract this wide range of genetic insults operate by the following basic mechanisms:[1-4]

(1) DNA damage reversal,

(2) excision of single bases, nucleotides or oligonucleotide segments,

(3) recombination.

As a consequence of this mechanistic diversity, mammalian cells depend on many different molecular strategies to recognize and process DNA damage. For example, DNA lesions may be detected directly by virtue of their unique chemical structure, or indirectly by their ability to distort the DNA double helix. After DNA damage recognition, DNA lesions may be simply reversed, leaving the structure of DNA intact, or may be cleaved out of the DNA molecule by excision repair processes. In DNA excision repair, lesions may be excised from DNA either as components of damaged or inappropriate bases (base excision repair), as intact nucleotides (for example mismatch correction) or as components of oligonucleotide segments (nucleotide excision repair). Depending on the particular excision pathway, these DNA metabolizing systems produce long, intermediate or short DNA repair patches, or no DNA repair patches at all.

All these repair processes must be initiated by specialized subunits that sense the presence of abnormal deoxyribonucleotide residues in DNA. Such damage recognition factors exploit alterations in the normal chemistry or conformation of DNA to discriminate between damaged and nondamaged substrates. Thus, recognition of DNA lesions involves specific binding of a protein or protein complex at or near sites of base or backbone anomalies although, in general, damage recognition factors also interact with at least some ele-

Mechanisms of DNA Damage Recognition in Mammalian Cells, by Hanspeter Naegeli.
© 1997 R.G. Landes Company.

ments of nondamaged DNA around the defective site.

BASIC RECOGNITION STRATEGIES

DNA damage recognition is effected by two basic strategies. A first mechanism (reviewed in this chapter) is based on a simple motif of one damage-one enzyme. The monomerization of cyclobutane pyrimidine dimers by photoreactivating enzyme (DNA photolyase)[5] or the excision of uracil bases by uracil-DNA glycosylase[6] provide typical examples for this simple strategy. These enzymes have a limited substrate range because their damage recognition domains interact with modified DNA in a very specific manner by exploiting complementary surfaces. Additionally, these repair proteins and their respective substrates remain in intimate contact throughout the catalytic reaction cycle.[7] In some cases, this binary recognition system is more relaxed, such that the same repair factor recognizes a family of chemically related lesions instead of just a single type of damage. For example, thymine glycol-DNA glycosylases are able to recognize and process a series of chemically related oxidation products derived from pyrimidine bases.[8] Similarly, most N-alkylpurine-DNA glycosylases are active on a family of closely related base alkylation products.[2,4]

A completely different DNA damage recognition strategy is exemplified by nucleotide excision repair and will be reviewed in chapters 5 and 6. In the nucleotide excision mode of DNA repair, a biochemically complex system endowed with multiple DNA binding and DNA metabolizing activities recognizes a broad range of chemically or structurally unrelated forms of base damage. An important prerequisite for this molecular versatility is that nucleotide excision repair processes defective DNA strands at some distance from the site of damage, i.e., it initiates repair by hydrolyzing two phosphodiester bonds, one on either side of the lesion, generating oligonucleotide excision products containing the defective base.[9]

DNA REPAIR BY DIRECT REVERSAL

Direct reversal is the simplest mechanism for repair of DNA lesions. This strategy is used to remove pyrimidine dimers (by DNA photolyases)[5] or O^6-alkylguanines (by O^6-methylguanine-DNA methyltransferases).[10] In both cases, damage is eliminated without removal or replacement of DNA nucleotides. Additional examples for this direct repair mechanism are the reversal of spore photoproducts in *Bacillus subtilis* by SP lyase,[11] and the reversal of anthramycin-DNA adducts by UvrA and UvrB (in the absence of UvrC).[12] In general, direct repair by reversal has a narrow substrate range.

REVERSAL OF O^6-ALKYLGUANINE

The principle of direct DNA damage reversal is important in mammals primarily because it serves to eliminate O^6-alkylguanine residues from DNA and, hence, protects the genome from a class of highly cytotoxic and mutagenic lesions induced by alkylating agents (summarized in chapter 3).

O^6-methylguanine-DNA methyltransferase (MGMT) binds to alkylated DNA and catalyzes the stoichiometric transfer of an alkyl group from O^6-guanine (or O^4-thymine) to a cysteine residue residing within the protein sequence, thereby returning the alkylated base to its nondamaged state.[13-15] The same MGMT polypeptide acts as both transferase and alkyl acceptor. Although this alkyl transfer reaction overcomes a large energy barrier, MGMT is not a true enzyme as it acts only once and the resulting alkylcysteine residue is not converted back to cysteine. The term "suicide enzyme" is frequently used to denote this unique mechanism. Because the cysteine acceptor site is not regenerated, the number of O^6-alkylguanines that can be reversed in a given cell is equal to the number of active MGMT molecules. In addition to its preferred substrate (O^6-methylguanine), MGMT also repairs other alkylguanines such as O^6-ethylguanine and O^6-butylguanine, as well as O^4-alkylthymines, but at a considerably lower efficiency.[16]

RECOGNITION OF O^6-ALKYLGUANINE BY MGMT

The structure of a prokaryotic MGMT (C-terminal domain of the Ada protein from *Escherichia coli*) has been solved by X-ray crystallography.[17] This 178-amino acid polypeptide spans residues 176-314 of the full-length MGMT. The polypeptide shows a high degree of amino acid sequence homology with related domains of other methyltransferases, suggesting that the structural features observed for this C-terminal domain is common to a wide range of O^6-methylguanine-DNA methyltransferases, including mammalian MGMT.

The protein contains a fold that resembles part of RNase H and a helix-turn-helix domain that may mediate binding to DNA. These structural components provide a basis for the known preference of MGMT for duplex DNA over single-stranded DNA, and for the observation that free O^6-methylguanine base derivatives are much poorer substrates than O^6-alkylguanine in double-stranded DNA.[17-19] Surprisingly, the cysteine acceptor site is found buried in the protein and apparently only becomes accessible when MGMT is in contact with DNA. Were the active site cysteine not already known, the obtained crystal structure would suggest that it plays no role in the alkyl transfer chemistry. Thus, MGMT must undergo a significant conformational change upon binding to DNA in order to provide access of O^6-alkylguanine residues to the active site. Moore et al[17] proposed that this conformational change is brought about by rupturing the noncovalent interaction between a glutamic acid side chain and the histidine adjacent to the cysteine acceptor site. Evidence that MGMT protein does in fact undergo major conformational changes upon binding DNA has been provided by studies using circular dichroism and fluorescence anisotropy.[20,21] How exactly MGMT interacts with its substrates remains unknown and additional biophysical investigations are required to understand the mechanism by which MGMT discriminates between nondamaged bases and O^6- or O^4-alkylated

bases. In any case, the available studies indicate that this molecular recognition process is accompanied by significant substrate-dependent conformational changes of the MGMT protein.

PHOTOREACTIVATION

Photoreactivation is one of several mechanism by which cells protect themselves from the formation of cyclobutane pyrimidine dimers or pyrimidine(6-4)pyrimidone photoproducts (see Fig. 3.6 in chapter 3). Photoreactivating enzyme or DNA photolyase utilizes the energy of visible light to reverse cyclobutane pyrimidine dimers into monomers. To accomplish this repair reaction, DNA photolyase binds to pyrimidine dimers and, upon excitation by blue light, splits the cyclobutane ring and restores the native bases. The activity of DNA photolyase may be summarized as follows:[22]

$$\text{Enzyme + Pyrimidine<>Pyrimidine} \xrightarrow{\text{Light (400 nm)}} \text{Enzyme + Pyrimidine-Pyrimidine}$$

The enzyme consists of a single polypeptide and two noncovalently bound prosthetic groups, or chromophores. The first prosthetic group (5,10-methenyltetrahydrofolylpolyglutamate, in some species deazaflavin) is the antenna that captures photons at visible wavelengths. The trapped energy is then transferred to the second cofactor (flavin adenine dinucleotide in the $FADH^-$ form), generating an excited state that is used to cleave the dimers into original monomeric bases.[5,22]

PHOTOREACTIVATION IN MAMMALS

Whether human cells possess an active DNA photolyase is a controversial issue. Interestingly, this enzyme is widespread in nature but is unpredictably missing in many species.[23] For example, the enteric bacterium *Escherichia coli*, which is normally not exposed to UV light, contains photoreactivating enzyme but the soil bacterium *Bacillus subtilis*, which is frequently exposed to sunlight, exhibits no photoreactivating activity. Also, the budding yeast *Saccharomyces cerevisiae* is subject to photoreactivation but the fission yeast *Schizosaccharomyces pombe* is not.[23] When mammalian cells were tested, early results indicated that marsupials are capable of photoreactivation, while placental animals are not (reviewed in Ref. 24). In species which do express active DNA photolyase, no correlation exists between the level of this enzyme in various tissues and the potential for formation of UV damage in these locations. For example, it has been reported that brain displays the highest photolyase activity of all tissues examined in chicken or opossum.[25] A possible function of DNA photolyase in organisms or tissues with essentially no cyclobutane pyrimidine dimer or (6-4) photoproduct formation is discussed in chapter 8.

Sancar and coworkers used defined DNA substrates containing a single cyclobutane pyrimidine dimer to confirm the presence of photoreactivating enzyme in *Escherichia coli*, *Saccharomyces cerevisiae*, and the rattlesnake *Crotalus horridus*.[23] However, they were unable to detect photolyase activity in extracts prepared from cultured human cells or from white blood cells collected from volunteers. On the basis of these results, they concluded that human cells possess no DNA photoreactivating enzyme in an active form. More recently, Sutherland and Bennett[26] provided new evidence supporting the presence of DNA photolyase activity in extracts from human white blood cells. Their assay involved incubation of UV-irradiated plasmid or bacteriophage λ DNA with a human cell-free extract, followed by analysis of the reaction products using *Micrococcus luteus* pyrimidine dimer-DNA glycosylase. Under their conditions, human cell extracts promoted cyclobutane pyrimidine dimer removal but only when the reaction mixtures were incubated under visible light. These results were confirmed using linear DNA fragments containing a site-specific cyclobutane pyrimidine dimer, and in vivo studies with human white blood cells supported the existence of an active photoreactivating enzyme operating on genomic DNA.[26] Sutherland and Bennett argue that the nega-

tive results obtained in other comparable studies may be due to the exquisite sensitivity of human photolyase to freezing during storage. They also pointed out that the putative human DNA photolyase activity is readily suppressed by high ionic strength conditions in vitro.

RECOGNITION OF CYCLOBUTANE DIMERS BY DNA PHOTOLYASE

Escherichia coli DNA photolyase binds to its target (cyclobutane pyrimidine dimers) both in double-stranded and single-stranded DNA. The dissociation constants are approximately $K_D = 10^{-9}$ M and $K_D = 10^{-8}$ M for double-stranded DNA and single-stranded DNA containing a cyclobutane dimer, respectively.[24,27,28] Photolyase also binds to cyclobutane pyrimidine dimers in the dinucleotide form, but with considerably lower affinity ($K_D = 10^{-4}$ M).[29] These quantitative binding studies indicate that DNA photolyase interacts with both the pyrimidine dimer and the surrounding native DNA sequence.

The structural basis for photolyase-DNA recognition is provided by X-ray crystallography.[5] The three-dimensional structure of DNA photolyase from *Escherichia coli* revealed a flat surface that promotes electrostatic interactions with phosphate residues of the DNA backbone. Mutational analysis confirmed that at least three arginine side chains in this flat domain are required for substrate binding.[5,30] The flat DNA binding surface is interrupted by a hole that has precisely the right dimensions to accommodate a single cyclobutane pyrimidine dimer. In addition, the residues lining this hole are hydrophobic on one side and polar on the other. It was noted that this asymmetry fits exactly with the polarity of cyclobutane pyrimidine dimers, as the cyclobutyl ring is hydrophobic but the opposite edges of the modified bases display polar groups capable of forming hydrogen bonds. Taken together, these structural features suggest that the cyclobutane pyrimidine dimer is brought near the protein active site by insertion into this hole.

This hypothesis involving insertion of the modified bases into a narrow active site hole of photoreactivating enzyme implies that the cyclobutane dimer is flipped out of the DNA double helix. Perhaps, extrahelical displacement of the damaged bases is facilitated by the resulting destabilization of the hydrogen bonding interactions between complementary bases upon cyclobutane dimer formation.[5] Importantly, the proposed model of substrate binding also implies that the physical properties of the active site hole (its dimensions and polarity) exclude nondamaged bases from entering the active site and being processed.

An activity related to cyclobutane dimer photoreactivation has been characterized in extracts from *Drosophila melanogaster* and the silkworm *Bombyx mori*. These organisms contain a photoreactivating enzyme that is capable of using visible light to reverse the formation of pyrimidine(6-4)-pyrimidone photoproducts.[31] The gene encoding the *Drosophila* (6-4) reactivating enzyme has been cloned,[32] but the molecular details of this novel repair mechanism and the strategy by which it recognizes its substrate remain to be elucidated.

BASE EXCISION REPAIR

Characteristic of excision repair is the removal of DNA damage by mechanisms that involve endonucleolytic cleavage of the deoxyribose-phosphate backbone (see Fig. 1.2). The correct nucleotide sequence and DNA continuity are restored by DNA synthesis and DNA ligation. Excision repair may be effected by removing the offending lesion as a base (base excision repair) or as a nucleotide, usually as a component of an oligonucleotide fragment (nucleotide excision repair). Nucleotide excision repair appears to be more active on bulky adducts, whereas base excision repair eliminates preferentially more subtle lesions that do not greatly distort the DNA double helix, although there is substantial overlap in the substrate ranges of these two repair systems.[1,4]

The process of base excision repair is initiated by a class of enzymes, termed DNA glycosylases, that hydrolyze the base-sugar *N*-glycosylic bond of modified, inappropriate or incorrect bases.[1,2,4] The reaction catalyzed by DNA glycosylases generates abasic (apurinic-apyrimidinic, AP) deoxyribose residues.

According to a widely accepted scheme,[33] base excision repair functions by five sequential steps (Fig. 4.1):

(1) removal of the targeted base from the sugar-phosphate backbone by a specific *N*-glycosylase to leave an AP site;

(2) incision of the AP residue on its 5' side by an AP endonuclease that hydrolyzes the phosphodiester bond, generating 5'-deoxyribose phosphate and 3'-hydroxyl termini;

(3) excision of the 5'-terminal deoxyribose phosphate by deoxyribose phosphatase to produce a single-nucleotide gap;

(4) DNA synthesis by DNA polymerase β to fill the gap;

(5) sealing of the remaining nick by a DNA ligase.

DNA GLYCOSYLASES

As indicated before, initiation of base excision repair depends on a family of specific glycosylases, each targeted to a defined type of damaged or inappropriate base. Another category of DNA glycosylases are those which recognize incorrect bases in the context of a particular mismatch (for example G·T base pairing).[35] As DNA glycosylases have a narrow substrate range, living cells must express multiple such enzymes. They are normally designated according to their specific targets. For example, uracil-DNA glycosylases initiate base excision repair of uracil residues in DNA and pyrimidine dimer-DNA glycosylases are directed to cyclobutane pyrimidine dimers. These small monomeric enzymes of typically 20-30 kDa perform several functions, i.e., binding to DNA, recognition of the target base and cleavage of the *N*-glycosyl bond to produce AP sites without any requirement for specific cofactors. Such base excision activities catalyzed by DNA glycosylases are found throughout phylogeny.[7,36,37]

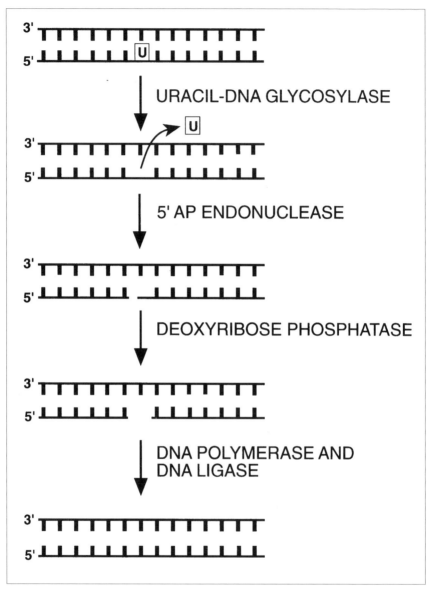

Fig. 4.1. Example of base excision repair: excision of uracil from DNA initiated by uracil-DNA glycosylase.

There are at least two types of DNA glycosylases, distinguished by their mechanism of action. One series, illustrated by the first DNA glycosylase activity that was discovered (uracil-DNA glycosylase), includes those enzymes that are limited in their activity to the cleavage of *N*-glycosylic bonds.[38,39] The resulting abasic site is then incised by a 5' AP endonuclease. The second type of DNA glycosylase includes bifunctional enzymes that carry out the glycosylase reaction and also catalyze subsequent AP lyase activity at a rate approximately equal to their glycosylase activity. This additional AP lyase function is effected by the same polypeptide and leads to polynucleotide strand scission 3' to AP sites by a β-elimination mechanism, leaving 3'-terminal unsaturated deoxyribose derivatives. Examples for this class of enzymes include formamidopyrimidine-DNA glycosylase,[40] thymine glycol-DNA glycosylase[41] and pyrimidine dimer-DNA glycosylase.[42] The AP lyase activity of these enzymes appears to be of limited biological significance.[4] Before elaborating on the mechanism of base damage recognition by various DNA glycosylases, the next two short sections will briefly describe the mechanistic steps that follow formation of AP sites in base excision repair.

HYDROLYTIC AP ENDONUCLEASES

Apurinic-apyrimidinic (AP) sites are potentially lethal and mutagenic.[43] As a major defense system against these lesions, cells possess AP endonucleases that cleave on the 5' side of AP sites by a hydrolytic mechanism to produce 3'-hydroxyl groups and 5'-deoxyribose phosphate residues (Fig. 4.1).[1,4,7] Hydrolytic AP endonucleases include *Escherichia coli* exonuclease III and the predominant human AP endonuclease designated Ape protein.[44] The same human enzyme is also called Hap1,[45] Apex[46] or Ref1 protein.[47] These AP endonucleases perform two functions. First, they eliminate AP sites generated by DNA glycosylases or by spontaneous base loss. Second, they process the 3' ends of strand breaks generated by free radicals. Such breaks often have 3'-phosphoglycolate or 3'-phosphate residues and these modified termini are trimmed by hydrolytic AP endonucleases to generate 3'-hydroxyl ends that are suitable for elongation by DNA polymerases.[48]

SUBSEQUENT ENZYMES IN THE BASE EXCISION REPAIR PATHWAY

As indicated, DNA incision by hydrolytic AP endonucleases leaves 3'-hydroxyl ends that constitute appropriate primer termini for gap filling by DNA polymerases. However, the 5'-deoxyribose phosphate residue must be removed before base excision re-

pair can be completed (Fig. 4.1). In prokaryotes, two proteins (Fpg and RecJ) have been reported to possess deoxyribose phosphatase activity.[49,50] In mammalian cells this deoxyribose phosphatase function may reside with DNA polymerase β. Recent reports demonstrated that polymerase β catalyzes DNA synthesis during base excision repair and, in this specialized function, cannot be replaced by DNA polymerase α, DNA polymerase δ or DNA polymerase ε.[51] Additionally, DNA polymerase β also catalyzes the release of 5'-terminal deoxyribose phosphate residues from incised AP sites.[52] This physical association of two enzymatic activities within the same DNA polymerase β polypeptide may enhance the efficiency of repair and minimize the possibility that incised reaction intermediates become a substrate for nonspecific nucleases or aberrant recombinational mechanisms.

DNA polymerase β leaves a single strand nick that remains to be ligated. This final step of the base excision repair pathway seems to be effected primarily by DNA ligase I.[53] Alternatively, DNA ligase III has been implicated in base excision repair. This enzyme forms a tight complex with XRCC1,[54] a protein that participates in the cellular DNA repair response to alkylating agents and ionizing radiation. The concerted action of these base excision repair enzymes result in narrow repair patches that may be as short as a single nucleotide (Fig. 4.1).[33]

URACIL-DNA GLYCOSYLASE

As outlined in chapter 3, uracil is produced in DNA by deamination of cytosine or by misincorporation of dUMP during replication. Uracil-DNA glycosylases are highly specialized but small enzymes that carry out the first catalytic step in the base excision repair of uracil residues in DNA. This enzymatic activity has been identified in virtually all prokaryotic and eukaryotic organisms. In addition, uracil-DNA glycosylases are encoded by viruses of the pox and herpes families.[55-58] These enzymes are extraordinarily specific, as they are able to discriminate between uracil, which is excised, and thymine, which differs from uracil by a single methyl group and is not excised (see Fig. 3.2 in chapter 3). Of course, uracil-DNA glycosylases also discriminate between uracil and cytosine, which differ only in the hydrogen bonding pattern at the C^4 and N^3 positions of the pyrimidine ring. Also, uracil-DNA glycosylases discriminate between uracil residues in DNA and RNA. The limited substrate range of uracil-DNA glycosylases includes, in addition to uracil, a few uracil derivatives such as isodialuric acid, 5-hydroxyuracil or alloxan that are generated from cytosine by oxidative processes.[59-61]

RECOGNITION OF URACIL BY URACIL-DNA GLYCOSYLASES

Biophysical analysis of the human[6] and virally encoded uracil-DNA glycosylase from herpes simplex type-1 (HSV-1)[62] revealed the molecular basis for the unique substrate specificity of these enzymes. The human enzyme was cocrystallized with an inhibitor, 6-aminouracil, bound in its active site. The HSV-1 enzyme, on the other hand, was either cocrystallized with a thymine trinucleotide or, alternatively, the protein crystals were soaked with free uracil to identify the active site. The resulting X-ray structures support a mechanism in which uracil-DNA glycosylase binds double-stranded DNA in a positively charged groove, and causes the target uracil base to swing out into an extrahelical position to fit into the active site of the enzyme. The protein does not appear to undergo major conformational changes upon binding its substrate.

The two X-ray studies identified a DNA binding groove on the surface of uracil-DNA glycosylase that has exactly the correct shape, size (diameter of 2.1 nm) and charge (basic amino acids) to accommodate duplex DNA. At the base of this groove, the uracil moiety is inserted into a rigid, preformed pocket. Uracil fits tightly into this narrow active site pocket and undergoes a set of hydrogen bonding interactions that involve Watson-Crick base-pairing groups and are unique to uracil (Fig. 4.2). This selective recognition mode excludes all other natural bases from the active site pocket. The bulkier adenine and guanine are excluded because their size does not fit into the rigid pocket. Thymine has identical dimensions and base pairing groups as uracil, but is excluded because its 5-methyl group clashes with the aromatic side chain of a tyrosine residue lining the recognition pocket. Interestingly, replacement of this tyrosine by alanine, cysteine or serine resulted in an enzyme that has thymine-DNA glycosylase activity.[63] Cytosine is excluded because its hydrogen bonding donor and acceptor sites are not compatible with binding into the uracil recognition pocket. However, the simple replacement of a single active site asparagine residue by aspartic acid produced an enzyme with cytosine-DNA glycosylase activity.[63] Effective discrimination against uracil bases located on RNA is thought to occur because the C2' hydroxyl group of the ribose moiety in RNA would clash with bulky aromatic side chains in the protein.[6,62]

The proposed model of recognition by uracil-DNA glycosylase assumes that the target base is flipped out from the DNA helix to reach the binding pocket within the active site.[64] Extrahelical dislocation of a substrate base has been observed before in the crystal structure of *Hha*I-methyltransferase in complex with its cognate DNA. Also, insertion of an extrahelical uracil into an active site pocket of the enzyme is consistent with several previous biochemical observations. First, uracil is excised three times faster from single-stranded DNA than from double-stranded DNA.[65] Second, the DNA sequence surrounding the uracil is expected to influence the rate of cleavage because uracil accessibility to the active site pocket of the enzyme may be reduced by stabilizing the duplex with increasing GC content. In fact, the rate of uracil excision decreases with the following order of nucleotide sequences: 5'-CAUAA > 5'-CGUAA > 5'-TGUGA > 5'-GGUGG.[65] Third, uracil DNA glycosylases excise uracil preferentially from U·G mispairs relative to the thermodynamically more stable U·A pairs,[66] indicating that extrahelical displacement of

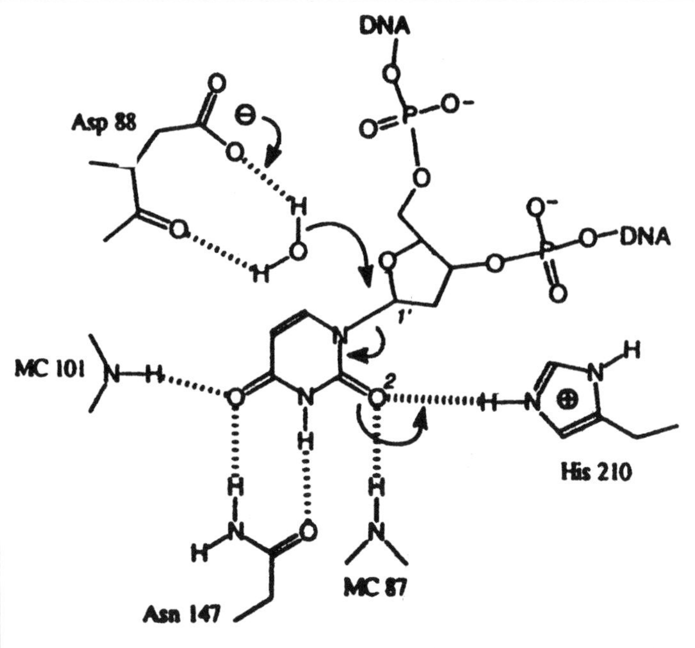

Fig. 4.2. Active site pocket of HSV-1 uracil-DNA glycosylase. The uracil makes specific hydrogen bonds with the side chain of Asn-147 and with the main chain atoms of Gln-87 and Phe-101. The arrows indicate the catalytic mechanism of uracil excision, which is initiated with protonation of O^2 by His-210 and concomitant hydrolytic attack of the *N*-glycosyl bond. Reprinted with permission from: Savva R, McAuley Hecht K, Brown T et al. The structural basis of specific base-excision repair by uracil-DNA glycosylase. Nature 1995; 373:487-493. © 1995 Macmillan Magazines Limited.

uracil from duplex DNA may constitute a rate-limiting step when the enzyme functions on double-stranded substrates. Careful inspection of the environment in the active site indicates that a leucine residue may facilitate this base displacement step by inserting its hydrophobic side chain between the bases.[64] A similar role for a leucine residue was previously found in glutaminyl-tRNA synthetase complexed with tRNA.[67]

PYRIMIDINE DIMER-DNA GLYCOSYLASE

T4 endonuclease V is a DNA repair enzyme encoded by the bacteriophage T4 *denV* gene. It was the first DNA repair enzyme ever to be isolated. Today, it is widely used as an enzymatic probe to detect and quan-tify UV radiation damage in DNA. Despite its small size (16 kDa), T4 endonuclease V has two distinct catalytic functions, i.e., a cyclobutane pyrimidine dimer-specific DNA glycosylase and an associated AP lyase activity. These enzymatic functions are used to initiate excision repair of cyclobutane pyrimidine dimers.[68-70] Specifically, T4 endonuclease V catalyzes the cleavage of the *N*-glycosyl bond on the 5' side of cyclobutane pyrimidine dimers, and subsequently incises the phosphodiester bond at the resulting apyrimidinic site through a β-elimination reaction.[71] A similar activity is also found in *Micrococcus luteus* and perhaps *Saccharomyces cerevisiae*,[72] but not in mammalian cells.[1,2,4] Knowledge of how this enzyme recognizes its substrate may nevertheless be useful to delineate possible mechanisms of cyclobutane pyrimidine dimer recognition in mammalian cells.

RECOGNITION OF CYCLOBUTANE PYRIMIDINE DIMERS BY T4 ENDONUCLEASE V

T4 endonuclease V has been extensively studied by biochemical, molecular and crystallographic methods.[73-78] The recently reported cocrystal structure of T4 endonuclease V in complex with a DNA duplex containing a cyclobutane pyrimidine dimer revealed the following striking features.[79]

(1) The enzyme interacts with the cyclobutane dimer in the minor groove of the double helix, as predicted from previous results based on phosphate-modified substrates or DNA methylation analysis of enzyme-DNA complexes.[80] This interaction in the minor groove accounts, in part, for the observation that T4 endonuclease V binds to cyclobutane pyrimidine dimers in double-stranded DNA with an affinity that is at least 1,000-times higher compared to pyrimidine dimers in single-stranded DNA.[81]

(2) When bound to the enzyme, the DNA duplex is sharply kinked by 60° (Fig. 4.3).[79] This considerable distortion appears to be protein-induced, as nuclear magnetic resonance analysis demonstrated that the normal B-DNA conformation remains largely intact at sites of cyclobutane pyrimidine dimer formation.[82,83] However, the weakened stacking and hydrogen bonding interactions at cyclobutane dimers may facilitate protein-induced deformation. Alternatively, one may also propose that the kink is already present in the absence of the enzyme as a minor conformer, and that upon interaction with the protein all the substrate may be driven into this kinked conformation.

(3) No evidence for an extrahelical cyclobutane pyrimidine dimer, analogous to the mechanism proposed for uracil-DNA glycosylase, has been found.[79] Instead, the nonmodified adenine base opposite the 5' side base of the dimer is completely rotated

out of the helix and trapped in a cavity located on the enzyme surface (Fig. 4.3). This conformational change generates a large hole in the double helix that is occupied by catalytically active side chains of T4 endonuclease V. Thus, the flipping out of a nondamaged complementary base enables the active amino acid residues to gain access to the lesion. Interestingly, the extrahelical localization of the nonmodified base within the cavity of the enzyme is stabilized by nonspecific van der Waal's interactions and, in principle, any other aromatic base should be accommodated in a similar manner. This structural observation is consistent with the finding that replacement of the complementary adenine by guanine, cytosine or thymine does not affect T4 endonuclease V activity.[79]

(4) Surprisingly, the enzyme does not form intimate contacts with the crosslinked bases of the cyclobutane pyrimidine dimer, as was observed in the case of photoreactivating enzyme (see previous section on DNA damage reversal). Instead, Vassylyev et al[79] proposed that the enzyme may recognize a specific protein-induced distortion of the deoxyribose-phosphate backbone in the vicinity of the cyclobutane pyrimidine dimer. A possibly unique feature of this deformed backbone is the spacing between the two phosphates of the dimer, which is shortened by 1.5 Å relative to normal B-DNA (Fig. 4.3). Also, recognition of backbone deformation is likely to involve hydrogen bonds between the two phosphates of the dimer and Arg-3, His-16, Arg-117 and Lys-121 of the glycosylase. The importance of this cluster of basic amino acid residues in substrate binding is supported by mutational analysis.[76]

In summary, these studies on T4 endonuclease V illustrate the principle of "indirect readout" by which conformational backbone distortion rather than the base damage itself is used for recognition. Although this enzyme is found only in phage T4-infected bacteria, its mode of recognition by "indirect readout" is of broader significance as it may also be exploited by DNA

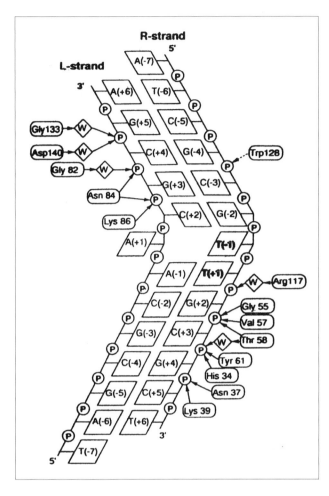

Fig. 4.3. Schematic drawing showing the DNA substrate deformation induced by T4 endonuclease V. The thymine-thymine cyclobutane dimer, marked by the open letters T(+1) and T(-1), is located in the center of a short DNA duplex. Upon interaction with T4 endonuclease V, the double helix is kinked by 60° and the adenine A(+1) opposite to the 5' base of the cyclobutane dimer is forced into an extrahelical localization. Reprinted with permission from: Vassylyev DG, Kashiwagi T, Mikami Y et al. Atomic model of a pyrimidine dimer excision repair enzyme complexed with a DNA substrate: structural basis for damaged DNA recognition. Cell 1995; 83:773-782. © 1995 Cell Press.

repair systems operating in mammalian cells (for example thymine glycol-DNA glycosylase or, perhaps, nucleotide excision repair).

THYMINE GLYCOL-DNA GLYCOSYLASE (ENDONUCLEASE III)

Endonuclease III, encoded by the *nth* gene of *Escherichia coli*, is the prototype thymine glycol-DNA glycosylase (see Fig. 3.4 for the chemical structure of thymine glycol). Despite its misleading name, endonuclease III is a DNA glycosylase with associated AP lyase activity. This enzyme is characterized by a rather relaxed substrate selectivity and initiates base excision repair of numerous forms of ring-saturated, ring-rearranged and ring-contracted pyrimidines in DNA, among which thymine glycol is the paradigm substrate. These thymine or cytosine lesions are generated upon exposure of DNA to UV radiation, ionizing radiation, active oxygen species, osmium tetroxide or acid pH.[8,43,84-89] The associated lyase activity of endonuclease III additionally cleaves the DNA phosphodiester backbone at the resulting apyrimidinic sites via a β-elimination reaction.[90-92] The enzymatic function exerted by endonuclease III is conserved from bacteria to eukaryotes, and enzymes with similar substrate specificity as *Escherichia coli* endonuclease III have also been found in mammalian cell extracts.[8]

RECOGNITION OF PYRIMIDINE LESIONS BY ENDONUCLEASE III

The crystal structure of endonuclease III, in combination with mutational analysis, disclosed the presence in this enzyme of two separate DNA binding motifs designated helix-hairpin-helix and iron-sulfur [4Fe-4S] cluster loop.[93-95] The helix-hairpin-helix structure is similar to the previously known helix-turn-helix DNA binding domain (Fig. 4.4).[96] The iron-sulfur cluster loop is a novel DNA binding domain that involves four cysteine residues in the sequence Cys-X_6-Cys-X_2-Cys-X_5-Cys anchoring an iron-sulfur [4Fe-4S] complex (Fig. 4.5). Between these two structural domains, the enzyme possesses a deep active site pocket that is lined by polar amino acids and is filled with water molecules. Also, the active site pocket is located in the proximity of the catalytically important amino acid side chains Lys-120 and Asp-138.

On the basis of these structural findings, Thayer et al[95] proposed a model for substrate recognition that involves binding of double-stranded DNA by the helix-hairpin-helix and iron-sulfur cluster loop domains. An important aspect of their model is that the target base becomes dislocated into an extrahelical position and is protruded into the solvent-filled active site pocket. This putative endonuclease III-substrate DNA complex is reminiscent of the crystallographic structure of *HhaI*-methyltransferase in association with its cognate DNA.[64] As mentioned before, flipping out of the target base has also been postulated for uracil-DNA glycosylase to explain how uracil residues may be accommodated into a rigid active site pocket.

Despite this apparent similarity in their mode of substrate binding, there is an important difference between uracil-DNA glycosylase and endonuclease III. As pointed out before, endonuclease III belongs to a class of DNA glycosylases that recognize a relatively broad range of modified bases. Thus, a recognition mechanism based on a rigid pocket, such as that found in the uracil-DNA glycosylase structure, would not be appropriate. In the literature, two distinct mechanisms have been proposed to accommodate the relaxed substrate specificity of endonuclease III. According to Thayer et al,[95] a large number of different thymine and cytosine derivatives may be accommodated within the active site pocket by the variable replacement of water molecules, thereby mediating specific interactions with a family of related target bases. This model should be further elaborated to understand how nondamaged pyrimidine bases are excluded from binding into the same active site pocket. Alternatively, it has been proposed that endonuclease III may recognize its target bases by "indirect readout", as proposed for T4 endonuclease V. This hypothesis is supported by footprinting data suggesting that endonuclease III binds DNA by interacting with the sugar-phosphate backbone.[97] Of course, these two putative recognition mechanisms must not be mutually exclusive.

Whether mammalian enzymes with thymine glycol-DNA glycosylase activity are structurally and functionally related to endonuclease III remains to be unequivocally established. Hilbert et al[98] were able to isolate a 31 kDa protein from calf thymus which operates through a *N*-acylimine (Shiff base) enzyme-substrate intermediate and, hence, displays exactly the same reaction mechanism as endonuclease III. Four different peptides derived from a proteolytic digest of this 31-kDa species were microsequenced and aligned with the amino acid sequences of endonuclease III and a homologous protein from *Caenorhabditis elegans*. This comparison indicated significant sequence homologies that extended into the iron-sulfur cluster loop region of these enzymes. The same authors also found that several 3' complementary DNA sequences from *Homo sapiens* and *Rattus sp.* that have been submitted to databases show a striking homology with the iron-sulfur cluster loop of endonuclease III.[98] Thus, there might a family of endonuclease III-like DNA repair glycosylases throughout phylogeny that share a common mode of DNA binding involving this iron-sulfur cluster loop domain.

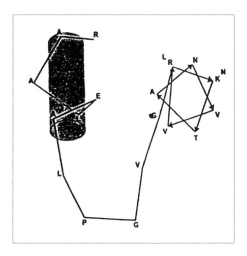

Fig. 4.4. A schematic diagram showing the helix-hairpin-helix motif of endonuclease III. The residues are indicated by one letter amino acid codes. Reprinted with permission from: Thayer MM, Ahern H, Xing D et al. Novel DNA binding motifs in the DNA repair enzyme endonuclease III crystal structure. EMBO J 1995; 14:4108-4120. © 1995 Oxford University Press.

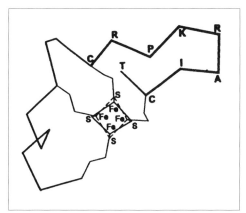

Fig. 4.5. Scheme of the iron-sulfur cluster loop found in endonuclease III. Fe, iron atoms; S, sulfur atoms. In this DNA binding domain, the iron-sulfur cluster is anchored by four cysteine residues in the sequence Cys-X_6-Cys-X_2-Cys-X_5-Cys. The cysteine side chains are indicated by thin lines. Reprinted with permission from: Thayer MM, Ahern H, Xing D et al. Novel DNA binding motifs in the DNA repair enzyme endonuclease III crystal structure. EMBO J 1995; 14:4108-4120. © 1995 Oxford University Press.

N-ALKYLPURINE-DNA GLYCOSYLASES

N-alkylpurine-DNA glycosylases provide another example of base excision repair enzymes with a rather broad substrate selectivity. For example, the *Escherichia coli* AlkA protein excises 3-methyladenine, 7-methylguanine and several minor alkylation products including those involving pyrimidine bases (O^2-methylcytosine, O^2-methylthymine).[99-101] *N*-alkylpurine-DNA glycosylases from eukaryotic sources have a broad substrate specificity similar to AlkA,[101,102] supporting the view that the bacterial enzyme constitutes a valuable model system for excision repair of alkylated bases in higher organisms. In contrast, a second *N*-alkylpurine-DNA glycosylase of *Escherichia coli* (Tag protein) has been found to excise only 3-methyladenine residues.[103]

RECOGNITION OF METHYLATED BASES BY ALKA PROTEIN

The crystal structure of AlkA protein revealed several structural motifs for DNA binding. AlkA is composed of three domains: an α-β sandwich structure (domain I) and two α-helical structures (domains II and III).[104,105] Domain I consists of a five-stranded antiparallel β sheet accompanied by two α helices. This folding scheme is classified as an α-β sandwich structure and is topologically identical to one half of the

DNA binding domain of the TATA-binding protein. Surprisingly, domains II and III of AlkA have a folding topology that is essentially identical to parts of endonuclease III, although the two enzymes share very little amino acid sequence homology. This region of the protein includes the previously described helix-hairpin-helix DNA binding motif but, in contrast to endonuclease III, AlkA has no iron-sulfur cluster loop.

Between the two α-helical domains II and III, AlkA protein contains a striking hydrophobic cleft lined with the electron-rich aromatic side chains of Phe-18, Trp-218, Tyr-222, Trp-272 and Tyr-273.[104,105] A common property of the alkylated bases processed by AlkA is the presence of a positive charge which is distributed throughout the entire π-system of the base, rendering it electron-deficient (see for example N^7-methylguanine and N^3-methyladenine in Fig. 3.5). Such electron-deficient aromatic rings are known to form strong noncovalent stacking interactions with electron-rich aromates.[106] Thus, the aromatic cleft in AlkA could serve as a binding pocket for electron-deficient alkylated bases. This model assumes that AlkA exploits π-electron complementarity to recognize its specific substrates. AlkA is capable of a hinge-like motion between the different domains, thereby conferring flexibility to the aromatic cleft.[104,105] Widening or narrowing of this recognition cleft may serve to accommodate diverse forms of electron-deficient alkylated bases, including both pyrimidine and the bulkier purine derivatives. Finally, recognition of alkylated bases in the aromatic cleft implies flipping out of the target base from DNA duplexes, as has been observed with DNA methyltransferase and uracil-DNA glycosylase,[64] or postulated for endonuclease III.[95]

FORMAMIDOPYRIMIDINE-DNA GLYCOSYLASE

A DNA glycosylase that catalyzes the excision of purines containing an open imidazole ring (formamidopyrimidine derivatives, Fig. 3.3) has been identified in *Escherichia coli* and in extracts of mammalian cells.[107-109] Both methylated and non-methylated formamidopyrimidine lesions are released by the action of this glycosylase. Subsequent studies showed that the same enzyme also processes 8-hydroxyguanine lesions (Fig. 3.3).[110] Interestingly, formamidopyrimidine-DNA glycosylase also excises imidazole ring-opened purines that are modified with bulky carcinogen adducts such as aflatoxin B_1-2,6-diamino-4-hydroxy-5-formamidopyrimidine,[111] and the same activity also processes certain oxidative pyrimidine products such as 5-hydroxy-2'-deoxycytidine and 5-hydroxy-2'-deoxyuridine.[59] Like endonuclease III or AlkA, formamidopyrimidine-DNA glycosylase is a repair enzyme with rather broad substrate specificity.

RECOGNITION OF DAMAGED BASES BY FORMAMIDOPYRIMIDINE-DNA GLYCOSYLASE

Escherichia coli Fpg is a 30 kDa enzyme with formamidopyrimidine-DNA glycosylase and associated AP lyase activity.[40,108,112,113] In the literature, Fpg is sometimes referred to as MutM protein or 8-hydroxyguanine-DNA glycosylase. In addition to its *N*-glycosylase and AP lyase function, Fpg protein also catalyzes a δ-elimination reaction that removes fragmented deoxyribose moieties from incised AP sites. Unlike other DNA glycosylases/AP lyases, the incision products of the Fpg reaction are 3'- and 5'-phosphate termini.

The mechanism of damage recognition by Fpg is unknown, but its interaction with DNA involves a zinc finger motif (Cys-X_2-Cys-X_{16}-Cys-X_2-Cys) located near its carboxyl terminus. Using this zinc finger domain, Fpg is thought to bind substrate DNA spanning a region of five nucleotides around the damaged site.[114-116] To understand how Fpg protein discriminates between non-damaged and damaged DNA, a structure-function analysis was performed using binding and kinetic parameters on a series of synthetic duplexes containing specific base lesions. This study led to the conclusion that Fpg protein recognizes duplex DNA containing unusual C=O moieties in the major groove.[117] For example, 8-hydroxyguanine may rearrange to form 8-oxoguanine with a C^8 keto group, and formamidopyrimidines contain a carbonyl group at exactly the same position. This model involving recognition of C=O groups in the major groove accommodates the observation that 8-hydroxyguanine lesions mispaired with adenine are not a substrate for Fpg protein.[117] In this case, the C^8 keto group is situated in the minor groove, where it apparently looses the capacity to trigger recognition by Fpg protein.

MOLECULAR SEARCH MECHANISMS

Early work on the *Escherichia coli* lac repressor suggested that random three-dimensional diffusion processes could not account for the efficient location of target sequences by DNA-interactive proteins.[118] Instead, specific protein-DNA interactions may be facilitated by limiting the diffusion to a single dimension along the DNA, and many proteins including transcription factors, restriction endonucleases, methylases, DNA or RNA polymerases and DNA helicases have been shown to utilize such facilitated (or one-dimensional) target site location mechanisms.[119-121]

The question of target sequence location is also pertinent to DNA repair enzymes, as these factors must be able to detect trace amounts of damaged bases in the context of a vast excess of nondamaged DNA. Several studies performed with a model repair enzyme (T4 endonuclease V) support a mechanism by which T4 endonuclease V binds nontarget DNA electrostatically and then translocates along the substrate molecule by linear diffusion until it reaches specific targets, i.e., cyclobutane pyrimidine dimers.[122-124] Similarly, cyclobutane pyrimidine dimer-DNA glycosylase of *Micrococcus luteus*[125] and uracil-DNA glycosylases[126] also appear to interact with DNA nonspecifically and to exploit linear diffusion to scan large template molecules for the presence of damaged bases. These few examples promulgated the notion that many (or all) DNA

repair enzymes may utilize facilitated (one-dimensional) target site location to search for defective sites in the genome.

Different mechanisms have been proposed for one-dimensional target site location.[127] For example, proteins may translocate along DNA by a "hopping" mechanism involving a series of microscopic association/dissociation events (Fig. 4.6; pathway 2). Microscopic hopping is believed to allow DNA binding proteins to translocate by as much as 10 base pairs per collision.[127] Alternatively, DNA binding proteins may translocate by "sliding", i.e., a linear diffusion process occurring while the proteins remain nonspecifically bound to the nucleic acid substrate (Fig. 4.6; pathway 4). Also, protein translocation along DNA may be facilitated by "intersegment transfer", i.e., by transient association with two different DNA segments to form a DNA loop in between (Fig. 4.6; pathway 3). Presumably, such unique nucleoprotein complexes are unstable, and when they dissociate the protein would have an equal probability of remaining associated with the first or the second segment. Electron microscopic studies suggest that T4 endonuclease V may exploit such an intersegment transfer mode of translocation while searching for cyclobutane pyrimidine dimers.[128] Hopping, sliding and intersegment transfer are an energy-independent, bidirectional diffusion process. Tracking, on the other hand, defines an energy-dependent and unidirectional mechanism that is tightly coupled to the hydrolysis of a nucleoside 5'-triphosphate. This latter mechanism is exploited by the prokaryotic (A)BC nucleotide excision repair system (see chapter 5).[129]

CONCLUSIONS

To be repairable, damage must be recognized by a protein (or a protein complex) that initiates a sequence of biochemical reactions leading to the elimination of DNA damage. To date, there are only few biophysical measurements that characterize the initial binding of a DNA repair factor to its target. However, these few examples

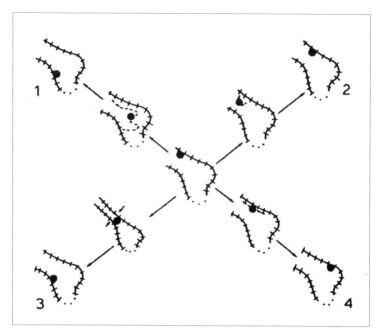

Fig. 4.6. Target site location of DNA-interactive proteins by facilitated (one-dimensional) transfer. Pathway 2: microscopic hopping; pathway 3: intersegment transfer; pathway 4: sliding. Pathway 1 shows an example of ineffective random (three-dimensional) search mechanism. Reprinted with permission from: Berg OG, Winter RB, von Hippel PH. Diffusion-driven mechanisms of protein translocation on nucleic acids. Biochemistry 1981; 20:6929-6948. © 1981 American Chemical Society.

illustrate that living cells utilize multiple strategies to discriminate covalent DNA modifications as lesions to be repaired. For example, damaged bases are accommodated into a rigid recognition pocket (DNA photolyase, uracil-DNA glycosylase) or into a flexible recognition pocket (endonuclease III, AlkA protein). The target structures in duplex DNA are made accessible by extra-helical displacement of the damaged base (DNA photolyase, uracil-DNA glycosylase, endonuclease III, AlkA protein) or by extrahelical displacement of the non-damaged complementary base, leaving the damaged residue inside the DNA duplex (T4 endonuclease V). Damaged bases are recognized directly (DNA photolyase, uracil-DNA glycosylase) or indirectly through conformational anomalies of the DNA backbone (T4 endonuclease V). Recognition

of DNA damage also involves various binding forces including specific hydrogen bonding (uracil-DNA glycosylase), hydrophobic interactions (DNA photolyase), electrostatic interactions (T4 endonuclease V) or π-electron complementarity between aromatic surfaces (AlkA protein). Finally, substrate recognition involves extensive conformational changes of the protein (O^6-methyl-guanine-DNA methyltransferase, AlkA protein) or no conformational changes at all (uracil-DNA glycosylase).

The present chapter is focused on DNA repair pathways that function by a principle of one damage-one enzyme. This mechanistically simple strategy achieves a remarkable level of genome protection in both prokaryotic and eukaryotic organisms. However, living cells were apparently not able to encode specific alkyltransferases, photolyases or

DNA glycosylases for every type of base damage that accompanied evolution. As a consequence, cells also display a more general strategy of damage recognition characterized by enhanced versatility. The next chapters are devoted to this general recognition strategy and are intended to discuss possible mechanisms by which a single repair pathway, referred to as nucleotide excision repair, is able to detect a large repertoire of structurally or chemically unrelated lesions.

REFERENCES

1. Sancar A. Mechanisms of DNA excision repair. Science 1994; 266:1954-1956.
2. Wood RD. DNA repair in eukaryotes. Annu Rev Biochem 1996; 65:135-167.
3. Modrich P. Mechanisms and biological effects of mismatch repair. Annu Rev Genet 1991; 25:229-253.
4. Friedberg EC, Walker GC, Siede W. DNA Repair and Mutagenesis. Washington, D.C.: American Society for Microbiology, 1995.
5. Park H-W, Kim S-T, Sancar A et al. Crystal structure of DNA photolyase from *Escherichia coli*. Science 1995; 268:1866-1872.
6. Mol CD, Arvai AS, Slupphaug G et al. Crystal structure and mutational analysis of human uracil-DNA glycosylase: structural basis for specificity and catalysis. Cell 1995; 80:869-878.
7. Sun B, Latham KA, Dodson ML et al. Studies on the catalytic mechanism of five DNA glycosylases. J Biol Chem 1995; 270:19501-19508.
8. Demple B, Harrison L. Repair of oxidative damage to DNA: enzymology and biology. Annu Rev Biochem 1994; 63:915-948.
9. Sancar A, Sancar GB. DNA repair enzymes. Annu Rev Biochem 1988; 57:29-67.
10. Pegg AE, Byers TL. Repair of DNA containing O^6-alkylguanine. FASEB J 1992; 6:2302-2310.
11. Setlow P. I will survive: protecting and repairing spore DNA. J Bacteriol 1992; 174:2737-2741.
12. Tang M-s, Nazimiec ME, Doisy RP et al. Repair of helix-stabilizing anthra-mycin-N2 guanine DNA adducts by UVRA and UVRB proteins. J Mol Biol 1991; 220:855-866.
13. Lindahl T, Sedgwick B, Sekiguchi M et al. Regulation and expression of the adaptive response to alkylating agents. Annu Rev Biochem 1988; 57:133-157.
14. Mitra S, Kaina B. Regulation of repair of alkylation damage in mammalian genomes. Progr Nucleic Acid Res Mol Biol 1993; 44:109-142.
15. Pegg AE, Dolan ME, Moschel RC. Structure, function, and inhibition of O^6-alkylguanine-DNA alkyltransferase. Progr Nucleic Acid Res Mol Biol 1995; 51:167-223.
16. Zak P, Kleibl K, Laval F. Repair of O^6-methylguanine and O^4-methylthymine by the human and rat O^6-methylguanine-DNA methyltransferase. J Biol Chem 1994; 269:730-733.
17. Moore MH, Gulbis JM, Dodson EJ et al. Crystal structure of a suicidal DNA repair protein: the Ada O^6-methylguanine-DNA methyltransferase from *E. coli*. EMBO J 1994; 13:1495-1501.
18. Kanugula S, Goodtzova K, Edara S et al. Alteration of arginine-128 to alanine abolishes the ability of human O^6-alkylguanine-DNA alkyltransferase to repair methylated DNA but has no effect on its reaction with O^6-benzylguanine. Biochemistry 1995; 34:7113-7119.
19. Goodtzova K, Crone TM, Pegg AE. Activation of O^6-alkyltransferase by DNA. Biochemistry 1994; 33:8385-8390.
20. Takahashi, Sakumi K, Sekiguchi M. Interaction of Ada protein with DNA examined by fluorescence anisotropy of the protein. Biochemistry 1990; 29:3431-3436.
21. Chan CL, Wu Z, Ciardelli T et al. Kinetic and DNA-binding properties of recombinant human O^6-methylguanine-DNA methyltransferase. Arch Biochem Biophys 1993; 300:193-200.
22. Hearst JE. The structure of photolyase: using photon energy for DNA repair. Science 1995; 268:1859-1859.
23. Li YF, Kim ST, Sancar A. Evidence for lack of DNA photoreactivating enzyme in humans. Proc Natl Acad Sci USA 1993; 90:4389-4393.

24. Kim ST, Sancar A. Photochemistry, photophysics, and mechanism of pyrimidine dimer repair by DNA photolyase. Photochem Photobiol 1993; 57:895-904.

25. Cook, J.S. Photoreactivation in animal cells. Photophysiology 1970; 5:191-213.

26. Sutherland BM, Bennett PV. Human white blood cells contain cyclobutyl pyrimidine dimer photolyase. Proc Natl Acad Sci USA 1995; 92:9732-9736.

27. Sancar GB, Smith FW, Sancar A. Binding of *Escherichia coli* DNA photolyase to UV-irradiated DNA. Biochemistry 1985; 24:1849-1855.

28. Husain I, Sancar A. Binding of *E. coli* DNA photolyase to a defined substrate containing a single T<>T dimer. Nuclei Acids Res 1987; 15:1109-1120.

29. Kim ST, Sancar A. Effect of base, pentose, and phosphodiester backbone structures on binding and repair of pyrimidine dimers by *Escherichia coli* DNA photolyase. Biochemistry 1991; 30:8623-8630.

30. Baer ME, Sancar GB. The role of conserved amino acids in substrate binding and discrimination by photolyase. J Biol Chem 1993; 268:16717-16724.

31. Todo T, Takemori H, Ryo H et al. A new photoreactivating enzyme that specifically repairs ultraviolet light-induced (6-4) photoproducts. Nature 1993; 361:371-374.

32. Todo T, Ryo H, Yamamoto K et al. Similarity among the Drosophila (6-4)photolyase, a human photolyase homolog, and the DNA photolyase-blue-light photoreceptor family. Science 1996; 272:109-112.

33. Dianov G, Price A, Lindahl T. Generation of single-nucleotide repair patches following excision of uracil residues from DNA. Mol Cell Biol 1992; 12:1605-1612.

34. Lindahl T. New class of enzymes acting on damaged DNA. Nature 1976; 259:64-66.

35. Jiricny J. Colon cancer and DNA repair: have mismatches met their match? Trends Genet 1994; 10:164-168.

36. Lindahl T. DNA repair enzymes. Annu Rev Biochem 1982; 51:61-87.

37. Kow YW, Wallace SS. Mechanism of action of *Escherichia coli* endonuclease III. Biochemistry 1987; 26:8200-8206.

38. Lindahl T. An *N*-glycosidase from *Escherichia coli* that releases free uracil from DNA containing deaminated cytosine residues. Proc Natl Acad Sci USA 1974; 71:3649-3653.

39. Friedberg EC, Ganesan AK, Minton K. *N*-Glycosidase activity in extracts of *Bacillus subtilis* and its inhibition after infection with bacteriophage PBS2. J Virol 1975; 16:315-321.

40. O'Connor TR, Laval J. Physical association of the 2,6-diamino-4-hydroxy-5N-formamidopyrimidine-DNA glycosylase of *Escherichia coli* and an activity nicking DNA at apurinic/apyrimidinic sites. Proc Natl Acad Sci USA 1989; 86:5222-5226.

41. Radman, M. An endonuclease from *Escherichia coli* that introduces single polynucleotide chain scissions in ultraviolet-irradiated DNA. J Biol Chem 1976; 251:1438-1445.

42. Friedberg EC, King JJ. Endonucleolytic cleavage of UV-irradiated DNA controlled by the V^+ gene in phage T4. Biochem Biophys Res Commun 1969; 37:646-651.

43. Loeb LA, Preston BD. Mutagenesis by apurinic/apyrimidinic sites. Annu Rev Genet 1986; 20:201-230.

44. Demple B, Herman T, Chem D. Cloning and expression of APE, the cDNA encoding the major human apurinic endonuclease: definition of a family of DNA repair enzymes. Proc Natl Acad Sci USA 1991; 88:11450-11454.

45. Robson CN, Hickson ID. Isolation of cDNA clones encoding a human apurinic/apyrimidinic endonuclease that corrects DNA repair and mutagenesis defects in *Escherichia coli* xth (exonuclease III) mutants. Nucleic Acids Res 1991; 19:5519-5523.

46. Seki S, Akiyama K, Watanabe S et al. A mouse DNA repair enzyme (APEX nuclease) having exonuclease and apurinic/apyrimidinic endonuclease activities: purification and characterization. Biochim Biophys Acta 1992; 1079:57-64.

47. Xanthoudakis S, Miao G, Wang F et al. Redox activation of Fos-Jun DNA binding activity is mediated by a DNA repair enzyme. EMBO J 1992; 11:3323-3335.

48. Demple B, Johnson A, Fung D. Exonu-

clease III and endonuclease IV remove 3' blocks from DNA synthesis primers in H₂O₂-damaged *Escherichia coli*. Proc Natl Acad Sci USA 1986; 83:7731-7735.

49. Graves RJ, Felzenszwalk I, Laval J et al. Excision of 5'-terminal deoxyribose phosphate from damaged DNA is catalyzed by the Fpg protein of *Escherichia coli*. J Biol Chem 1992; 267:1442914435.

50. Dianov G, Sedgwick B, Graham D et al. Release of 5'-terminal deoxyribose-phosphate residues from incised abasic sites in DNA by the Escherichia coli RecJ protein. Nucleic Acids Res 1994; 22:993-998.

51. Sobol RB, Horton JK, Kühn R et al. Requirement of mammalian DNA polymerase-β in base-excision repair. Nature 1996; 379:183-186.

52. Matsumoto Y, Kim K. Excision of deoxyribose phosphate residues by DNA polymerase β during DNA repair. Science 1995; 269:699-702.

53. Prigent C, Satoh MS, Daly G et al. Aberrant DNA repair and DNA replication due to an inherited enzymatic defect in human DNA ligase I. Mol Cell Biol 1994; 14:310-317.

54. Caldecott KW, Tucker JD, Stanker LH et al. Characterization of the XRCC1-DNA ligase III complex in vitro and its absence from mutant hamster cells. Nucleic Acids Res 1995; 23:4836-4843.

55. Lindahl T. An *N*-glycosydase from *Escherichia coli* that releases free uracil from DNA containing deaminated cytosine residues. Proc Natl Acad Sci USA 1974; 71:3649-3653.

56. Olsen LC, Aasland R, Wittwer CU et al. Molecular cloning of a human uracil-DNA glycosylase, a highly conserved DNA repair enzyme. EMBO J 1989; 8:3121-3125.

57. Upton C, Stuart DT, McFadden G. Identification of a poxvirus gene encoding a uracil DNA glycosylase. Proc Natl Acad Sci USA 1993; 90:4518-4522.

58. Gallinari P, Jiricny J. A new class of uracil-DNA glycosylases related to human thymine-DNA glycosylases. Nature 1996; 383:735-738.

59. Hatahet Z, Kow YW, Purmal AA et al. New substrates for old enzymes. 5-Hydroxy-2'-deoxycytidine and 5-hydroxy-2'-deoxyuridine are substrates for *Escherichia coli* endonuclease III and formamidopyrimidine DNA *N*-glycosylase, while 5-hydroxy-2'-deoxyuridine is a substrate for uracil DNA *N*-glycosylase. J Biol Chem 1994; 269:18814-18820.

60. Zastawny TH, Doetsch PW, Dizdaroglu M. A novel activity of *E. coli* uracil DNA *N*-glycosylase: excision of isodialuric acid (5,6-dihydroxyuracil), a major product of oxidative DNA damage, from DNA. FEBS Lett 1995; 364:255-258.

61. Dizdaroglu M, Karakaya A, Jaruga P et al. Novel activities of human uracil-DNA *N*-glycosylase for cytosine-derived products of oxidative DNA damage. Nucleic Acids Res 1996; 24:418-422.

62. Savva R, McAuley-Hecht K, Brown T et al. The structural basis of specific base-excision repair by uracil-DNA glycosylase. Nature 1995; 373:487-493.

63. Kavli B, Slupphaug G, Mol CD et al. Excision of cytosine and thymine from DNA by mutants of human uracil-DNA glycosylase. EMBO J 1996; 15:3442-3447.

64. Slupphaug G, Mol CD, Kavli B et al. A nucleotide-flipping mechanism from the structure of human uracil-DNA glycosylase bound to DNA. Nature 1996; 386:87-92.

65. Slupphaug G, Eftedal I, Kavli B et al. Properties of a recombinant human uracil-DNA glycosylase from the *UNG*-gene and evidence that the *UNG*-gene encodes the major uracil-DNA glycosylase. Biochemistry 1995; 34:128-138.

66. Verri A, Mazzarello P, Spadari S et al. Uracil-DNA glycosylases preferentially excise mispaired uracil. Biochem J 1992; 287:1007-1010.

67. Rould MA, Perona JJ, Soll D et al. Structure of an *E. coli* glutamyl-tRNA synthetase complexed with tRNAGln and ATP at 2.8 Å resolution: implications for tRNA discrimination. Science 1989; 246:1135-1142.

68. Yasuda S, Sekiguchi M. T4 endonuclease involved in repair of DNA. Proc Natl Acad USA 1970; 67:1839-1845.

69. McMillan S, Edenberg HJ, Radany EH et al. *denV* gene of bacteriophage T4 codes for both pyrimidine dimer-DNA glycosylase and apyrimidinic endonuclease

activities. J Virol 1981; 40:211-223.

70. Dodson ML, Michaels ML, Lloyd RS. Unified catalytic mechanism for DNA glycosylase. J Biol Chem 1994; 269:32709-32712.

71. Grafstrom RH, Park L, Grossman L. Enzymatic repair of pyrimidine dimer-containing DNA. A 5' dimer DNA glycosylase: 3'-apyrimidinic endonuclease mechanism from *Micrococcus luteus.* J Biol Chem 1982; 257:13465-13474.

72. Hamilton KK, Kim PM, Doetsch PW. A eukaryotic DNA glycosylase/lyase recognizing ultraviolet light-induced pyrimidine dimers. Nature 1992; 356:725-728.

73. Morikava K, Matsumoto O, Tsujimoto M et al. X-ray structure of T4 endonuclease V: an excision repair enzyme specific for a pyrimidine dimer. Science 1992; 256:523-526.

74. Morikava K. DNA repair enzymes. Curr Opin Struct Biol 1993; 3:17-24.

75. Morikava K, Ariyoshi M, Vassylyev D et al. Crystal structure of a pyrimidine dimer-specific excision repair enzyme from bacteriophage T4: refinement at 1.45 Å resolution and X-ray analysis of the three active site mutants. J Mol Biol 1995; 249:360-375.

76. Doi T, Recktenwald A, Karaki Y et al. Role of the basic amino acid cluster and Glu-23 in pyrimidine dimer glycosylase activity of T4 endonuclease V. Proc Natl Acad Sci USA 1992; 89:9420-9424.

77. Schrock RD, Lloyd RS. Reductive methylation of the amino terminus of endonuclease V eradicates catalytic activities. J Biol Chem 1991; 266:17631-17639.

78. Schrock RD, Lloyd RS. Site-directed mutagenesis of the NH2 terminus of T4 endonuclease V. J Biol Chem 1993; 268:880-886.

79. Vassylyev DG, Kashiwagi T, Mikami Y et al. Atomic model of a pyrimidine dimer excision repair enzyme complexed with a DNA substrate: structural basis for damaged DNA recognition. Cell 1995; 83:773-782.

80. Iwai S, Maeda M, Shimada Y et al. Endonuclease V from bacteriophage T4 interacts with its substrate in the minor groove. Biochemistry 1994; 33:5581-5588.

81. Inaoka T, Ishida M, Ohtsuka E. Affinity of single- or double-stranded oligodeoxyribonucleotides containing a thymine photodimer for T4 endonuclease V. J Biol Chem 1989; 264:2609-2614.

82. Kemmink J, Boelens R, Koning T et al. ¹H NMR study of the exchangeable protons of the duplex d(GCGTTGCG)-·d(CGCAACGC) containing a thymine photodimer. Nucleic Acids Res 1987; 15:4645-4653.

83. Taylor J-S, Garrett DS, Brockie IR et al. ¹H NMR assignment and melting temperature study of *cis-syn* and *trans-syn* thymine dimer containing duplexes of d(CGTATTATGC)·d(GCATAATACG). Biochemistry 1990; 29:8858-8866.

84. Breimer L, Lindahl T. A DNA glycosylase from *Escherichia coli* that releases free urea from a polydeoxyribonucleotide containing fragments of base residues. Nucleic Acids Res 1980; 8:6199-6211.

85. Katcher HL, Wallace SS. Characterization of the *Escherichia coli* X-ray endonuclease, endonuclease III. Biochemistry 1983; 22:4071-4081.

86. Boorstein RJ, Hilbert TP, Cadet J et al. UV-induced pyrimidine hydrates in DNA are repaired by bacterial and mammalian DNA glycosylase activity. Biochemistry 1989; 28:6164-6170.

87. Dizdaroglu M, Laval J, Boiteux S. Substrate specificity of the *Escherichia coli* endonuclease III: excision of thymine- and cytosine-derived lesions in DNA produced by radiation-generated free radicals. Biochemistry 1993; 32:12105-12111.

88. Strniste GF, Wallace SS. Endonucleolytic incision of x-irradiated deoxyribonucleic acid by extracts of *Escherichia coli.* Proc Natl Acad Sci USA 1975; 72:1997-2001.

89. Gates FT, Linn S. Endonuclease from *Escherichia coli* that acts specifically upon duplex DNA damaged by ultraviolet light, osmium tetroxide, acid, or x-rays. J Biol Chem 1977; 252:2802-2807.

90. Bailly V, Verly WG. *Escherichia coli* endonuclease III is not an endonuclease but a β-elimination catalyst. Biochem J 1987; 242:565-572:

91. Kim J, Linn S. The mechanism of action of *E. coli* endonuclease III and T4 UV endonuclease (endonuclease V) at AP sites. Nucleic Acids Res 1988;

16:1135-1141.

92. Mazumder A, Gerlt JA, Absalon MJ et al. Stereochemical studies of the β-elimination reactions at aldehydic abasic sites in DNA: endonuclease III from *Escherichia coli*, sodium hydroxide and Lys-Trp-Lys. Biochemistry 1991; 30:1119-1126.

93. Kuo C-F, McRee DE, Fisher CL et al. Atomic structure of the DNA repair [4Fe-4S] enzyme endonuclease III. Science 1992; 258:434-440.

94. Kuo C-F, McRee DE, Cunningham RP et al. Crystallization and crystallographic characterization of the iron-sulfur-containing enzyme endonuclease III from *Escherichia coli*. J Mol Biol 1992; 227:347-351.

95. Thayer MM, Ahern H, Xing D et al. Novel DNA binding motifs in the DNA repair enzyme endonuclease III crystal structure. EMBO J 1995; 14:4108-4120.

96. Ohlendorf DH, Anderson WF, Fisher RG et al. The molecular basis of DNA-protein recognition inferred from the structure of cro repressor. Nature 1982; 298:718-723.

97. O'Handley S, Scholes CP, Cunningham RP. Endonuclease III interactions with DNA substrates. 1. Binding and footprinting studies with oligonucleotides containing a reduced apyrimidinic site. Biochemistry 1995; 34:2528-2536.

98. Hilbert TP, Boorstein RJ, Kung HC et al. Purification of a mammalian homologue of *Escherichia coli* endonuclease III: identification of a bovine pyrimidine hydrate-thymine glycol DNA-glycosylase/AP lyase by irreversible cross linking to a thymine glycol-containing oligodeoxynucleotide. Biochemistry 1996; 35:2505-2511.

99. Mattes WB, Lee C-S, Laval J et al. Excision of DNA adducts of nitrogen mustards by bacterial and mammalian 3-methyladenine-DNA glycosylases. Carcinogenesis 1996; 17:643-648.

100. Nakabeppu Y, Kondo H, Sekiguchi M. Cloning and characterization of the AlkA gene of Escherichia coli that encodes 3-methyladenine DNA glycosylase II. J Biol Chem 1984; 259:13723-13729.

101. Saparbaev M, Kleibl K, Laval J. *Escherichia coli, Saccharomyces cerevisiae*, rat, and human 3-methyladenine DNA glycosylases repair 1,N^6-ethenoadenine when present in DNA. Nucleic Acids Res 1995; 23:3750-3755.

102. Seeberg E, Eide L, Bjoras M. The base-excision repair pathway. Trends Biochem Sci 1995; 20:391-397.

103. Thomas L, Yang C-H, Goldthwait DA. Two DNA glycosylases in Escherichia coli which release primarily 3-methyladenine. Biochemistry 1982; 21:1162-1169.

104. Yamagata Y, Kato M, Odawara et al. Three-dimensional structure of a DNA repair enzyme, 3-methyladenine DNA glycosylase II, from *Escherichia coli*. Cell 1996; 86:311-319.

105. Labahn J, Schärer OD, Long A et al. Structural basis for the excision repair of alkylation-damaged DNA. Cell 1996; 86:321-329.

106. Ishida T, Doi M, Ueda H et al. Specific ring stacking interaction on the tryptophan-7-methylguanine system: comparative crystallographic studies of indole derivatives-7-methylguanine base, nucleoside, and nucleotide complexes. J Am Chem Soc 1988; 110:2286-2294.

107. Boiteux S, Bichara M, Fuchs RP et al. Excision of the imidazole ring-opened form of N-2-aminofluorene-C(8)-guanine adduct in poly(dG-dC) by *Escherichia coli* formamidopyrimidine-DNA glycosylase. Carcinogenesis 1989; 10:1905-1909.

108. Chetsanga CJ, Lindahl T. Release of 7-methylguanine residues whose imidazole rings have been opened from damaged DNA by a DNA glycosylase from *Escherichia coli*. Nucleic Acids Res 1979; 6:3673-3683.

109. Bessho T, Roy R, Yamamoto K et al. Repair of 8-hydroxyguanine in DNA by mammalian N-methylpurine-DNA glycosylase. Proc Natl Acad Sci USA 1993; 90:8901-8904.

110. Tchou J, Kasai H, Shibutani S et al. 8-Oxoguanine (8-hydroxyguanine) DNA glycosylase and its substrate specificity. Proc Natl Acad Sci USA 1991; 88:4690-4694.

111. Chetsanga CJ, Frenette GP. Excision of aflatoxin B_1-imidazole ring opened guanine adducts from DNA by formamidopyrimidine-DNA glycosylase. Carcinogenesis 1983; 4:997-1000.

112. Laval J, Boiteux S, O'Connor TR. Physiological properties and repair of apurinic/apyrimidinic sites and imidazole ring-opened guanines in DNA. Mutat Res 1990; 233:73-79.

113. Tchou J, Grollman AP. Repair of DNA containing the oxidatively-damaged base, 8-oxoguanine. Mutat Res 1993; 299:277-287.

114. O'Connor TR, Graves RJ, de Murcia G et al. Fpg protein of *Escherichia coli* is a zinc finger protein whose cysteine residues have a structural and/or functional role. J Biol Chem 1993; 268:9063-9070.

115. Tchou J, Michaels ML, Miller JH et al. Function of the zinc finger in *Escherichia coli* Fpg protein. J Biol Chem 1993; 268:26738-26744.

116. Castaing B, Boiteux S, Zelwer C. DNA containing a chemically reduced apurinic site is a high affinity ligand for the *Escherichia coli* formamidopyrimidine-DNA glycosylase. Nucleic Acids Res 1993; 21:2899-2905.

117. Tchou J, Bodepudi V, Shibutani S et al. Substrate specificity of Fpg protein. J Biol Chem 1994; 269:15318-15324.

118. Riggs AD, Bourgeois S, Cohn M. The lac repressor-operator interaction. J Mol Biol 1970; 53:401-417.

119. Berg OG, Winter RB, von Hippel PH. How do genome-regulatory proteins locate their DNA target sites? Trends Biochem Sci 1982; 7:52-55.

120. Ehrbrecht H-J, Pingoud A, Urbanke C et al. Linear diffusion of restriction endonucleases on DNA. J Biol Chem 1985; 260:6160-6166.

121. Singer P, Wu C-W. Promoter search by *Escherichia coli* RNA polymerase on a circular DNA template. J Biol Chem 1987; 262:14178-14189.

122. Ganesan AK, Seawell P, Lewis RJ et al. Processivity of T4 endonuclease V is sensitive to NaCl concentration. Biochemistry 1986; 25:5751-5755.

123. Gruskin EA, Lloyd RS. The DNA scanning mechanism of T4 endonuclease V: effect of NaCl concentration on processive nicking activity. J Biol Chem 1986; 261:9607-9613.

124. Gruskin EA, Lloyd RS. Molecular analysis of plasmid DNA repair within UV-irradiated *Escherichia coli*. I: T4 endonuclease V-initiated excision repair. J Biol Chem 1988; 263:12728-12737.

125. Hamilton RW, Lloyd RS. Modulation of the DNA scanning activity of the *Micrococcus luteus* UV endonuclease. J Biol Chem 1989; 264:17422-17427.

126. Higley M, Lloyd RS. Processivity of uracil DNA glycosylase. Mutat Res 1993; 294:109-116.

127. Berg OG, Winter RB, von Hippel PH. Diffusion-driven mechanisms of protein translocation on nucleic acids. Biochemistry 1981; 20:6929-6948.

128. Lloyd RS, Dodson ML, Gruskin EA et al. T4 endonuclease V promotes the formation of multimeric DNA structures. Mutat Res 1987; 183:109-115.

129. Grossman L, Thiagalingam S. Nucleotide excision repair, a tracking mechanism in search of damage. J Biol Chem 1993; 268:16871-16874.

MOLECULAR RECOGNITION STRATEGIES II: (A)BC EXCINUCLEASE

All DNA repair mechanisms illustrated in the previous chapter (operating either by DNA damage reversal or by base excision repair) achieve their selectivity for damaged sites through noncovalent interactions between complementary surfaces. These binary DNA repair systems are initiated by specific enzymes (DNA photolyases, alkyltransferases, glycosylases) that bind a narrow range of lesions, i.e., a particular type of base damage, thereby excluding nondamaged DNA from being processed. For example, the substrate binding pocket of uracil-DNA glycosylase accommodates uracil and a few uracil derivatives, but efficiently rejects adenine, cytosine or thymine. Thus, the recognition strategy used in such repair systems is highly selective for damaged DNA, is efficient and (with exception of the "suicidal" O^6-methylguanine-DNA methyltransferase) of low energetic cost,[1] but limits dramatically the spectrum of lesions that can be recognized and processed by a given pathway.

As outlined in chapter 3, DNA is permanently damaged by a multiplicity of genotoxic agents deriving from endogenous reactions or from environmental sources. It is not possible, however, to encode a specific photoreactivating enzyme, alkyltransferase or DNA glycosylase for every type of damage that accompanied evolution of life. As a consequence, cells must be capable of dealing with a wide diversity of chemical modifications in DNA with only a limited number of repair enzymes. To compensate for the narrow substrate range of DNA damage reversal and base excision repair, an excision repair mode has evolved that accommodates to virtually all types of covalent damage to DNA bases. This versatile repair mechanism is referred to as nucleotide excision repair because, in all organisms, it releases DNA damage as a component of oligonucleotide fragments.[2-6]

Schemes based on the mechanism of nucleotide excision repair in *Escherichia coli* have shaped ideas about the analogous process in eukaryotic cells for many years. The purpose of this short chapter is to introduce the prokaryotic (A)BC excinuclease as a model system to study the biochemically more complex reactions involved in DNA damage recognition during nucleotide excision repair in mammals.

GENERAL OUTLINE OF NUCLEOTIDE EXCISION REPAIR

Nucleotide excision repair operates by a cut and patch mechanism that consists of five steps (Fig. 5.1):[7]

Mechanisms of DNA Damage Recognition in Mammalian Cells, by Hanspeter Naegeli.
© 1997 R.G. Landes Company.

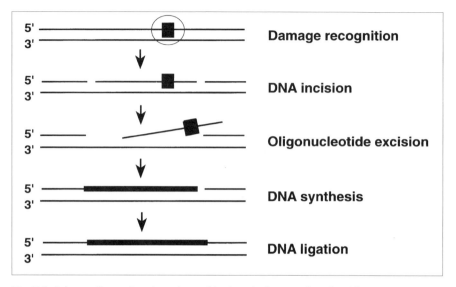

Fig. 5.1. Scheme illustrating the primary biochemical steps of nucleotide excision repair. This general mechanism based on oligonucleotide excision and DNA repair synthesis is functionally conserved throughout phylogenic evolution.

(1) damage recognition,
(2) incision of damaged strands on both sides of a lesion and at some distance from it,
(3) excision of oligonucleotides containing the damaged site,
(4) DNA synthesis using the complementary nondamaged strand as a template
(5) DNA ligation.

This basic mechanism is conserved throughout evolution. In most prokaryotic and eukaryotic cells, nucleotide excision repair is the principal pathway for removal of the major UV radiation products, including cyclobutane pyrimidine dimers and pyrimidine(6-4)pyrimidone photoproducts or their Dewar isomers. Also, nucleotide excision repair is, in most cases, the sole mechanism that eliminates bulky carcinogen-DNA adducts such as those induced by aromatic amines, polycyclic aromatic hydrocarbons, natural toxins (for example aflatoxin, anthramycin or CC-1065) and mono- or diadducts of crosslinking agents (furocoumarins, mitomycins, nitrosoureas, platinum derivatives). The repertoire of DNA lesions recognized and processed by nucleotide excision repair systems comprises even more subtle DNA modifications such as

methylation products, thymine glycols or abasic sites.[3,6-8] To explain the striking capacity of nucleotide excision repair to detect a broad range of chemically diverse lesions, it has been suggested that its molecular recognition subunits may be able to sense damage-induced conformational distortion of the DNA double helix rather than the precise chemical structure of defective nucleotides.[3,5] The problem of DNA damage recognition in prokaryotic nucleotide excision repair is discussed in the second part of this chapter. DNA damage recognition in mammalian nucleotide excision repair is the topic of chapter 6. Additionally, chapter 7 elaborates on the hypothesis of conformational recognition and examines how the secondary structure of DNA influences mammalian nucleotide excision repair activity.

(A)BC EXCINUCLEASE AS A MODEL FOR MAMMALIAN EXCISION REPAIR

In the prokaryote *Escherichia coli*, nucleotide excision repair is initiated by (A)BC excinuclease, a dynamic enzyme consisting of three distinct gene products in variable molecular associations: UvrA, UvrB and UvrC.[8] (A)BC excinuclease shares with mammalian nucleotide excision repair the ability to recognize and process an extremely large repertoire of DNA adducts. However, the two systems diverge at the mechanistic level. For example, the length of the oligonucleotide excision product in prokaryotes (12-13 nucleotides) is considerably smaller than that found in mammals, where nucleotide excision repair releases oligomers of 27 to 29 residues in length.[9] In addition, very little sequence homology is found between bacterial and mammalian nucleotide excision repair proteins.[3,7] While only three gene products are required for DNA incision by prokaryotic nucleotide excision repair, the mammalian pathway is initiated by several more proteins that may be classified into three categories. The first group includes XPA and XPC, and consists of specialized factors that, like UvrABC, function uniquely in nucleotide excision repair (reviewed in chapter 6).[10,11] The second group of proteins is exemplified by RPA (for *replication protein A*), a three-subunit factor that is required for initiation of nucleotide excision repair as well as for DNA replication,[12] recombination[13] and, possibly, other chromosomal processes.[14] The third category comprises XPB and XPD (subunits of transcription factor TFIIH), i.e., proteins that are shared between nucleotide excision repair and RNA polymerase II-dependent transcription.[15] Despite the considerable genetic and biochemical differences between prokaryotic and eukaryotic nucleotide excision repair, it is nevertheless believed that deciphering the mechanisms of this process in *Escherichia coli* is useful to understand the molecular details of its mammalian counterpart.[1,3,6-8]

Enzymology of Prokaryotic Nucleotide Excision Repair

Nucleotide excision repair in *Escherichia coli* is effected by the coordinated action of a total of six gene products: UvrA, UvrB and UvrC are responsible for damage recognition and damage-specific incision, while UvrD, DNA polymerase I and DNA ligase accomplish the postincisional reactions.[3-8] Two auxiliary proteins, DNA photolyase[16]

and the *mfd* gene product (for *m*utational *f*requency *d*ecline)[17] modulate this mechanism by targeting the nucleotide excision repair system to specific sites in the genome. As stated before, in reconstituted in vitro systems three gene products (UvrA, UvrB and UvrC) are necessary and sufficient to incise damaged double-stranded DNA.[18]

Several detailed descriptions of the properties of the (A)BC excinuclease have been provided during the last few years.[3,6-8] UvrA is a protein of 104 kDa containing structural motifs for ATP binding, ATP hydrolysis and zinc DNA binding fingers. UvrB is a highly hydrophobic protein of 76 kDa that displays an ATP binding motif and short regions of homology that are highly conserved among DNA helicases, i.e., enzymes that separate the complementary strands of duplex DNA in a reaction that is coupled to ATP hydrolysis (Fig. 5.2). On the other hand, no specific protein domain or motif was apparent in the predicted amino acid sequence of UvrC (a polypeptide of 66 kDa).[1,3,7] These three factors constitute the subunits of (A)BC excinuclease, an enzyme that removes damaged residues from DNA in oligomeric segments by hydrolyzing mainly the 8th phosphodiester bond 5' and the 5th phosphodiester bond 3' to the lesion. All other prokaryotes tested remove DNA damage by the same excision pattern.[7]

Substrate Binding

UvrA, UvrB and UvrC operate in a cascade-like mechanism that may be summarized as follows:[3,8]

(1) UvrA+UvrA→UvrA$_2$
(2) UvrA$_2$+DNA→UvrA$_2$-DNA
(3) UvrA$_2$-DNA+UvrB→UvrA$_2$B-DNA
(4) UvrA$_2$B-DNA→UvrB-DNA+UvrA$_2$
(5) UvrB-DNA+UvrC→UvrBC-DNA

As indicated by Lin and Sancar,[8] discrimination between target and nontarget DNA is achieved by a "selectivity cascade", i.e., by the succession of a series of sequential biochemical reactions in which each single step potentiates the overall selectivity of the (A)BC system for damaged sites.

Step 1: UvrA+UvrA→UvrA$_2$

UvrA is an ATP-binding and ATP-hydrolyzing factor that exists in equilibrium between a monomeric and a dimeric form in solution.[19] An early event in the nucleotide excision repair pathway involves dimerization of UvrA to generate UvrA$_2$ homodimers, the active form of UvrA.[20] ATP binding drives UvrA-UvrA association, while ATP hydrolysis induces dissociation of this protein-protein complex.

Step 2: UvrA$_2$+DNA→UvrA$_2$-DNA

UvrA dimers are capable of binding double-stranded DNA with some selectivity for damaged sites, protecting 33 base pairs from DNase I digestion. UvrA$_2$ provides about 3 orders of magnitude of discrimination between nontarget and target DNA. ATP promotes the binding of UvrA$_2$ to DNA, while ATP hydrolysis results in the release of UvrA monomers from DNA.[6-8,20]

Step 3: UvrA$_2$-DNA+UvrB→UvrA$_2$B-DNA

UvrA dimers on DNA associate with UvrB to form UvrA$_2$B heterotrimers characterized by DNA binding, DNA-dependent ATPase and 5'→3' DNA helicase activity (Fig. 5.2).[21-23] Alternatively, the UvrA$_2$B heterotrimer may also form by interaction of UvrA$_2$ with UvrB in solution. The UvrA$_2$B trimer is thought to constitute a damage recognition complex that moves along DNA in search for lesions.[6]

Step 4: UvrA$_2$B-DNA→UvrB-DNA+UvrA$_2$

In the presence of DNA damage, a UvrA$_2$B-DNA complex is formed that yields a 45-base pair DNase I footprint.[6,23-25] At this stage, UvrA$_2$B uses energy from ATP hydrolysis and its 5'→3' helicase activity to bend and unwind the DNA substrate.[26,27] These ATP-driven structural changes force DNA into a conformation that is suitable for interaction with UvrB, thereby forming a tight nucleoprotein complex at the lesion site and promoting the release of UvrA.[22,28]

The conversion from a dynamic UvrA$_2$B-DNA complex to the stable UvrB-DNA intermediate is considered the rate limiting step of the (A)BC nucleotide excision repair pathway.[3,8] These ATP-dependent allosteric reactions increase the overall selectivity of the system for damaged bases to approximately 10^{12} (reviewed in Ref 8).

Step 5: UvrB-DNA+UvrC→UvrBC-DNA

The resulting UvrB-DNA complex has a smaller 19-base pair DNase I footprint.[29,30] The DNA in this nucleoprotein structure is locally denatured by about five base pairs around the lesion and kinked by 130°.[31,32] Also, the UvrB-DNA complex is extremely stable, having a half-life of approximately 2-3 h.[19,32] UvrC binds specifically and with high affinity to this UvrB-DNA complex.[28,32] In contrast, UvrC has no affinity for the UvrA$_2$B complex neither in solution nor on DNA, and only binds to the UvrB-DNA complex after release of UvrA from the DNA substrate.[30,32] On the basis of these findings, Sancar proposed the term (A)BC excinuclease to indicate the role of UvrA in the assembly of incision complexes without being physically present during subsequent endonucleolytic cleavage reactions.[8]

DUAL DNA INCISION

Binding of UvrC to the UvrB-DNA complex induces further conformational changes that are required for incision of DNA on both sides of the lesion.[33] Only damaged strands are targeted by the dual incision reaction. These damage-specific endonucleolytic cleavages are effected in an ordered fashion, first on the 3' side and then on the 5' side of the lesion.[7,18] Incision on the 3' side may be effected at the 5th, 6th or 7th phosphodiester bond 3' to the lesion.[34] In contrast, the 5' incision reactions occurs predominantly at the 8th phosphate 5' to the lesion. Following double incision of DNA, UvrB, UvrC and the excised oligomer of 12-13 residues remain bound to the substrate in a post-incision complex.[28]

It was postulated for some time that UvrB makes the 3' incision, while UvrC performs the 5' incision.[8,32] This hypothesis was mainly derived from the observation that a UvrB mutant that was thought to carry a glutamic acid (E) to alanine (A) substitution at amino acid position 640 was unable to promote DNA incision on the 3' side.[32] However, when the same E640A mutant was produced in an independent laboratory, this substitution resulted in a UvrB protein with identical properties to those of the wild type protein.[35] The original "E640A" mutant turned out to contain a frameshift resulting in a truncated UvrB that lacks the C-terminal 33 amino acids. This region contains a sequence that is predicted to form coiled-coil structures[36,37] for interactions with UvrC, but is not directly involved in the 3' incision. Thus, it appears more likely that both incisions are effected by UvrC, although it remains to be determined whether a single UvrC molecule or, alternatively, two UvrC molecules are required for the dual incision reaction.

POSTINCISIONAL STEPS IN PROKARYOTIC NUCLEOTIDE EXCISION REPAIR

The 3'→5' DNA helicase activity of UvrD (Fig. 5.2) is required to release both UvrC and the excised oligomer from the post-incision complex.[28] The resulting UvrB-gapped DNA complex is a substrate for strand elongation by DNA polymerase I. In the presence of deoxyribonucleoside triphosphates, polymerase I binds to the 3' terminus, carries out gap filling and, simultaneously, displaces UvrB from DNA.[28] The remaining nick is ligated by DNA ligase. It is believed that the nucleotide excision gaps are filled without nick translation, thereby generating repair patches of exactly the same size as the excised oligomers.[7]

THE MOLECULAR MATCHMAKER CONCEPT

Sancar and Tang[1] used the following sentence to describe the early events that initiate prokaryotic nucleotide excision repair: "UvrA delivers UvrB delivers UvrC". The UvrA dimer plays a central role in this mechanism because it induces conformational changes in both UvrB and DNA to load UvrB onto damaged DNA, and then dissociates from the UvrB-damaged DNA complex. In view of this ATP-dependent activity, UvrA has been referred to as a molecular matchmaker.[38]

Molecular matchmakers are a unique class of enzymes that use the energy released from ATP hydrolysis to promote the association of two macromolecules that have no affinity to one another in their native conformation. After promoting stable complex formation, the molecular matchmaker dissociates to allow specific biochemical reactions between the matched macromolecules.[38] Specifically, UvrB has little affinity for DNA, but UvrA is able to convert UvrB to a DNA binding protein presumably by exposing aliphatic and aromatic amino acid residues that serve as partners for interaction with DNA bases.[8] The structure of DNA is also changed by UvrA, as the double helix becomes kinked and unwound to expose its aromatic components and generate a structure that constitutes a suitable binding substrate for UvrB.[31,32]

DAMAGE RECOGNITION BY (A)BC EXCINUCLEASE

Although DNA damage excision by the (A)BC complex appears to be a relatively simple biochemical reaction involving only four or five polypeptides (UvrA$_2$, UvrB, 1 or 2 UvrC molecules) and one cofactor (ATP), the mechanism by which this system discriminates damaged nucleotides as substrates for dual DNA incision remains a subject of intense debate. Several different models of damage recognition by (A)BC excinuclease have been proposed,[1,3,6,8,23,39-41] and a selection of the most relevant scenarios of substrate discrimination is provided below. Central to most models is the conception that the molecular matchmaker function of UvrA$_2$ is essential to load UvrB onto DNA at sites of damage. Also, the different models of damage recognition share the idea that UvrB constitutes the ultimate

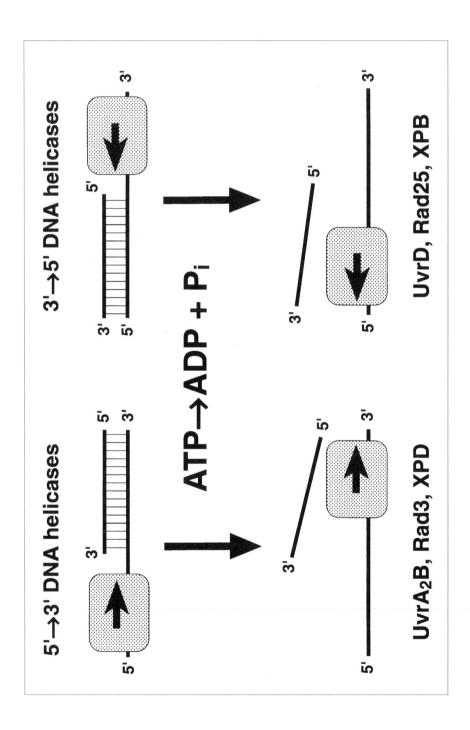

recognition subunit, thereby accommodating the observation that loading of UvrB is the rate-limiting step of prokaryotic nucleotide excision repair. Moreover, there is agreement on the idea that the stability of UvrB-DNA complexes largely correlates with the subsequent rate of double DNA incision.[42-44] However, the different models disagree on the question of how exactly the $UvrA_2B$ complex distinguishes damaged nucleotides from nondamaged DNA constituents.

THE "ATP-FUELLED TRACKING" HYPOTHESIS

Grossman and coworkers proposed that the DNA helicase activity of $UvrA_2B$ may serve to scan duplex DNA for damaged nucleotides in a tracking reaction that is tightly coupled to the hydrolysis of ATP as an energy cofactor. By tracking it is meant that a protein or an assembly of proteins binds to a site "a" and translocates to a site "b" along the DNA.[45]

These authors found that the $UvrA_2B$ heterotrimer displays a limited $5' \rightarrow 3'$ helicase activity, i.e., the complex is able to unwind short stretches of DNA duplexes of 17-25 residues in length using ATP hydrolysis as an energy source.[23,46] This ATPase/ DNA helicase activity drives translocation of the heterotrimer on DNA with $5' \rightarrow 3'$ polarity (Fig. 5.2).[23,24] They also found that the DNA helicase activity of $UvrA_2B$ is strongly inhibited by DNA lesions such as those induced by UV radiation.[23,46] These properties are consistent with a $UvrA_2B$ tracking mechanism that functions to search for damaged nucleotides on DNA. Protein translocation, ATPase and DNA helicase activity are terminated at sites of damage,

where $UvrA_2$ dissociates and highly stable UvrB-DNA complexes are formed. This model involves inhibition of DNA helicase activity as the principal determinant of damage recognition and accommodates the observation that (A)BC excinuclease acts on a large diversity of chemically distinct lesions. Presumably, the ATPase/DNA helicase activity of $UvrA_2B$ is sensitive to a broad range of defective nucleotides independently of the type of modification. The "ATP-fuelled tracking" model exploits the potential ability of $UvrA_2B$ to read the chemistry of DNA and arrest $5' \rightarrow 3'$ protein translocation at sites where DNA integrity is disrupted.

THE "DYNAMIC BACKBONE DISTORTION" HYPOTHESIS

Several authors proposed that the increased susceptibility of damaged DNA to be kinked may constitute a decisive factor in discriminating between damaged and nondamaged sites.[1,47,48] Indirect readout is the main feature of this hypothesis because the recognition signal is thought to reside in the sugar-phosphate backbone and not in the damaged bases.

A damage-induced kink in the DNA helix is not necessarily static, as sites of abnormally increased flexibility in the backbone, or hinge points, may adequately serve as determinants for recognition. For example, monoadducts and interstrand crosslinks formed by psoralens destabilize the deoxyribose-phosphate conformation in the DNA backbone. As a consequence, the DNA helix is able to adopt an abnormally large range of rapidly interconverting conformers as compared to nonmodified DNA.[47,48] The $UvrA_2B$ complex may target DNA lesions by probing the duplex for conformational

Fig. 5.2. (Opposite) Several ATPases/DNA helicases participate in prokaryotic or eukaryotic nucleotide excision repair. These enzymes exploit the energy resulting from ATP hydrolysis to separate the two strands of duplex DNA. In *Escherichia coli*, $UvrA_2B$ (a DNA helicase with $5' \rightarrow 3'$ polarity) is required for damage recognition and damage-specific DNA incision, whereas UvrD (a DNA helicase with $3' \rightarrow 5'$ polarity) is involved in postincisional reactions. The mammalian enzymes XPB and XPD (and their yeast *Saccharomyces cerevisiae* homologs Rad25 and Rad3) are components of the multisubunit nucleotide excision repair/transcription factor TFIIH. These eukaryotic enzymes have opposite reaction polarities, and their possible function in DNA damage recognition is discussed in chapter 6.

flexibility of the sugar-phosphate backbone arising at sites of base damage. According to this model, UvrB is preferentially loaded at sites where the backbone displays elevated flexibility and can be more easily deformed by the ATPase/DNA helicase function of the UvrA$_2$B complex. In this "dynamic backbone distortion" model, abnormally enhanced backbone flexibility is the major determinant of damage recognition.

THE "HYDROPHOBIC BINDING POCKET" HYPOTHESIS

Van Houten and Snowden[39] elaborated on the UvrA$_2$B tracking mechanism and came to the conclusion that UvrB may use hydrophobic stacking interactions between its aromatic amino acid side chains and DNA bases to promote selectivity for damaged sites. In their model, the ATPase/DNA helicase activity of the UvrA$_2$B complex serves to actively separate the two strands of duplex DNA and provide greater accessibility to the bases. During this unwinding reaction, the hydrophobic components of bases that are normally located on the inside of the double helix are exposed, allowing for intercalation of tyrosine or phenylalanine side chains. UvrB is specifically loaded at sites that are unable to undergo appropriate base stacking interactions. Thus, the main recognition signal in this model resides in the disruption of regular base stacking interactions, rather than in backbone distortion.

This hypothesis predicts that UvrB makes intimate contact with the aromatic surface of the damaged bases. Experimental evidence that supports this hypothesis has been presented. For example, UvrB could be covalently crosslinked to a psoralen monoadduct or to a bis(platinum)-DNA adduct, indicating close interactions between UvrB and damaged bases.[28,49] Also, recent studies showed that UvrB is able to bind to single-stranded DNA in the absence of UvrA in vitro, although only at very high protein concentrations.[50] Under experimental conditions that favor these interactions with single-stranded DNA, UvrB bound to psoralen or cisplatin-damaged DNA with higher affinity ($K_d = 5 \times 10^{-6}$ M) than to nonmodified DNA ($K_d > 5 \times 10^{-5}$ M). Moreover, by a combination of tryptophan replacements and fluorescence quenching it was shown that two phenylalanine residues of UvrB, Phe-365 and Phe-496, are in direct contact with DNA bases, presumably through stacking interactions. These and other observations led Hsu et al[50] to postulate that UvrB possesses a lesion binding pocket. This pocket appears to be hydrophobic and, hence, may accommodate a large number of different damaged base derivatives. In fact, the lack of requirements for more specific binding forces (hydrogen bonds, salt bridges) is an important prerequisite to explain the striking substrate versatility associated with prokaryotic nucleotide excision repair. Conceivably, a damaged base is more susceptible to entering this hydrophobic recognition pocket if its stacking interactions with neighboring bases are diminished or disrupted.

CONCLUSIONS

As early as 1965 it was discovered that a protein complex encoded by the *uvrA*, *uvrB* and *uvrC* genes is involved in the repair of several chemically unrelated lesions.[51] Based on this finding, Hanawalt and Hayes suggested that the excision repair system catalyzed by these proteins may be able to detect DNA damage by a mechanism that is formally equivalent to threading the DNA through a close-fitting "sleeve". They suggested that the sleeve may function to gauge the closeness-of-fit to the Watson-Crick structure, and all deviations from the normal structure would be perceived as damaged DNA to be processed.[5,51] During the subsequent 30 years, the protein products of *uvrA*, *uvrB* and *uvrC* have been studied in great detail using genetic, molecular, biochemical and biophysical methods. Their mode of binding to DNA (damaged or nondamaged), their ability to undergo protein-protein interactions and their enzymatic activities have also been investigated. In addition, structure-function relationships have been established with regard to their capacity to excise structurally or chemically distinct lesions. Excellent and comprehensive reviews have been provided to summarize this wealth of information on prokaryotic nucleotide excision repair.[1-3,6-8]

Several lines of research converge on the concept that the remarkable selectivity of (A)BC excinuclease for damaged sites results from a biochemical cascade during which UvrA$_2$B binds to DNA and utilizes the energy released from ATP hydrolysis to probe its secondary structure and chemistry. This probing mechanism has been shown to involve specific conformational changes in the vicinity of damaged sites. Various scenarios have been proposed to describe how "gaugeing the closeness-of-fit to the Watson-Crick structure" by (A)BC excinuclease might occur at the molecular level. More research needs to be done to understand in detail how bacteria patrol their DNA for potential products of genotoxic reactions, but these models provide useful working hypotheses for further experimentation. Undoubtedly, molecular insights gained from the prokaryotic system will also provide useful models to study the analogous process in mammalian cells. This view is supported by the finding that at least two essential elements of (A)BC excinuclease (the participation of zinc finger domains and enzymes with ATPase/DNA helicase activity) are reiterated in the DNA damage recognition step of mammalian nucleotide excision repair.

REFERENCES

1. Sancar A, Tang M-s. Nucleotide excision repair. Photochem Photobiol 1993; 57:905-921.
2. Friedberg EC, Walker GC, Siede W. DNA Repair and Mutagenesis. Washington, D.C.: American Society for Microbiology, 1995.
3. Van Houten B. Nucleotide excision repair in *Escherichia coli*. Microbiol Rev 1990; 54:18-51.
4. Hanawalt PC, Cooper PK, Ganesan AK et al. DNA repair in bacteria and mammalian cells. Annu Rev Biochem 1979; 48:783-836.
5. Hanawalt PC. "Close-fitting sleeves"-Recognition of structural defects in duplex DNA. Mutat Res 1993; 289:7-15.
6. Grossman L, Thiangalingam S. Nucleotide excision repair, a tracking mechanism in search of damage. J Biol Chem 1993; 268:16871-16874.
7. Sancar A. DNA excision repair. Annu Rev Biochem 1996; 65:43-81.
8. Lin J-J, Sancar A. (A)BC excinuclease: the *Escherichia coli* nucleotide excision repair enzyme. Mol Microbiol 1992; 6:2219-2224.
9. Huang J-C, Svoboda DL, Reardon JT et al. Human nucleotide excision nuclease removes thymine dimers from DNA by incising the 22nd phosphodiester bond 5' and the 6th phosphodiester bond 3' to the photodimer. Proc Natl Acad Sci USA 1992; 89:3664-3668.
10. Tanaka K, Wood RD. Xeroderma pigmentosum and nucleotide excision repair. Trends Biochem Sci 1994; 19:83-86.
11. Hoeijmakers JHJ. Nucleotide excision repair II: from yeast to mammals. Trends Genet Sci 1993; 9:211-217.
12. Coverley D, Kenny MK, Munn M et al. Requirement for the replication protein SSB in human DNA excision repair. Nature 1991; 349:538-541.
13. Heyer W-D, Rao MRS, Erdile LF et al. An essential *Saccharomyces cerevisiae* single-stranded DNA binding protein is homologous to the large subunit of human R-PA. EMBO J 1990; 9:2321-2329.
14. Li R, Botchan M. The acidic transcriptional activation domain of VP16 and p53 bind the cellular RPA and stimulate in vitro BPV-1 DNA replication. Cell 1993; 73:1207-1221.
15. Schaeffer L, Roy R, Humbert S et al. DNA repair helicase: a component of BTF2 (TFIIH) basic transcription factor. Science 1993; 260:58-63.
16. Sancar A, Franklin KA, Sancar GB. *Escherichia coli* DNA photolyase stimulates UvrABC excision nuclease in vitro. Proc Natl Acad Sci USA 1984; 81:7397-7401.
17. Selby CP, Witkin EM, Sancar A. *Escherichia coli mfd* mutant deficient in "mutation frequency decline" lacks strand-specific repair: in vitro complementation with purified coupling factor. Proc Natl Acad Sci USA 1991; 88:11574-11578.

18. Sancar A, Rupp WD. A novel repair enzyme: UVRABC excision nuclease of *Escherichia coli* cuts a DNA strand on both sides of the damaged region. Cell 1983; 33:249-260.

19. Oh EY, Grossman L. The effect of *Escherichia coli* Uvr protein binding on the topology of supercoiled DNA. Nucleic Acids Res 1986; 14:8557-8571.

20. Mazur SJ, Grossman L. Dimerization of *Escherichia coli* UvrA and its binding to undamaged and ultraviolet light damaged DNA. Biochemistry 1991; 30:4432-4443.

21. Orren DK, Sancar A. The (A)BC excinuclease of *Escherichia coli* has only the UvrB and UvrC subunits in the incision complex. Proc Natl Acad Sci USA 1989; 86:5237-5241.

22. Orren DK, Sancar A. Formation and enzymatic properties of the UvrB DNA complex. J Biol Chem 1990; 265:15796-15803.

23. Oh EY, Grossman L. Characterization of the helicase activity of the *Escherichia coli* Uvr AB protein complex. J Biol Chem 1989; 264:1336-1343.

24. Koo H-S, Classen L, Grossman L et al. ATP-dependent partitioning of the DNA template into supercoiled domains by *Escherichia coli* UvrAB. Proc Natl Acad Sci USA 1991; 88:1212-1216.

25. Moolenar GF, Visse R, Ortiz-Buysse M et al. Helicase motifs V and VI of the *Escherichia coli* UvrB protein of the UvrABC endonuclease are essential for the formation of the preincision complex. J Mol Biol 1994; 240:294-307.

26. Backendorf C, Spaink H, Barbiero AP et al. Structure of the *uvrB* gene of E. *coli*, homology with DNA repair enzymes and characterization of *uvrB5* mutation. Nucleic Acids Res 1986; 14:2877-2890.

27. Seeley TW, Grossman L. The role of the E. *coli* UvrB in nucleotide excision repair. J Biol Chem 1990; 265:7158-7165.

28. Orren DK, Selby CP, Hearst JE et al. Post-incision steps of nucleotide excision repair in E. *coli:* disassembly of the UvrBC-DNA complex by helicase II and DNA polymerase. J Biol Chem 1992; 267:780-788.

29. Van Houten B, Gamper H, Sancar A et al. DNase I footprint of ABC excinuclease. J Biol Chem 1987; 262:13180-13187.

30. Bertrand-Burggraf E, Selby CP, Hearst JE et al. Identification of the different intermediates in the interaction of (A)BC excinuclease with its substrate by DNase I footprinting on two uniquely modified oligonucleotides. J Mol Biol 1991; 218:27-36.

31. Shi Q, Thresher R, Sancar A et al. An electron microscopic study of (A)BC excinuclease. J Mol Biol 1992; 226:425-432.

32. Lin J-J, Phillips AM, Hearst JE et al. Active site of (A)BC excinuclease II. Binding, bending, and catalysis mutants of UvrB reveal a direct role in 3' and an indirect role in 5' incision. J Biol Chem 1992; 267:17693-17700.

33. Lin J-J, Sancar A. Active site of (A)BC excinuclease I. Evidence for 5' incision by UvrC through a catalytic site involving Asp[399], Asp[438], Asp[466], and His[538] residues. J Biol Chem 1992; 267:17688-17692.

34. Zou Y, Liu T-M, Geacintov NE et al. Interaction of the UvrABC nuclease system with a DNA duplex containing a single stereoisomer of dG-(+)- or dG-(-)-anti-BPDE. Biochemistry 1995; 34:13582-13593.

35. Moolenar GF, Franken KLMC, Dijkstra DM et al. The C-terminal region of UvrB protein of *Escherichia coli* contains an important determinant for UvrC binding to the preincision complex but not the catalytic site for 3'-incision. J Biol Chem 1995; 270:30508-30515.

36. Cohen C and Parry DA. Alpha-helical coiled coils and bundles: how to design an alpha-helical protein. Proteins 1990; 7:1-15.

37. Glover JN, Harrison SC. Crystal structure of the heterodimeric bZIP transcription factor c-Fos-c-Jun bound to DNA. Nature 1995; 373:257-261.

38. Sancar A, Hearst JE. Molecular matchmakers. Science 1993; 259:1415-1420.

39. Van Houten B, Snowden A. Mechanism of action of the *Escherichia coli* UvrABC nuclease: clues to the damage recognition problem. BioEssays 1993; 15:51-59.

40. Grossman L, Yeung AT. The UvrABC endonuclease system of *Escherichia coli*-a view from Baltimore. Mutat Res 1990; 236:213-221.

41. Walter RB, Pierce J, Case R et al. Rec-

ognition of the DNA helix stabilizing anthramycin-N2-guanine adduct by UVRABC nuclease. J Mol Biol 1988; 203:939-947.

42. Ramaswamy M, Yeung JE. Sequence-specific interactions of UvrABC endonuclease with psoralen interstrand crosslinks. J Biol Chem 1994; 269:485-492.

43. Snowden A, Van Houten B. Initiation of the UvrABC nuclease cleavage reaction: extent of incisions is not correlated with UvrA binding affinity. J Mol Biol 1991; 220:19-33.

44. Visse R, van Gool AJ, Moolenaar GF et al. The actual incision determines the efficiency of repair of cisplatin damaged DNA by the *Escherichia coli* UvrABC endonuclease. Biochemistry 1994; 33:1804-1811.

45. Wang JC, Giaever GN. Action at a distance along a DNA. Science 1988; 240:300-304.

46. Oh EY, Grossman L. Helicase properties of the *Escherichia coli* UvrAB protein complex. Proc Natl Acad Sci USA 1987; 84:3638-3642.

47. Spielmann HP, Dwyer TJ, Sastry SS et al. DNA structural reorganization upon conversion of a psoralen furan-side monoadduct to an interstrand cross-link: implications for DNA repair. Proc Natl Acad Sci USA 1995; 92:2345-2349.

48. Spielmann HP, Dwyer TJ, Hearst JE et al. Solution structure of psoralen monoadducted and cross-linked DNA oligomers by NMR spectroscopy and restrained molecular dynamics. Biochemistry 1995; 34:12937-12953.

49. Van Houten B, Illenye S, Qu Y et al. Homodinuclear (Pt, Pt) and heterodinuclear (Ru, Pt) metal compounds as DNA-protein cross-linking agents: potential suicide DNA lesions. Biochemistry 1993; 32:11794-11801.

50. Hsu DS, Kim S-T, Sun Q et al. Structure and function of UvrB. J Biol Chem 1995; 270:8319-8327.

51. Hanawalt PC, Haynes R. Repair replication of DNA in bacteria: irrelevance of chemical nature of base defect. Biochem Biophys Res Commun 1965; 19:462-467.

CHAPTER 6

MAMMALIAN NUCLEOTIDE EXCISION REPAIR

Mammalian nucleotide excision repair and the analog process initiated in bacteria by (A)BC excinuclease share many basic biochemical steps including DNA damage recognition, dual incision, oligonucleotide excision, repair synthesis and ligation (see Fig. 5.1). The similarities of mammalian nucleotide excision repair to the *Escherichia coli* system include the following:[1-7]

(1) one multisubunit excision nuclease removes all types of base adducts,

(2) DNA damage is released as a component of oligonucleotide segments,

(3) the nucleotide excision repair reaction is strictly ATP-dependent,

(4) transcribed strands of active genes are preferentially repaired (see chapter 9).

On the other hand, there are also important differences between (A)BC excinuclease in *Escherichia coli* and nucleotide excision repair in mammalian cells. The main differences are as follows:

(1) The number of mammalian nucleotide excision repair genes (and proteins) is significantly higher: a set of at least 30 polypeptides is required to catalyze the whole nucleotide excision repair process in mammals compared to only 6 polypeptides in bacteria.[3,5,8] A summary of all known factors that are directly involved in mammalian nucleotide excision repair is provided in Table 6.1,

(2) There is no extended sequence homology between the nucleotide excision repair proteins of mammals and prokaryotes,[1-4]

(3) The mammalian excision gap is 27-29 nucleotides in length compared to only 12-13 nucleotides in prokaryotes.[9,10] The main sites of DNA incision in mammalian nucleotide excision repair are the 22nd phosphodiester bond 5' and the 6th phosphodiester bond 3' to the lesion,[5,9] but some variability in this incision reaction has been observed,[11]

(4) Mammalian and prokaryotic nucleotide excision repair have distinctly different substrate preferences,[12]

(5) The mammalian nucleotide excision repair system recruits many of its factors from proteins involved in other basic pathways of DNA metabolism including replication (RPA, RFC, PCNA, DNA polymerase ε),[13] transcription (TFIIH),[14] and presumably recombination (XPF-ERCC1).[15] Proteins that are uniquely involved in nucleotide excision repair, primarily XPA and XPC, may serve to divert multifunctional factors from their normal activity in replication, transcription or recombination and assemble them into an excision repair-competent complex.

Mechanisms of DNA Damage Recognition in Mammalian Cells, by Hanspeter Naegeli.
© 1997 R.G. Landes Company.

Table 6.1. Factors required for mammalian nucleotide excision repair (see text for references)

Factor	Subunits	Main properties	Role in nucleotide excision repair
XPA	1 polypeptide	DNA binding with a preference for single-stranded DNA, interactions with RPA, TFIIH and ERCC1	DNA damage recognition? Anchor for nucleoprotein assembly?
RPA	p70 p34 p11	Single-stranded DNA binding Stimulation of XPF-ERCC1 and XPG endonucleases	DNA damage recognition? Stabilization of unwound DNA in the preincision complex? Stimulation of DNA repair synthesis?
TFIIH	p89/XPB/ERCC3 p80/XPD/ERCC2 p62/TFB1, p52 p44/SSL1 p38/cdk7/MO15 p37/cyclin H MAT1 p34	$3' \rightarrow 5'$ DNA helicase and ATPase $5' \rightarrow 3'$ DNA helicase and ATPase Zinc finger Cyclin-dependent protein kinase Regulatory partner of cdk7 Stimulation of protein kinase activity Zinc finger	DNA damage recognition? DNA unwinding in the preincision complex? Coupling nucleotide excision repair to transcription and cell cycle regulation?
XPC-HHR23B	XPC HHR23B	Single-stranded DNA binding Ubiquitin-like domain	DNA damage recognition? Molecular chaperone? Interaction with chromatin?
XPF-ERCC1	XPF, ERCC1	Single-stranded DNA endonuclease	5' Incision
XPG	1 polypeptide	Structure-specific DNA endonuclease	3' Incision
RFC	p140, p40, p38, p37, p36	DNA-dependent ATPase	Loading of PCNA onto DNA
PCNA	(p32)$_3$	DNA polymerase clamp	Processivity of DNA repair synthesis
DNA polymerase ε	p258 p55	PCNA-dependent DNA polymerase with proofreading activity	Synthesis of repair patches
DNA ligase I	DNA ligase I	DNA ligase	Ligation of repair patches

THE GENETIC FRAMEWORK

Research on mammalian nucleotide excision repair was greatly facilitated by the availability of human and Chinese hamster ovary cells deficient in this process. These cell lines provided a genetic framework to study the basic mechanisms involved in nucleotide excision repair.

XERODERMA PIGMENTOSUM COMPLEMENTATION GROUPS

As outlined in chapter 1, xeroderma pigmentosum is a rare autosomal recessive disease characterized by abnormal sensitivity to UV radiation, a strong predisposition to the development of skin cancer and, in some cases, neurological disorders.[2-4] A report published in 1968 established that the biochemical basis of xeroderma pigmentosum is a defect in nucleotide excision repair of UV radiation damage.[16] At that time, the most useful method to determine excision repair activity in mammalian cells was unscheduled DNA synthesis. This technique uses autoradiography to detect the incorporation of radioactive thymidine into nuclear DNA of individual cells treated with UV or bulky chemicals.[2,4,16] Replication of chromosomal DNA is restricted to the S phase of cell cycle. Thus, cells in S phase will be heavily labeled and can be distinguished from the lightly labeled nuclei containing nucleotide excision repair patches.

When unscheduled DNA synthesis was examined in the nuclei of heterokaryons formed by fusing cells from different xeroderma pigmentosum patients, certain combinations appeared normal in excision repair capacity. Such combinations were able to complement each other with respect to their nucleotide excision repair deficiency and, as a consequence, reflected mutations in distinct genes. These cell fusion studies defined seven xeroderma pigmentosum complementation groups, designated A-G. The respective genes were designated *XPA* through *XPG* (Table 6.1).[17] Strand break measurements in the chromatin of xeroderma pigmentosum cell lines, using alkaline sucrose gradients or the DNA unwinding technique, indicated that these complementa-tion groups are defective in the DNA incision step of nucleotide excision repair.[18] An eighth complementation group (XP-V or xeroderma pigmentosum variant) is defective in DNA replication (see chapter 10).[19]

HUMAN NUCLEOTIDE EXCISION REPAIR GENES

Screening of mutagenized cultures of rodent cell lines led to the isolation of UV-sensitive mutants that have subsequently been characterized and classified by cell fusion and unscheduled DNA synthesis.[20] To date, mutant rodent cells defective in nucleotide excision repair have been divided into 11 complementation groups designated with Arabic numerals.[17]

Because rodent cells are more easily transfected than human cells, these UV-sensitive mutants have been used as recipients in cloning human genes. Transfection of rodent cell lines with human genomic libraries, combined with phenotypic selection, was for many years the most successful strategy to isolate human nucleotide excision repair genes. This work led to the isolation of a number of human *ERCC* genes (for *excision repair cross complementing*). *ERCC1* has no *XP* equivalent;[21] *ERCC2* is identical to *XPD*;[22,23] *ERCC3* is the same as *XPB*;[24] *ERCC4* and *ERCC11* are equivalent to *XPF*;[25-27] *ERCC5* is the same as *XPG*;[28,29] *ERCC6* complements Cockayne syndrome complementation group B (CS-B),[30] *ERCC8* corresponds to *CSA*.[31]

XPA and *XPC* remain the only nucleotide excision repair genes that were cloned directly by complementation of the repair defect in xeroderma pigmentosum cell lines. *XPA* was originally isolated by transfection of XP-A cells with mouse genomic DNA,[32] and *XPC* was cloned using a human complementary DNA library engineered into an extrachromosomally replicating Epstein-Barr virus-based vector.[33]

THE BIOCHEMICAL FRAMEWORK

The identification of mammalian nucleotide excision repair genes stimulated overproduction, purification and biochemical

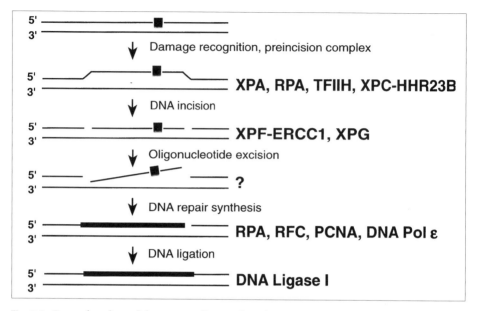

Fig. 6.1. General outline of the mammalian nucleotide excision repair pathway and its known factors (see also Table 6.1). The rectangle symbolizes a base adduct.

analysis of the respective gene products (summarized in Table 6.1). Studies performed in many laboratories established that most mammalian nucleotide excision repair proteins are active as components of multiprotein complexes. As schematically shown in Figure 6.1, three multisubunit factors (RPA, TFIIH and XPC-HHR23B) in combination with XPA are thought to promote DNA damage recognition and subsequent conformational changes of the DNA substrate required for damage-specific endonucleolytic cleavage. XPG then performs DNA incision on the 3' side to the lesion, followed by 5' incision through the action of the XPF-ERCC1 heterodimer (Fig. 6.1). After excision of an oligonucleotide containing the damage, synthesis of repair patches involves the coordinated action of the multisubunit replication factors RFC, PCNA, RPA and DNA polymerase ε (Fig. 6.1). Finally, continuity of the repaired strand is reestablished by DNA ligase I.[1-9]

XPA

XPA is a protein of 273 amino acids containing a zinc finger motif, a nuclear localization signal and at least four domains for spe-cific protein-protein interactions (Fig. 6.2). XPA binds DNA with a preference for single-stranded over double-stranded DNA.[34,35] It also has a preference for duplex DNA containing pyrimidine(6-4)pyrimidone photoproducts or other base lesions that disrupt the native conformation of the double helix and enhance its single strand character.[34,36] Purified XPA protein binds to all these forms of DNA rather weakly with association constants ranging between $6 \times 10^5 \, M^{-1}$ and $3 \times 10^6 \, M^{-1}$ (Ref 34). However, its striking ability to undergo multiple protein-protein interactions suggests that XPA may function as a nucleation center for the assembly of a damage recognition or incision complex on DNA.[5]

The zinc finger domain of XPA displays the motif $Cys-X_2-Cys-X_{17}-Cys-X_2-Cys$,[32,37] and is indispensable for normal XPA protein structure and function. A mutant XPA protein, in which the zinc finger was disrupted by replacing one of the cysteine residues with serine, has a vastly different protein conformation and results in the complete loss of nucleotide excision repair activity.[37] The requirement for zinc finger motifs in DNA repair is a common theme as there are several other examples of zinc

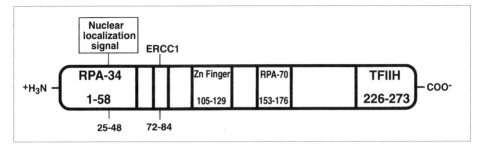

Fig. 6.2. Domain structure of XPA, a zinc finger protein of 273 amino acids that interacts with RPA, ERCC1 and TFIIH. Residues 1-58 and 153-176 are required for the interactions with the 34 and 70 kDa subunits of RPA. Residues 72-84 are involved in the binding to ERCC1, while residues 226-273 have been implicated in interactions with TFIIH (see refs. 42-49).

fingers implicated in the interaction with damaged DNA substrates, including UvrA[38] (a subunit of bacterial nucleotide excision repair; see chapter 5), Fpg protein[39] (a DNA glycosylase; see chapter 4), yeast Rad18 protein[40] (a putative DNA helicase involved in replication of damaged DNA; see chapter 10), and poly(ADP-ribose)polymerase[41] (a unique eukaryotic enzyme that synthesizes ADP-ribose polymers in response to DNA strand breaks).

As mentioned before, XPA has several separate domains for specific interactions with other nucleotide excision repair factors (Fig. 6.2). XPA makes a tight complex with two subunits (p34 and p70) of replication protein A (RPA).[42-44] The resulting tetrameric XPA-RPA complex displays a preference for damaged over nondamaged double-stranded DNA and, hence, constitutes a putative damage recognition subunit of mammalian nucleotide excision repair (see section on multiprotein assembly below).[42,44] XPA also interacts with XPF-ERCC1,[45-48] a dimeric protein complex with endonuclease activity that is responsible for DNA incision 5' to the lesion. Finally, the C-terminal domain of XPA is able to interact with the multisubunit transcription factor TFIIH.[49]

RPA

Mammalian single-stranded DNA binding protein, designated replication protein A (RPA) is an essential factor for multiple processes in DNA metabolism, including DNA replication,[50-52] nucleotide excision repair,[52,53] recombination[52,54] and transcription.[55] RPA is a tight complex made up of three polypeptides of 70 (RPA-70), 34 (RPA-34) and 14 kDa (RPA-14). The p70 subunit mediates binding to single-stranded DNA with an association constant of ~10^9 M^{-1}, but it has a low affinity for native double-stranded DNA. The other two subunits are thought to provide interfaces for the interaction with other cellular proteins.[56-59] RPA-70 also displays a putative zinc finger domain near its C terminus.[60]

By virtue of its single-stranded DNA binding activity, RPA may adopt several specific functions in mammalian nucleotide excision repair. As an essential replication accessory factor, it may stimulate DNA repair synthesis following dual incision and formation of single strand gaps in the substrate.[50,51,53] Additionally, RPA may act at biochemical steps that precede dual incision, for example by stabilizing a preincision complex in which DNA is partially denatured.[61]

Early work by Toulmé and coworkers on the single-stranded DNA binding protein of bacteriophage T4 (gene 32 protein) predicted that factors with affinity for DNA single strands may contribute to DNA damage recognition.[62] These authors used fluorescence spectroscopy to investigate the interaction between gene 32 protein and DNA substrates that had been treated with different genotoxic agents. Their results indicate that gene 32 protein binds more efficiently to DNA modified with either cis-diamminedichloroplatinum(II) or various aminofluorene derivatives than to native double-stranded DNA. The DNA binding domain of gene 32 protein may include a tryptophan residue.[50] Interestingly, the short peptide Lys-Trp-Lys also interacts preferentially with damaged double-stranded DNA,[63] presumably through stacking interactions between the indole ring of tryptophan and DNA bases. These studies led to the conclusion that certain single-stranded DNA binding proteins may serve to probe DNA conformation and recognize sites at which the double helix is partially denatured by base adducts. The increased affinity of the XPA-RPA complex for damaged DNA relative to nondamaged substrates[42,44] seems to confirm the prediction by Toulmé et al that single-stranded DNA binding proteins may be directly involved in DNA damage recognition.

Transcription Factor TFIIH

The involvement of a general transcription factor (TFIIH) during nucleotide excision repair in mammals[14,64,65] and yeast[66] was reported independently by several laboratories. While searching for a factor that promotes transcription by RNA polymerase II, these groups stumbled upon a multiprotein complex that contains, among other polypeptides, the products of the nucleotide excision repair genes *XPB* and *XPD*.

TFIIH is a multifunctional complex made up of 7-9 polypeptides, depending on the purification scheme.[67-70] Its subunits include p89/XPB/ERCC3, p80/XPD/ERCC2, p62/TFB1, p52/TFB2, p44/SSL1, p38/cdk7/MO15, p37/cyclin H and p32/MAT1 (Table 6.1).[69] This complex possesses DNA-dependent ATPase, 5'→3' DNA helicase, 3'→5' DNA helicase (see Fig. 5.2 for the polarity of DNA helicases) and protein kinase activity, and participates in a late step of transcription initiation,[14,64-66,69] in nucleotide excision repair[14,64-66] and, presumably, cell cycle regulation.[68,71] These different functions appear to be associated with several distinct TFIIH complexes that differ qualitatively and/or quantitatively in their polypeptide composition.[67,71,72]

In transcription, TFIIH is one of several factors required for initiation of RNA synthesis by RNA polymerase II. In the absence of TFIIH, RNA polymerase II stalls immediately downstream of the transcription start site and produces aborted transcripts of less than 50 residues in length, indicating that TFIIH relieves a block that prevents the polymerase from entering the elongation phase of RNA synthesis.[73] The term "promoter clearance" has been coined to indicate that TFIIH functions to overcome a functional block to the initial translocation of RNA polymerase II along DNA.[69,73] It is reasonable to assume that TFIIH may catalyze a similar reaction in the context of nucleotide excision repair, perhaps by using its 5'→3' and 3'→5' DNA helicase activities to unwind the duplex and promote translocation of a recognition complex at or near sites of damage (see section on DNA helicases below).

The presence of a protein kinase activity in TFIIH led to the attractive idea that this multifunctional complex may provide a link between transcription, repair and cell cycle regulation.[68,71] The protein kinase component of TFIIH is strictly required in transcription, where it phosphorylates the C-terminal domain of the large subunit of RNA polymerase II.[74] Conversely, this TFIIH-associated protein kinase seems to be dispensable for nucleotide excision repair.[6] However, it may play a role in cell cycle regulation, as several groups found that the protein kinase complexed with TFIIH is identical to cdk7/MO15, a polypeptide that was previously identified as the catalytic subunit of a cyclin H-dependent kinase.[71,75-77] Subsequent studies showed that TFIIH also contains cyclin H, the regulatory partner of cdk7/MO15, and MAT1 (for *ménage a trois*), a factor that stimulates the protein kinase activity of cdk7.[68,71] Since the complex formed by cdk7, cyclin H and MAT1 is believed to play a role in cell cycle regulation,[78] it was postulated that TFIIH may respond to DNA damage by initiating (or interrupting) a signaling cascade, thereby leading to arrest of cell cycle progression.[71] Recently, Egly and coworkers reported that the TFIIH-associated cdk7 kinase activity is reduced in cells that have been exposed to UV light.[68] This finding argues in favor of a regulatory circuit that is responsive to phosphorylation signals emanating from TFIIH.

XPC-HHR23B

The XPC-HHR23B heterodimer was purified to homogeneity by in vitro complementation of the XP complementation group C excision repair defect in a cell-free repair system.[79] The isolated factor consists of two tightly associated proteins of 125 and 58 kDa. The 125 kDa protein represents the *XPC* gene product,[33] while the 58 kDa was designated HHR23B because it is a human homolog of the yeast nucleotide excision repair protein Rad23. A second human homolog of yeast *RAD23* was also identified and designated *HHR23A*, but only the *HHR23B* gene product is found complexed with XPC.[79,80]

The XPC-HHR23B heterodimer binds single-stranded and double-stranded DNA with similar affinity (binding constants of $\sim 2 \cdot 10^8$ M^{-1}),[81] but no enzymatic activity could be associated with this factor. The only striking domain recognizable in the amino acid sequence of its two subunits is an ubiquitin-like N-terminus in the Rad23 homolog.[79] It has been suggested that ubiquitin moieties may act as a chaperone enabling proper folding and assembly in multiprotein complexes.[82] Support for this hypothesis is provided by the recent observation that Rad23 promotes association of Rad14 protein (the yeast homolog of XPA) with TFIIH in vitro.[83]

XPC protein appears to be selectively implicated in the repair of nontranscribed bulk DNA. In fact, XP-C cell lines are defective in the genome overall repair but maintain normal capacity to process the transcribed strand of active genes.[84] Several models have been proposed to accommodate this unique phenotype. For example, XPC-HHR23B may exploit its affinity for DNA to facilitate the assembly of a nucleoprotein complex (possibly involving XPA and TFIIH) and initiate nucleotide excision repair on nontranscribed DNA.[85] In this model, the repair of transcribed sequences does not require XPC-HHR23B because XPA and TFIIH may be recruited directly by the stalled transcription elongation complex (see chapter 9). Alternatively, XPC-HHR23B may function to uncouple TFIIH from the basal transcription initiation machinery and make it available for nucleotide excision repair of inactive genomic regions.[79] Another scenario predicts that XPC-HHR23B may be involved in the repair of transcriptionally silent DNA segments by altering chromatin structure and providing accessibility of nucleotide excision repair to the DNA substrate.[86] The possible function of the XPC-HHR23B complex was also investigated by reconstituting the human nucleotide excision repair system in vitro.[6] These biochemical studies indicate that XPC-HHR23B may be important to target the nuclease subunits (XPF-ERCC1 and XPG) to the proper sites of dual incision, thereby protecting nondamaged DNA near the lesion from nonspecific degradation.[6] Following this model, XPC-HHR23B may be dispensable during the repair of transcribed strands because RNA polymerase II provides sufficient protection from nonspecific incision reactions.

XPF-ERCC1 AND XPG: DUAL DNA INCISION

Two distinct nuclease activities are involved in double DNA incision during mammalian nucleotide excision repair. XPF-ERCC1 incises DNA on the 5' side of a lesion, while XPG incises DNA on the 3' side.

The products of *XPF* and *ERCC1* genes form a tight heterodimer with 1:1 stoichiometry[25,26] that displays single strand-specific endonuclease activity.[87] The XPF-ERCC1 complex acts preferentially on single-stranded DNA or the single-stranded region of "bubble" substrates, i.e., duplex DNA containing a noncomplementary sequence of 30 nucleotides in the center.[87] In the presence of RPA, XPF-ERCC1 stops to cut single-stranded DNA indiscriminately and adopts a specific double-stranded/single-stranded DNA junction cutting activity, indicating that RPA confers structure specificity to the XPF-ERCC1 endonuclease.[61] Under these conditions, only the strand that undergoes the transition from double-stranded to single-stranded DNA in the 5' to 3' direction is cleaved by XPF-ERCC1 (Fig. 6.3).[87] In addition to conferring structure specificity, RPA also greatly stimulates the DNA incision activity of XPF-ERCC1.[61]

XPG incises "Y-shaped" DNA substrates (consisting of a duplex region with two single-stranded arms) by cutting at the boundary between double-stranded and single-stranded DNA.[88] The incision is made at the branch point, a few bases into the duplex region (Fig. 6.3). When incubated with a "bubble" structure, XPG also cleaves at the border between double-stranded and single-stranded DNA, but only at the 3' side of the noncomplementary region (Fig. 6.3).[88] XPG is active as a solitary protein but, as was shown for the XPF-ERCC1 endonuclease, its structure-specific endonuclease function is stimulated by RPA.[61] Also, in vitro-translated XPG has been shown to interact with several subunits of TFIIH and with Cockayne syndrome group B protein.[89]

In summary, studies using model "bubble" or "Y-shaped" substrates are consistent with a model of damage excision where RPA stabilizes a region of unwound DNA around the damage and recruits the two endonucleases, XPF-ERCC1 on the 5' side and XPG on the 3' side.[1,3,61] This polarity of dual incision was confirmed by Matsunaga et al,[90] who showed that anti-ERCC1 antibodies inhibit the 5' incision without significantly affecting the 3' incision. Similarly, anti-XPG antibodies affected the 3' incision, confirming that XPF-ERCC1 is responsible for the 5' cut and XPG for the 3' cut. Subsequent kinetic studies using a reconstituted nucleotide excision repair system showed that 3' incision by XPG precedes 5' incision by XPF-ERCC1.[6]

POLYMERASES AND ACCESSORY FACTORS: DNA REPAIR SYNTHESIS

Dual incision generates a single strand gap of 27-29 residues in length that is filled in by DNA repair synthesis. The resulting repair patch was found to match exactly the gap generated by oligonucleotide excision without enlargement in either the 3' or the 5' direction.[9,10] DNA synthesis during nucleotide excision repair requires replication protein A (RPA), replication factor C (RFC), proliferating cell nuclear antigen (PCNA) and a PCNA-dependent DNA polymerase (Table 6.1).[13,53,91,92]

In DNA replication, the multisubunit factor RFC loads the ring-shaped homotrimeric PCNA onto DNA to form a polymerase clamp that enhances processivity of DNA synthesis.[93] In nucleotide excision repair, RFC and PCNA may have a dual function by first dissociating incision proteins from the substrate and then promoting the formation of a polymerase clamp.[6,13,94] Of the five mammalian DNA polymerases, only polymerase δ and polymerase ε require RFC/PCNA indicating that DNA repair synthesis in nucleotide excision repair is carried out by one of these two enzymes. In reconstituted in vitro systems, both polymerase δ and polymerase ε are able to promote DNA repair synthesis, but DNA polymerase ε is more efficient in generating DNA repair patches that are suitable substrates for subsequent ligation by DNA ligase I.[92]

FIDELITY OF DNA REPAIR SYNTHESIS

DNA repair synthesis in nucleotide excision repair should be essentially error-free for the following reasons. First, the nondamaged strand is used as a template to fill in the gap produced by oligonucleotide excision. Second, the DNA polymerase involved in this step (most likely DNA polymerase ε) is endowed with a proofreading activity that enhances its fidelity.[91,92]

In at least one example, however, DNA synthesis during nucleotide excision repair appears to be highly error-prone.[95] Simian virus 40-base shuttle vectors were site-specifically damaged with either O^4-methylthymine, O^4-ethylthymine or O^4-propylthymine. These alkyl adducts at O^4 of thymine are highly mutagenic lesions.[96-98] When the modified vectors were transfected into nucleotide excision repair-proficient

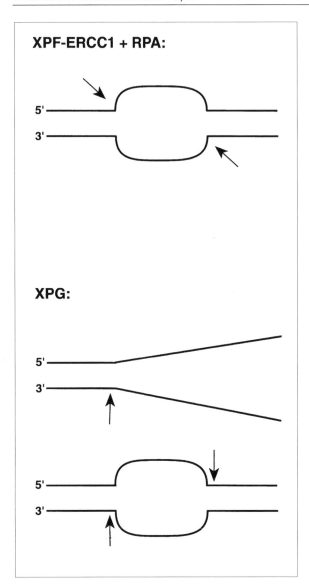

Fig. 6.3. Structure-specific endonuclease activity of the XPF-ERCC1-RPA complex and of XPG. The sites of DNA incision are indicated by the arrows. In the presence of RPA, XPF-ERCC1 incises only the strands that undergo a transition from double-stranded to single-stranded DNA in the 5' to 3' direction. XPG is active on the same junctions but with opposite polarity.[61,87,88]

human cells, all three types of O^4-alkyl-thymines exhibited mutagenic properties.[95] In nucleotide excision repair-deficient XPA cells, on the other hand, the vectors were poorly replicated but the tested O^4-alkyl lesions lost their mutagenicity. These surprising results suggest that processing of O^4-alkylthymines by nucleotide excision repair correlates with a high frequency of mutation induction. The authors proposed, as a possible explanation for this phenomenon, that the nucleotide excision repair complex may be unable to correctly discriminate between nondamaged and O^4-alkyl-thymine-damaged strands. As a consequence, nondamaged strands are erroneously incised, and subsequent DNA repair synthesis using damaged strands as templates generates mutations opposite O^4-alkylthymine residues.[95] Alternatively, these studies may hint to an unexpected function of XPA in facilitating the mutagenic bypass of DNA lesions during replication.

DAMAGE RECOGNITION IN MAMMALIAN NUCLEOTIDE EXCISION REPAIR

What proteins are implicated in the damage recognition step of mammalian nucleotide excision repair? After eliminating those factors that are unequivocally involved in dual DNA incision, DNA repair synthesis and DNA ligation, we remain with a minimum of four known factors (and at least 15 polypeptides) that may potentially operate during early, preincisional reactions: XPA, RPA, XPC-HHR23B and TFIIH (Table 6.1). However, the mechanism by which three factors with affinity for single-stranded DNA (XPA, RPA, XPC-HHR23B) or double-stranded DNA (XPC-HHR23B), in combination with a multisubunit complex containing ATPase and DNA helicase activity (TFIIH), may cooperate to recognize and locate a broad range of DNA lesions in the genome is poorly understood.[5] Therefore, the second part of this chapter summarizes current problems associated with the unresolved question of DNA damage recognition in mammalian nucleotide excision repair.

DIFFERENTIAL RECOGNITION OF UV RADIATION PRODUCTS

Although mammalian nucleotide excision repair processes essentially all forms of base adducts, its response is highly variable with a preference for particular types of lesions. This heterogeneity of mammalian nucleotide excision repair is illustrated by kinetic differences in the removal of cyclobutane pyrimidine dimers and pyrimidine(6-4)pyrimidone photoproducts.

When DNA is exposed to UV light with peak output at 254 nm, cyclobutane pyrimidine dimers and pyrimidine(6-4)pyrimidone photoproducts are formed in a ratio of about 3:1 (see Fig. 3.6).[99] Both major forms of UV radiation damage are processed efficiently in mammalian cells, but with dif-

ferent rates. Human cells, for example, remove 50% of (6-4) photoproducts in only 1 h after UV irradiation, but 50% of cyclobutane dimers in 8 h.[100] Kinetic differences in the excision of these UV radiation products have also been observed in rodent systems (see Table 2.1). Several studies have implicated (6-4) photoproducts as the main UV-induced cytotoxic and mutagenic lesion.[100-102] Thus, the up to 10-fold higher repair rate of (6-4) photoproducts may serve to protect cells from the more deleterious consequences of this particular lesion.

DAMAGED DNA BINDING PROTEINS

Several proteins that bind specifically to damaged DNA have been identified, but not all of these factors are involved in DNA repair. The main technique used to search for damaged DNA binding (DDB) proteins, various forms of electrophoretic gel mobility shift assays, has been extremely useful to identify and characterize sequence-specific factors. However, this method has yielded disappointing results with regard to the isolation of DNA damage recognition proteins from crude cell extracts. For example, XPA and RPA, which are now believed to constitute damage recognition subunits of mammalian nucleotide excision repair were completely missed. On the other hand, many proteins that were isolated on the basis of their specific interaction with damaged DNA, as assayed by mobility shift analysis, are not involved in DNA excision repair at all. On the contrary, some of these proteins may potentiate the cytotoxic effects of DNA damaging agents by inhibiting rather than facilitating DNA repair. For example, certain transcription factors such as hUBF (*h*uman *u*pstream *b*inding *f*actor) bind with high affinity to a specific class of DNA adducts generated by the antineoplastic drug cisplatin.[103] This type of interaction may titrate essential transcription factors away from their natural promoter or enhancer sequences and, additionally, protect the platinated sites from excision repair processes (see chapter 8). DDB proteins have also been reported for UV radiation products,[104] acetylaminofluorene adducts,[105] apurinic/apyrimidinic sites[106] or $1,N^6$-ethenoadenine in DNA.[107]

The first human DDB factor was identified 20 years ago by filter retention assay using UV irradiated DNA probes,[108] and partially purified from human placenta as a polypeptide of 120 kDa.[109] The same factor was rediscovered about 12 years later, when Chu and Chang observed that cells from some XP-E patients were missing a DDB activity.[110] This finding indicated that DDB protein may function as a DNA damage recognition factor in nucleotide excision repair, as xeroderma pigmentosum patients are specifically defective in this repair process.

The purification of DDB/XPE protein was reported almost simultaneously by two laboratories with slightly different results. Hwang and Chu[111] obtained a 125 kDa polypeptide which migrated as a monomer on gel filtration and glycerol gradient sedimentation. Keeney et al[112] isolated a heterodimeric protein consisting of two polypeptides of 124 and 41 kDa. Microinjection experiments have subsequently demonstrated that this heterodimeric factor corrects the DNA repair defect in those XP-E cells lacking DDB activity.[113]

Using synthetic DNA substrates containing one of the major UV radiation products at a single position, Sancar and collaborators analyzed the DNA binding affinity of the heterodimeric DDB/XPE factor by mobility shift assays.[114] These authors found that DDB/XPE discriminates only modestly between cyclobutane pyrimidine dimer-containing DNA and nondamaged DNA while the factor binds to pyrimidine-(6-4)pyrimidone photoproducts with much higher affinity. The binding constants are 5.5×10^8 M^{-1} for nondamaged DNA fragments, 1.7×10^9 M^{-1} for fragments containing a site-directed cyclobutane dimer and 1.6×10^{10} M^{-1} for fragments containing a site-directed (6-4) photoproduct. Thus, the reaction constants for binding of DDB/XPE to DNA are ~3 and ~30 times higher in the presence of cyclobutane dimers and (6-4) photoproducts, respectively, than in the absence of

DNA damage. These results are consistent with DDB/XPE being a damage recognition component of mammalian nucleotide excision repair, as cyclobutane dimers are a relatively poor substrate for this process, whereas (6-4) photoproducts are rapidly removed in mammalian cells.

Despite these detailed studies, the precise function of DDB/XPE remains uncertain. In fact, a complete lack of DDB activity correlates with only ~50% reduction in nucleotide excision repair efficiency in the affected XP-E patients,[4] indicating that DDB/XPE cannot be the only factor responsible for targeting the nucleotide excision repair machinery to sites of DNA damage. This view is supported by the observation that DDB/XPE is not absolutely necessary to reconstitute human nucleotide excision repair from purified components in vitro.[6,8,91]

DDB/XPE is relatively abundant with nearly 10^5 molecules/cell and, on the basis of this observation, it has been proposed that it may constitute a structural element of mammalian chromatin that facilitates some steps of nucleotide excision repair.[112,114] For example, XPE/DDB may increase the efficiency of damage recognition by a mechanism similar to that proposed for DNA photolyase of *Escherichia coli*, which enhances the rate of DNA incision by (A)BC excinuclease (see chapter 8). According to this model, XPE would bind to UV photoproducts and act as an antenna to recruit the nucleotide excision repair complex. This function could be mediated directly by protein-protein interactions, or indirectly by inducing conformational changes of the DNA helix that promote recognition. Alternatively, it has been proposed that XPE/DDB may facilitate dissociation of nucleotide excision repair complexes from the substrate after dual DNA incision.[115] By increasing the turnover of postincision complexes, XPE/DDB may improve the overall rate of excision repair. Another model proposes that XPE/DDB is a molecular chaperone that promotes specific protein-protein interactions during the assembly of multimeric complexes.[116] A similar function has already been tentatively assigned to the XPC-HHR23B heterodimer.[82,83]

THE DNA BINDING FUNCTION OF XPA

The studies conducted on DDB/XPE protein indicated that mammalian cells must express additional damage recognition proteins. One such candidate is XPA, a zinc finger protein that was shown by mobility shift assays to bind to DNA fragments with a preference for UV-irradiated over non-irradiated DNA.[34] The binding constant for the irradiated fragment ($\sim 3 \cdot 10^6$ M^{-1}) was 5-fold higher than that for the nonirradiated counterpart ($\sim 6 \cdot 10^5$ M^{-1}). This interaction of purified XPA with DNA is evidently rather weak with a binding affinity approximately 5,000-fold lower than that of XPE for the same substrate (see previous section). Furthermore, removal of cyclobutane dimers by treatment with DNA photolyase did not detectably reduce binding of XPA to the irradiated fragment. Thus, the increased affinity of XPA protein for UV-irradiated DNA is mediated exclusively by pyrimidine(6-4)pyrimidone photoproducts.[34] Additional experiments showed that XPA also has a 5-fold greater affinity for single-stranded over double-stranded DNA. Collectively, these results are consistent with XPA being able to recognize the single strand character of DNA containing certain types of lesions which, like (6-4) photoproducts, greatly distort the double helix.[34,35]

Although the preference of XPA for (6-4) photoproducts seems to correlate again with the higher rate of repair of this particular lesion in mammalian cells, it was surprising to find that XPA protein does not bind to cyclobutane pyrimidine dimers at all. In fact, the repair of both major UV radiation products is strictly dependent on XPA protein, and cyclobutane dimer removal is essentially abolished in XP-A cells.[1-5] As a consequence, it appears that the interaction of purified XPA protein with damaged DNA fragments in vitro does not precisely reflect its function in the nucleotide excision repair pathway. Also, the mechanism by which

more subtle DNA lesions such as cyclobutane pyrimidine dimers are recognized in mammalian cells remains unknown.

MULTIPROTEIN ASSEMBLY IN DNA DAMAGE RECOGNITION

To discriminate various forms of damage among the 10^{10} base pairs of mammalian genomes, but avoid DNA repair at improper (nondamaged) sites, nucleotide excision repair requires extremely high levels of precision in the localization of modified nucleotides. This requirement is not unique for DNA repair, as other DNA transactions also depend on the ability to select highly specific sites in the genome. It has been pointed out that sequence-dependent events such as transcription, replication, or site-specific recombination are unlikely to be directed by the binding of a single protein to a single DNA site.[117] On the contrary, these complex DNA transactions are effected by multiple protein-DNA and protein-protein interactions. Such macromolecular interactions are thought to regulate the ordered assembly of DNA-bound proteins to generate highly organized nucleoprotein structures in which DNA becomes folded or wound, thereby enhancing the specificity of protein-DNA recognition.[117]

Mechanistic studies on the prokaryotic (A)BC nucleotide excision repair system[118] indicate, on the other hand, that DNA repair pathways may achieve selectivity for their specific targets by a sequence of transient interactions between individual factors, rather then by the assembly of large multiprotein complexes. Bertrand-Burggraf et al[118] introduced the term "selectivity cascades" to indicate that in the prokaryotic nucleotide excision pathway a sequence of partly overlapping steps of low selectivity eventually results in very high specificity for sites of DNA damage. Similarly, the in vitro finding that human nucleotide excision repair acts efficiently on relatively short DNA substrates of ~ 100 base pairs in length has been used as an argument to support the idea that the proteins involved in damage recognition and damage-specific incision in the mammalian system may function in a sequential manner, as it appears unlikely that such short DNA fragments would be able to accommodate large nucleoprotein structures.[119] This view is apparently contradicted by the observation that yeast nucleotide excision repair, which is highly homologous to the mammalian system, does indeed involve the assembly of large multiprotein complexes.[72,120]

In mammalian nucleotide excision repair, an increasing number of multiprotein complexes have been identified indicating that protein-protein interactions play a crucial role, although it remains to be elucidated whether these interactions occur transiently or result in progressively more complex structures at sites of damage. RPA is a stable complex composed of three subunits.[56] XPB, XPD and several additional gene products are subunits of TFIIH.[64-66] XPF and ERCC1 form a tight dimeric complex with single strand-specific endonuclease activity.[25,26] XPC forms a heterodimer with HHR23B.[79]

In addition to these stable complexes, weaker interactions of presumably transient nature have been identified, mainly using the two-hybrid system and coprecipitation with either specific antibodies or glutathione beads. For example, XPA protein associates with at least three different nucleotide excision repair partners, i.e., the 34 and 70 kDa subunits of RPA,[42-44] the XPF-ERCC1 heterodimer[45-48] and transcription factor TFIIH.[49] As shown in the map of Figure 6.2, specialized domains of XPA are involved in these interactions. The portion of XPA mediating binding to RPA-34 was identified within its first 58 residues. A second domain, located between XPA residues 153 and 176, mediates the interaction with RPA-70. Deletion mutants of XPA that fail to bind RPA-70 are deficient in nucleotide excision repair.[44] The interaction domain with ERCC1 has been mapped to XPA residues 72-84.[46,48] Again, mutations in XPA that prevent association with ERCC1 confer defective nucleotide excision repair.[46] Finally, binding to TFIIH has been shown to involve XPA residues 226-273. TFIIH may additionally be

loosely associated with XPC, XPG and the XPF-ERCC1 heterodimer.[8,72,89,121] In summary, the many interaction partners of XPA and TFIIH define a network of protein-protein interactions that may constitute the molecular basis for the assembly of a large excision repair factory.

Importantly, some of these interactions appear to have functional consequences with respect to DNA damage recognition. For example, DNA that has been UV- or acetylaminofluorene-damaged and subsequently immobilized on Dynabeads displays a 5- to 10-fold higher capacity to bind XPA-RPA complexes than either component (XPA or RPA) alone.[42,44] These results are suggestive of damage recognition by a mechanism that is based on the cooperative binding of XPA and RPA to DNA lesions. It is not clear whether the greater binding capacity of damaged DNA for XPA-RPA reflects a higher affinity of these proteins for damaged substrates or a greater stability of the resulting nucleoprotein complexes. Another caveat that has to be kept in mind in future experiments is that physical association between XPA and RPA also stimulates their binding to nondamaged DNA.[44,122]

Using filter retention assays, it has been shown that association of XPA with ERCC1 stimulates protein binding to UV-irradiated DNA up to 7-fold.[48] No increase in binding activity was observed in the presence of nondamaged DNA. Also, ERCC1 alone was completely unable to bind DNA, regardless of whether the nucleic acid substrate was damaged or not.[48] Collectively, these reports support the idea that the extraordinary selectivity of mammalian nucleotide excision repair for damaged DNA may be achieved, at least in part, by the assembly of multiprotein complexes at potential target sites.

THE POSSIBLE CONTRIBUTION OF DNA HELICASES

DNA helicases catalyze strand displacement by disrupting the hydrogen bonds that hold the two strands of duplex DNA together. This strand separation reaction strictly requires the hydrolysis of nucleoside

5'-triphosphates and involves unidirectional translocation of helicase enzymes along the DNA substrate (see Fig. 5.2).[123]

DNA helicases have been identified as major players in the sequence-specific initiation of prokaryotic replication.[124] Oh and Grossman were the first to propose that a DNA helicase may also be essential in the initiation of nucleotide excision repair.[125] As outlined in chapter 5, the (A)BC excinuclease performs damage recognition and double DNA incision in prokaryotic nucleotide excision repair.[126,127] UvrA and UvrB associate to form a UvrA$_2$B trimer that behaves like a DNA helicase on partial duplex DNA substrates, i.e., it catalyzes ATP hydrolysis coupled to DNA strand displacement.[125,127] The DNA helicase activity of UvrA$_2$B is inhibited by UV radiation damage, indicating that translocation of the protein complex is arrested by DNA lesions.[125,127] On the basis of these observations, Grossman and coworkers proposed that the DNA helicase activity of UvrA$_2$B may serve to scan short segments of DNA in search for damage to be processed by nucleotide excision repair.[125,127,128]

At least two DNA helicases, XPB and XPD, also participate during early steps of mammalian nucleotide excision repair.[14,64,65] These two enzymes are catalytic subunits of the multiprotein transcription/repair factor TFIIH. Their precise role in nucleotide excision repair is unknown, but it was noted that the DNA helicase activity of the highly conserved *Saccharomyces cerevisiae* homolog of XPD, Rad3 protein, is strongly inhibited by DNA damage. As illustrated in Figure 6.4, this inhibitory effect was observed when damage was located in the strand along which the enzyme translocates during the unwinding reaction, but no inhibition of Rad3 helicase activity was observed when damage was located in the complementary strand.[129-132] Using filter binding and competition assays, it was additionally shown that Rad3 protein forms stable complexes with damaged DNA but not with nondamaged DNA.[121,130,132] Thus, DNA lesions apparently block translocation of Rad3 pro-

tein in a strand-specific manner and induce the formation of abnormally stable Rad3 protein-DNA complexes. As a consequence, Rad3 protein becomes sequestered on the DNA template, presumably at sites of nucleotide damage. These results suggest that nucleotide excision repair may have adopted enzymes with DNA helicase activity (Rad3 in yeast and its homolog XPD in humans) to discriminate between nondamaged and damaged DNA constituents.[133]

FACILITATED DISTORTION ENHANCES RECOGNITION

A novel principle in protein-DNA recognition was indicated by studies on the mechanism by which the restriction enzyme *Eco*RI selects its recognition sequence 5'-GAATTC. Lesser et al[134] analyzed the energetics of site-specific contacts within the *Eco*RI-DNA complex using base analogs that change only a single hydrogen bonding position. Most base substitutions produced the expected losses of binding free energy. However, with one of the analogs tested, one that eliminates the N^6 amino group from a central A·T base pair, the binding free energy became more favorable. Thus, deletion of this particular hydrogen bonding position for protein-DNA interaction improved the binding of *Eco*RI to its recognition sequence.[134] Since previous work demonstrated that the *Eco*RI recognition sequence is accommodated into the *Eco*RI-DNA complex by a kink in the nucleic acid substrate,[135] it was concluded that the base substitution must facilitate DNA distortion into the unique conformation necessary for *Eco*RI-DNA interactions.[134] The loss of a protein-DNA hydrogen bond at the central base pairs was more than compensated for by this distortive effect on DNA structure, indicating that, in addition to hydrogen bonding, distortion of the DNA structure is a thermodynamically important feature of this interaction.

Protein-induced distortion seems to be a common theme in protein-DNA recognition, where it appears to enhance the selectivity for specific binding sites.[136] For ex-

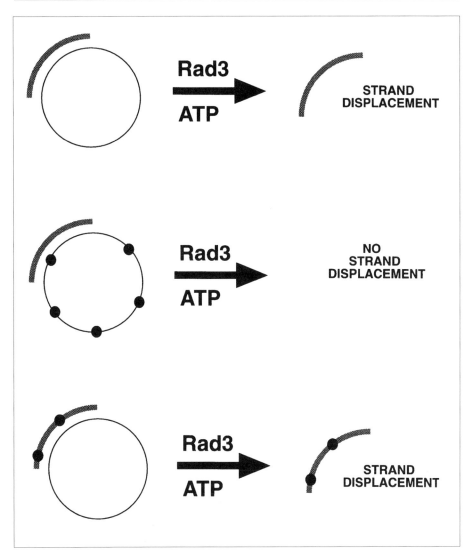

Fig. 6.4. Strand-specific inhibition of Rad3 ATPase/DNA helicase activity. Rad3 DNA helicase is blocked by damage located in the strand to which the enzyme binds, but not by damage located in the opposite complementary strand.[129-132] DNA damage is indicated by the filled circles.

ample, the first step in the formation of many transcription initiation complexes is the interaction of the TATA box binding protein (TBP), a component of transcription factor TFIID, to the TATA box element. Crystallographic analysis of this protein-DNA complex showed dramatic changes in the conformation of the double helix, in that

TBP causes DNA to bend by an angle of ~80°; in parallel, the TATA element is unwound by ~110°.[137] Both DNA bending and unwinding at the TATA element are likely to regulate the next levels of assembly of the multiprotein transcription initiation complex: DNA bending could bring other initiation factors closer together than on linear

DNA, while DNA unwinding could be the signal for further separation of double-stranded DNA.[137] Other transcription factors for which substantial DNA bending has been demonstrated include Cro protein or Sp1.[138,139]

With regard to DNA repair, a relevant example is provided by the UvrB-damaged DNA complex formed by prokaryotic (A)BC excinuclease. DNA is kinked and locally denatured by the $UvrA_2B$ heterotrimer; $UvrA_2$ then dissociates from the kinked DNA complex to induce subsequent steps required for damage-specific incision. Electron microscopy studies demonstrated that the remaining UvrB-DNA structure kinks the DNA molecule by an average of 127° (see chapter 5).[140] From these results, it can be expected that the mammalian nucleotide excision repair machinery also induces distortion of DNA at target sites.

A HYPOTHETICAL MODEL OF DAMAGE RECOGNITION

The purpose of the previous sections was to show that complex DNA transactions such as replication, recombination or transcription, achieve high levels of target site selectivity by the assembly of multiprotein complexes, often with significant contributions from ATPases/DNA helicases. In the resulting nucleoprotein structures, conformation of the nucleic acid substrate is normally distorted. These principles in protein-DNA recognition, together with the known biochemical properties of mammalian nucleotide excision repair factors, may be integrated in the following hypothetical model of DNA damage recognition (Fig. 6.5).

(1) Initiation. A damage recognition complex consisting of the multisubunit factor TFIIH, or only part of it, and other components is assembled on DNA. This initial step is facilitated by the DNA binding activity and the ubiquitin tail of XPC-HHR23B. XPA serves as a nucleation center for correct assembly of the recognition complex on DNA, and uses the single-stranded DNA binding domain of RPA to target the recognition complex preferentially to sites of re-duced helical stability. Recent experimental evidence underlying the impact of thermo-dynamic parameters during these early processes will be presented in chapter 7.

(2) Translocation. The DNA helicase activity of XPD and XPB drives unidirectional translocation of the recognition complex along DNA, in a reaction involving hydrolysis of ATP and transient strand displacement. The opposite polarity of the two DNA helicases (5'→3' and 3'→5') indicates that each enzyme translocates along one strand of the duplex. This translocation function may represent an extension of the activity of XPD and XPB during transcription initiation, where TFIIH acts to overcome functional blocks that prevent RNA polymerase II from entering the elongation phase of RNA synthesis.

(3) Strand-specific block. Translocation of the recognition complex is arrested at sites of covalent DNA damage, for example cyclobutane pyrimidine dimers, located in one of the two strands. The term strand-specific block is used here to indicate that only one of the two DNA helicases, either XPB or XPD, is blocked by the lesion.

(4) Distortion. Studies on Rad3 protein, the highly conserved yeast homolog of XPD, showed that inhibition of Rad3 DNA helicase activity by DNA damage is strictly DNA strand-specific (Fig. 6.4). Thus, after inhibition of one (either XPD or XPB) of the two DNA helicases by DNA damage, the opposite helicase bound to the nondamaged strand may retain the ability to proceed for at least one or a few additional nucleotides. To compensate for this asymmetric movement driven by ATP hydrolysis, DNA is distorted to form a unique structural intermediate. DNA bending and unwinding in this hypothetical recognition intermediate constitute critical parameters for subsequent protein-DNA and protein-protein interactions. For example, the particular orientation of DNA bending should eventually determine which strand is exposed to cleavage by endonucleases.

(5) DNA Incision. After these conformational changes, the resulting nucleoprotein

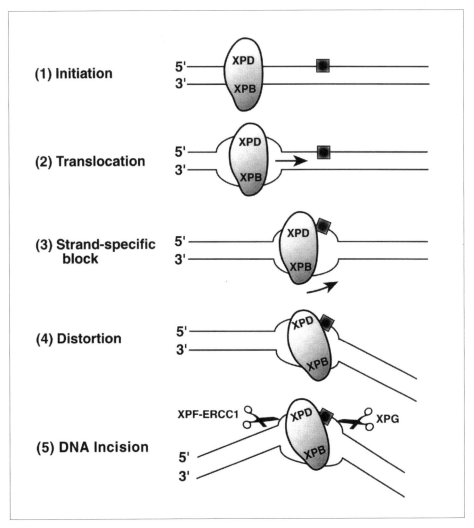

Fig. 6.5. Hypothetical model illustrating how a recognition complex may use the DNA helicase activity of XPB and XPD to discriminate between nondamaged and damaged substrates. The rectangle symbolizes a base adduct.

structure provides the substrate for dual incision by structure-specific DNA endonucleases, XPG protein on the one side and XPF-ERCC1 on the other side of the damage. XPA and RPA may be required throughout this sequence of reactions to anchor the damage recognition and incision complex to DNA.

The recognition model shown in Figure 6.5 accommodates the observation that nucleotide excision repair is active on a wide range of structurally or chemically dissimilar types of DNA damage. The model also accommodates the strand specificity associated with nucleotide excision repair, i.e., its ability to discriminate between the two strands of duplex DNA and target only the damaged strand for DNA incision and DNA repair synthesis. In addition, the concept of facilitated DNA distortion in protein-DNA recognition suggests that the model may also account for the enhanced repair rate of (6-4) photoproducts and some other specific DNA modifications.

FACILITATED DNA DISTORTION IN NUCLEOTIDE EXCISION REPAIR

XPA and XPE are not the only factors with a preference for pyrimidine(6-4)-pyrimidone photoproducts. RPA and Rad3 protein, the yeast homolog of XPD, also turned out to bind preferentially to DNA containing (6-4) photoproducts.[141,142] As was demonstrated for XPA and XPE, RPA and Rad3 protein have no affinity for cyclobutane pyrimidine dimers.[141,142] These results suggest that the (6-4) photoproduct binding activity of Rad3 may be conserved in XPD protein, its human homolog, and that XPD protein may mediate the binding of the whole TFIIH complex, of which it is a component, to the lesion.[142]

In summary, most nucleotide excision repair proteins that have been tested for interactions with (6-4) photoproducts actually have an affinity for this particular lesion. However, the temptation to conclude from these experiments that XPA, XPE, RPA and Rad3/XPD represent specific factors for the recognition of (6-4) photoproducts should be resisted. An alternative interpretation, suggested by the phenomenon of facilitated DNA distortion, is that DNA at (6-4) photoproducts contains structural elements which are also encountered during the formation of recognition complexes at any other type of DNA damage. The particular deformation introduced by (6-4) photoproducts may mimic a favorable DNA conformation that occurs during the assembly of a critical intermediate in damage recognition. Solution structure analysis of duplex DNA fragments containing a (6-4) photoproduct in the center showed a 44° bending in the helix and complete disruption of hydrogen bonding interactions between complementary bases at the 3' side of the photolesion.[143] Several nucleotide excision repair factors may have some affinity for this favorable conformation and are able

to bind to this structure. As a consequence, (6-4) photoproducts may promote the assembly of nucleotide excision repair recognition and incision complexes, just like a specific base modification was found to improve the binding of *Eco*RI to its recognition sequence.[134] Thus, recognition of (6-4) photoproducts and some other specific lesions may be initiated directly with step 5 of the model shown in Figure 6.5, thereby avoiding most rate limiting molecular events occurring during steps 1 to 4.

CONCLUSIONS

Damage recognition is a crucial step in any DNA repair or DNA damage processing pathway, but the versatility of nucleotide excision repair poses special problems with respect to the mechanism by which DNA modifications are discriminated as substrates for damage-specific incision and DNA repair synthesis. The molecular details involved in dual DNA incision, DNA repair synthesis or ligation have been elucidated to a considerable extent, but the most challenging aspect of the biochemistry of mammalian nucleotide excision repair appears to be its ability to recognize multiple forms of DNA damage, and efficiently reject nondamaged nucleotides as targets for DNA incision.

Interestingly, mammalian cells remove some of the most cytotoxic and mutagenic lesions with highest priority, and then patrol the genome to detect other types of damage, such as cyclobutane pyrimidine dimers, which are not recognized during the first rapid screening. As a consequence, any model of mammalian nucleotide excision repair must account for its ability to process a broad range of different DNA modifications and, simultaneously, explain its preference for specific lesions.

Many current models invoke XPA as the central player in DNA damage recognition during mammalian nucleotide excision repair. However, it appears difficult to conceive how a protein with poor DNA binding capacity (association constants of ~10^6 M^{-1}) and a moderate (~5-fold) preference for single-stranded DNA or substrates containing (6-4) photoproducts may promote, as a solitary protein or in complex with other subunits, the enormous level of molecular precision required for mammalian nucleotide excision repair. On the contrary, DNA damage recognition likely involves a series of highly coordinated molecular events dictated by the assembly of protein-DNA and protein-protein structures.

The sequential order and the manner by which the nucleotide excision repair machinery is assembled at sites of DNA damage remains to be determined. Figure 6.5 provides a hypothetical scenario in which selectivity for target sites is achieved through the assembly of a nucleoprotein complex that uses its ATPase/DNA helicase components to probe DNA chemistry. An important prediction in this alternative model is the formation of a specific intermediate in which DNA shares some conformational elements with the bent and unwound structure induced by (6-4) photoproducts, thereby facilitating the recognition of these specific lesions through their distortive effect on DNA. The proposed model is consistent with our recent finding that damage recognition in mammalian nucleotide excision repair is achieved by a bipartite strategy of substrate discrimination (summarized in chapter 7).

REFERENCES

1. Sancar A. DNA excision repair. Annu Rev Biochem 1996; 65:43-81.
2. Tanaka K, Wood RD. Xeroderma pigmentosum and nucleotide excision repair. Trends Biochem Sci 1994; 19:83-86.
3. Wood RD. DNA repair in eukaryotes. Annu Rev Biochem 1996; 65:135-167.
4. Hoeijmakers JHJ. Nucleotide excision repair II: from yeast to mammals. Trends Genet 1993; 9:211-217.
5. Sancar A. Excision repair in mammalian cells. J Biol Chem 1995; 270:15915-15918.
6. Mu D, Hsu DS, Sancar A. Reaction mechanism of human DNA repair excision nuclease. J Biol Chem 1996; 271:8285-8294.
7. Wood RD, Robins P, Lindahl T. Complementation of the xeroderma pigmentosum DNA repair defect in cell-free extracts. Cell 1988; 53:97-106.
8. Mu D, Park C-H, Matsunaga T et al. Reconstitution of human DNA repair excinuclease in a highly defined system. J Biol Chem 1995; 270:2415-2418.
9. Huang J-C, Svoboda DL, Reardon JT et al. Human nucleotide excision nuclease removes thymine dimers from DNA by the 22nd phosphodiester bond 5' and the 6th phosphodiester bond 3' to the photodimer. Proc Natl Acad Sci USA 1992; 89:3664-3668.
10. Hansson J, Munn M, Rupp WD et al. Localization of DNA repair synthesis by human cell extracts to a short region at the site of a lesion. J Biol Chem 1989; 264:21788-21792.
11. Moggs JG, Yarema KJ, Essigmann JM et al. Analysis of incision sites produced by human cell extracts and purified proteins during nucleotide excision repair of a 1,3-intrastrand d(GpTpG)-cisplatin adduct. J Biol Chem 1996; 271:7177-7186.
12. Huang J-C, Hsu DS, Kazantsev A et al. Substrate spectrum of human excinuclease: repair of abasic sites, methylated bases, mismatches, and bulky adducts. Proc Natl Acad Sci USA 1994; 91:12213-12217.
13. Shivji MKK, Kenny MK, Wood RD. Proliferating cell nuclear antigen is required for DNA excision repair. Cell 1992; 69:367-374.
14. Schaeffer L, Roy R, Humbert S et al. DNA repair helicase: a component of BTF2 (TFIIH) basic transcription factor. Science 1993; 260:58-63.
15. Schiestl RH, Prakash S. *RAD10*, an excision repair gene of Saccharomyces cerevisiae, is involved in the *RAD1* pathway of mitotic recombination. Mol Cell Biol 1990; 10:2485-2491.
16. Cleaver JE. Defective repair replication of DNA in xeroderma pigmentosum. Nature 1968; 218:652-656.
17. Lehmann AR, Bootsma D, Clarkson SG et al. Nomenclature of human DNA repair genes. Mutat Res 1994; 315:41-42.
18. Fornace AJ, Kohn KW, Kann HE. DNA single-strand breaks during repair of UV

damage in human fibroblasts and abnormalities or repair in xeroderma pigmentosum. Proc Natl Acad Sci USA 1976; 73:39-43.

19. Lehmann AR, Kirk-Bell S, Arlett CF et al. Xeroderma pigmentosum with normal levels of excision repair have a defect in DNA synthesis after UV-irradiation. Proc Natl Acad Sci USA 1975; 72:219-223.

20. Thompson LH. Somatic cell genetics approach to dissecting mammalian DNA repair. Environ Mol Mutagen 1989; 14:264-281.

21. Westerveld A, Hoeijmakers JH, van Duin M et al. Molecular cloning of a human DNA repair gene. Nature 1984; 310:425-429.

22. Weber CA, Salazar EP, Stewart SA et al. *ERCC2*: cDNA cloning and molecular characterization of a human nucleotide excision repair gene with high homology to yeast *RAD3*. EMBO J 1990; 9:1437-1447.

23. Flejter WL, McDaniel LD, Johns D et al. Correction of xeroderma pigmentosum complementation group D mutant cell phenotypes by chromosome and gene transfer: involvement of the human *ERCC2* DNA repair gene. Proc Natl Acad Sci USA 1992; 89:261-265.

24. Weeda G, Van Ham RCA, Vermeulen W et al. A presumed DNA helicases encoded by *ERCC-3* is involved in the human repair disorders xeroderma pigmentosum and Cockayne's syndrome. Cell 1990; 62:777-791.

25. Biggerstaff M, Szymkowski DE, Wood RD. Co-correction of the ERCC1, ERCC4 and xeroderma pigmentosum group F DNA repair defects in vitro. EMBO J 1993; 12:3685-3692.

26. van Vuuren AJ, Appeldoorn E, Odijk H et al. Evidence for a repair enzyme complex involving ERCC1 and complementing activities of ERCC4, ERCC11 and xeroderma pigmentosum group F. EMBO J 1993; 12:3693:3701.

27. Sijbers AM, de Laat WL, Ariza RR et al. Xeroderma pigmentosum group F caused by a defect in a structure-specific DNA repair endonuclease. Cell 1996; 86:811-822.

28. Scherly D, Nouspikel T, Corlet J et al.

Complementation of the DNA repair defect in xeroderma pigmentosum group G cells by a human cDNA related to yeast *RAD2*. Nature 1993: 363:182-185.

29. O'Donovan A, Wood RD. Identical defect in DNA repair in xeroderma pigmentosum group G and rodent ERCC group 5. Nature 1993; 363:185-188.

30. Troelstra C, Van Gool A, De Wit J et al. *ERCC6*, a member of a subfamily of putative helicases, is involved in Cockayne's syndrome and preferential repair of active genes. Cell 1992; 71:1-15.

31. Itoh T, Shiomi T, Harada Y et al. Rodent complementation group 8 (ERCC8) corresponds to Cockayne syndrome complementation group A. Mutat Res 1996; 362:167-174.

32. Tanaka K, Miura N, Satokata I et al. Analysis of a human DNA excision repair gene involved in group A xeroderma pigmentosum and containing a zinc finger domain. Nature 1990; 348:73-76.

33. Legerski R, Peterson C. Expression cloning of a human DNA repair gene involved in xeroderma pigmentosum group C. Nature 1992; 359:70-73.

34. Jones CJ, Wood RD. Preferential binding of the xeroderma pigmentosum group A complementing protein to damaged DNA. Biochemistry 1993; 32:12096-12104.

35. Eker APM, Vermeulen W, Miura N et al. Xeroderma pigmentosum group A correcting protein from calf thymus. Mutat Res 1992; 274:211-224.

36. Asashina H, Kuraoka I, Shirakawa M et al. The XPA protein is a zinc metalloprotein with an ability to recognize various kinds of DNA damage. Mutat Res 1994; 315:229-237.

37. Miyamoto I, Miura N, Niwa H et al. Mutational analysis of the structure and function of the xeroderma pigmentosum group A complementing protein: identification of essential domains for nuclear localization and DNA excision repair. J Biol Chem 1992; 267: 12182-12187.

38. Van Houten B. Nucleotide excision repair in Escherichia coli. Microbiol Rev 1990; 54:18-51.

39. Tchou J, Michaels ML, Miller JH et al. Function of the zinc finger in *Escherichia coli* Fpg protein. J Biol Chem 1993;

268:26738-26744.

40. Jones JS, Weber S, Prakash L. The Saccharomyces cerevisiae *RAD18* gene encodes a protein that contains potential zinc finger domains for nucleic acid binding and a putative nucleotide binding sequence. Nucleic Acids Res 1988; 16:7119-7131.

41. Gradwohl G, Menissier de Murcia JM, Molinete M et al. The second zinc-finger domain of poly(ADP-ribose) polymerase determines specificity for single-stranded breaks in DNA. Proc Natl Acad Sci USA 1990; 87:2990-2994.

42. He Z, Henricksen LA, Wold MS et al. RPA involvement in the damage-recognition and incision steps of nucleotide excision repair. Nature 1995; 374:566-568.

43. Matsuda T, Saijo M, Kuraoka I et al. DNA repair protein XPA binds replication protein A (RPA). J Biol Chem 1995; 270:4152-4157.

44. Li L, Lu X, Peterson CA et al. An interaction between the DNA repair factor XPA and replication protein A appears essential for nucleotide excision repair. Mol Cell Biol 1995; 15:5396-5402.

45. Li L, Elledge SJ, Peterson CA et al. Specific association between the human DNA repair proteins XPA and ERCC1. Proc Natl Acad Sci USA 1994; 91:5012-5016.

46. Li L, Peterson CA, Legerski RF. Mutations in XPA that prevent association with ERCC1 are defective in nucleotide excision repair. Mol Cell Biol 1995; 15:1993-1998.

47. Park CH, Sancar A. Formation of a ternary complex by human XPA, ERCC1 and ERCC4(XPF) excision repair proteins. Proc Natl Acad Sci USA 1994; 91:5017-5021.

48. Nagai A, Saijo M, Kuraoka I et al. Enhancement of damage-specific DNA binding of XPA by interaction with the ERCC1 DNA repair protein. Biochem Biophys Res Comm 1995; 211:960-966.

49. Park CH, Mu D, Reardon JT et al. The general transcription-repair factor TFIIH is recruited to the excision repair complex by the XPA protein independent of the TFIIE transcription factor. J Biol Chem 1995; 270:4896-4902.

50. Challberg MD, Kelly TJ. Animal virus DNA replication. Annu Rev Biochem 1989; 58:671-717.

51. Hurwitz J, Dean FB, Kwong AD et al. The in vitro replication of DNA containing the SV40 origin. J Biol Chem 1990; 265:18043-18046.

52. Longhese MP, Plevani P, Lucchini G. Replication factor A is required in vivo for DNA replication, repair, and recombination. Mol Cell Biol 1994; 14:7884-7890.

53. Coverley D, Kenny M, Munn M et al. Requirement for the replication protein SSB in human DNA excision repair. Nature 1991; 349:538-541.

54. Heyer W-D, Rao MR, Erdile LF et al. An essential *Saccharomyces cerevisiae* single-stranded DNA binding protein is homologous to the large subunit of human RP-A. EMBO J 1990; 9:2321-2329.

55. Li R, Botchan M. The acidic transcriptional activation domain of VP16 and p53 bind the cellular RPA and stimulate in vitro BPV-1 DNA replication. Cell 1993; 73:1207-1221.

56. Gomes XV, Wold MS. Structural analysis of human replication protein A. J Biol Chem 1995; 270:4534-4543.

57. Wobble CR, Weissbach L, Borowiec JA et al. Replication of simian virus 40 origin-containing DNA in vitro with purified proteins. Proc Natl Acad Sci USA 1987; 84:1834-1838.

58. Fairman MP, Stillman B. Cellular factors required for multiple stages of SV40 DNA replication in vitro. EMBO J. 1988; 7:1211-1218.

59. Wold MS, Kelly T. Purification and characterization of replication protein A, a cellular protein required for in vitro replication of simian virus 40 DNA. Proc Natl Acad Sci USA 1988; 85:2523-2527.

60. Lin Y-L, Clark C, Keshav KF et al. Dissection of functional domains of the human DNA replication protein complex replication protein A. J Biol Chem 1996; 271:17190-17198.

61. Matsunaga T, Park C-H, Bessho T et al. Replication protein A confers structure-specific endonuclease activities to the XPF-ERCC1 and XPG subunits of human DNA repair excision nuclease. J Biol Chem 1996; 271:11047-11050.

62. Toulmé JJ, Behmoaras T, Guignes M et

al. Recognition of chemically damaged DNA by the gene 32 protein from bacteriophage T4. EMBO J. 1983; 2:505-510.

63. Toulmé F, Hélène C, Fuchs RPP et al. Binding of a tryptophan-containing peptide (lysyltryptophyllysine) to deoxyribonucleic acid modified by 2(N-acetoxyacetylamino)fluorene. Biochemistry 1980; 19:870-875.

64. Drapkin R, Reardon JT, Ansari A et al. Dual role of TFIIH in DNA excision repair and in transcription by RNA polymerase II. Nature 1994; 368:769-772.

65. Hwang JR, Moncollin V, Vermeulen W et al. A 3'→5' XPB helicase defect in repair/transcription factor TFIIH of xeroderma pigmentosum group B affects both repair and transcription. J Biol Chem 1996; 271:15898-15904.

66. Feaver WJ, Svejstrup JQ, Bardwell L et al. Dual roles of a multiprotein complex from Saccharomyces cerevisiae in transcription and DNA repair. Cell 1993; 75:1379-1387.

67. Reardon JT, Ge H, Gibbs E et al. Isolation and characterization of two human transcription factor IIH (TFIIH)-related complexes: ERCC2/CAK and TFIIH*. Proc Natl Acad Sci USA 1996; 93:6482-6487.

68. Adamczewski JP, Rossignol M, Tassan J-P et al. MAT1, cdk7 and cyclin H form a kinase complex which is UV-light sensitive upon association with TFIIH. EMBO J 1996; 15:1877-1884.

69. Drapkin R, Reinberg D. The multifunctional TFIIH complex and transcriptional control. Trends Biochem Sci 1994; 19:504-508.

70. Serizawa H, Conaway RC, Conaway JW. Multifunctional RNA polymerase II initiation factor δ from rat liver. J Biol Chem 1993; 268:17300-17308.

71. Roy R, Adamczewski JP, Seroz T et al. The MO15 cell cycle kinase is associated with the TFIIH transcription-DNA repair factor. Cell 1994; 79:1093-1101.

72. Svejstrup JQ, Wang Z, Feaver WJ et al. Different forms of TFIIH for transcription and DNA repair: holo-TFIIH and a nucleotide excision repairosome. Cell 1995; 80:21-28.

73. Goodrich JA, Tjian R. Transcription factor IIE and IIH and ATP hydrolysis direct promoter clearance by RNA polymerase II. Cell 1994; 77:145-156.

74. Lu H, Zawel L, Fisher L et al. Human general transcription factor IIH phosphorylates the C-terminal domain of RNA polymerase II. Nature 1992; 358:641-645.

75. Shiekhattar R, Mermelstein F, Fisher R et al. Cdk-activating kinase (CAK) complex is a component of human transcription factor IIH. Nature 1995; 374:283-287.

76. Fesquet D, Labbé J-C, Derancourt J et al. The MO15 gene encodes the catalytic subunit of a protein kinase that inactivates cdc2 and other cyclin-dependent kinases (CDKs) through phosphorylation of Thr161 and its homologues. EMBO J 1993; 12:3111-3121.

77. Mäkelä TP, Tassan JP, Nigg EA et al. A cyclin associated with the CDK-activating kinase MO15. Nature 1994; 371:254-257.

78. Solomon MJ. The function(s) of CAK, the p34^{cdc2}-activating kinase. Trends Biochem Sci 1994; 19:496-500.

79. Masutani C, Sugusawa K, Yanagisawa J et al. Purification and cloning of a nucleotide excision repair complex involving the xeroderma pigmentosum group C protein and a human homolog of yeast RAD23. EMBO J 1994; 13:1831-1843.

80. van der Speck PJ, Eker A, Rademakers S et al. XPC and human homologs of RAD23: intracellular localization and relationship to other nucleotide excision repair complexes. Nucleic Acids Res 1996; 24:2551-2559.

81. Reardon JT, Mu D, Sancar A. Overproduction, purification, and characterization of the XPC subunit of the human DNA repair excision nuclease. J Biol Chem 1996; 271:19451-19456.

82. Finley D, Bartel B, Varshavsky A. The tails of ubiquitin precursors are ribosomal proteins whose fusion to ubiquitin facilitates ribosome biogenesis. Nature 1989; 338:394-401.

83. Guzder SN, Bailly V, Sung P et al. Yeast DNA repair protein RAD23 promotes complex formation between transcription factor TFIIH and DNA damage recognition factor RAD14. J Biol Chem 1995; 270:8385-8388.

84. Venema J, van Hoffen A, Karcagi V et

al. Xeroderma pigmentosum comple-mentation group C cells remove pyrimi-dine dimers selectively from the tran-scribed strand of active genes. Mol Cell Biol 1991;11:4128-4134.

85. Naegeli H. Mechanisms of DNA damage recognition in mammalian nucleotide ex-cision repair. FASEB J 1995; 9:1043-1050.

86. Hoeijmakers JHJ. Human nucleotide ex-cision repair syndromes: molecular clues to unexpected intricacies. Europ J Can-cer 1994; 30A:1912-1921.

87. Park CH, Bessho T, Matsunaga T et al. Purification and characterization of the XPF-ERCC1 complex of human DNA repair excision nuclease. J Biol Chem 1995; 270:22657-22660.

88. O'Donovan, Davies AA, Moggs JG et al. XPG endonuclease makes the 3' incision in human DNA nucleotide excision re-pair. Nature 1994; 371:432-435.

89. Iyer N, Reagan MS, Wu KJ et al. Inter-actions involving the human RNA poly-merase II transcription/nucleotide exci-sion repair complex TFIIH, the nucle-otide excision repair protein XPG, and Cockayne syndrome group B (CSB) pro-tein. Biochemistry 1996; 35:2157-2167.

90. Matsunaga T, Mu D, Park C-H. Human DNA repair excision nuclease. Analysis of the roles of the subunits involved in dual incisions by using anti-XPG and anti-ERCC1 antibodies. J Biol Chem 1995; 270:20862-20869.

91. Aboussekhra A, Biggerstaff M, Shivji MKK et al. Mammalian DNA nucleotide excision repair reconstituted with puri-fied components. Cell 1995; 80:859-.

92. Shivji MKK, Podust VN, Hübscher U et al. Nucleotide excision repair DNA syn-thesis by DNA polymerase ε in the pres-ence of PCNA, RFC, and RPA. Biochem-istry 1995; 34:5011-5017.

93. Sancar A, Hearst JE. Molecular match-makers. Science 1993; 259:1415-1420.

94. Nichols AF, Sancar A. Purification of PCNA as a nucleotide excision repair protein. Nucleic Acids Res 1992; 20:2441-2446.

95. Klein JC, Bleeker MJ, Roelen HCPF et al. Role of nucleotide excision repair in pro-cessing of O^4-alkylthymines in human cells. J Biol Chem 1994; 269:25521-25528.

96. Saffhill R, Margison GP, O'Connor PJ, Mechanisms of carcinogenesis induced by alkylating agents. Biochim Biophys Acta 1985; 823:111-145.

97. Saffhill R. In vitro miscoding of alkyl-thymines with DNA and RNA poly-merases. Chem Biol Interact 1985; 53:121-130.

98. Singer B. O-alkyl pyrimidines in mu-tagenesis and carcinogenesis: occur-rence and significance. Cancer Res 1986; 46:4879-4885.

99. Mitchell DL, Nairn RS. The biology of the (6-4) photoproduct. Photochem Photobiol 1989; 49:805-819.

100. Mitchell DL. The relative cytotoxicity of (6-4) photoproducts and cyclobutane pyrimidine dimers in mammalian cells. Photochem Photobiol 1988; 48:51-57.

101. Franklin WA, Haseltine WA. The role of the (6-4) photoproduct in ultraviolet light-induced transition mutations in E. coli. Mutat Res 1986; 165:1-7.

102. Zdzinicka MZ, Venema J, Mitchell DL et al. (6-4) Photoproducts and not cyclo-butane pyrimidine dimers are the main UV-induced mutagenic lesion in Chinese hamster cells. Mutat Res 1992; 273:73-83.

103. Treiber DK, Zhai X, Jantzen H-M et al. Cisplatin-DNA adducts are molecular decoys for the ribosomal RNA transcrip-tion factor hUBF (human upstream binding factor). Proc Natl Acad Sci USA 1994; 91:5672-5676.

104. McLenigan M, Levine AS, Protic M. Dif-ferential expression of pyrimidine dimer-binding proteins in normal and UV light-treated vertebrate cells. Photochem Photobiol 1993; 57:655-662.

105. Moranelli F, Lieberman MW. Recogni-tion of chemical carcinogen-modified DNA by a DNA-binding protein. Proc Natl Acad Sci USA 1980; 77:3201-3205.

106. Lenz J, Okenquist SA, LoSardo JE et al. Identification of a mammalian nuclear factor and human cDNA-encoded pro-teins that recognize DNA containing apurinic sites. Proc Natl Acad Sci USA 1990; 87:3396-3400.

107. Rydberg B, Dosanjh MK, Singer B. Hu-man cells contain protein specifically binding to a single 1,N^6-ethenoadenine in a DNA fragment. Proc Natl Acad Sci

USA 1991; 88:6839-6842.

108. Feldberg RS, Grossman L. A DNA binding protein from human placenta specific for ultraviolet-damaged DNA. Biochemistry 1976; 15:2402-2408.

109. Feldberg RS, Lucas JL, Dannenberg A. A damage-specific DNA binding protein. J Biol Chem 1982; 257:6394-6401.

110. Chu G, Chang E. Xeroderma pigmentosum group E cells lack a nuclear factor that binds to damaged DNA. Science 1988; 242:564-567.

111. Hwang BJ, Chu G. Purification and characterization of a human protein that binds to damaged DNA. Biochemistry 1993; 32:1657-1666.

112. Keeney S, Chang GJ, Linn S. Characterization of a human DNA damage binding protein implicated in xeroderma pigmentosum E. J Biol Chem 1993; 268:21293-21300.

113. Keeney S, Eker APM, Brody T et al. Correction of the DNA repair defect in xeroderma pigmentosum group E by injection of a DNA damage-binding protein. Proc Natl Acad Sci USA 1994; 91:4053-4056.

114. Reardon JT, Nichols AF, Keeney S et al. Comparative analysis of binding of human damaged DNA-binding protein (XPE) and *Escherichia coli* damage recognition protein (UvrA) to the major ultraviolet photoproducts: T[c,s]T, T[t,s]T, T[6-4]T, and T[Dewar]T. J Biol Chem 1993; 268:21301-21308.

115. Treiber D, Chen Z, Essigmann J. An ultraviolet light-damaged DNA recognition protein absent in xeroderma pigmentosum group E cells binds selectively to pyrimidine (6-4) pyrimidone photoproducts. Nucleic Acids Res 1992; 20:5805-5810.

116. Kazantsev A, Mu D, Nichols AF et al. Functional complementation of xeroderma pigmentosum complementation group E by replication protein A in an in vitro system. Proc Natl Acad Sci USA 1996; 93: 5014-5018.

117. Echols H. Multiple DNA-protein interactions governing high-precision DNA interactions. Science 1986; 233:1050-1055.

118. Bertrand-Burggraf E, Selby CP, Hearst JE et al. Identification of the different intermediates in the interaction of (A)BC excinuclease with its substrate by DNaseI footprinting on two uniquely modified oligonucleotides. J Mol Biol 1991; 219:27-36.

119. Huang J-C, Sancar A. Determination of minimum substrate size for human excinuclease. J Biol Chem 1994; 269:19034-19040.

120. Guzder SN, Sung P, Prakash L et al. Nucleotide excision repair in yeast is mediated by sequential assembly of repair factors and not by a pre-assembled repairosome. J Biol Chem 1996; 271:8903-8910.

121. Maldonado E, Shiekhattar R, Sheldon M et al. A human RNA polymerase II complex associated with SRB and DNA repair proteins. Nature 1996; 381:86-89.

122. Lee S-H, Kim D-K, Drissi R. Human xeroderma pigmentosum group A protein interacts with human replication protein A and inhibits DNA replication. J Biol Chem 1995; 270:21800-21805.

123. Matson SW, Kaiser-Rogers KA. DNA helicases. Annu Rev Biochem 1990; 59:289-329.

124. Arai K-I, Low R, Kobori J et al. Mechanism of *dnaB* protein action. J Biol Chem 1981; 256:5273-5280.

125. Oh EY, Grossman L. Helicase properties of the *Escherichia coli* UvrAB protein complex. Proc Natl Acad Sci USA 1987; 84:3638-3642.

126. Sancar A., Rupp WD. A novel repair enzyme: UVRABC excision of *Escherichia coli* cuts a DNA strand on both sides of the damaged region. Cell 1983; 33:249-260.

127. Grossman L, Thiagalingam S. Nucleotide excision repair, a tracking mechanism in search of damage. J Biol Chem 1993; 268:16871-16874.

128. Oh EY, Grossman L. Characterization of the helicase activity of the *Escherichia coli* UvrAB protein complex. J Biol Chem 1989; 264:1336-1343.

129. Naegeli H, Bardwell L, Friedberg EC. The DNA helicase and adenosine triphosphatase activities of yeast Rad3 protein are inhibited by DNA damage. J Biol Chem 1992; 267:392-398.

130. Naegeli H, Bardwell L, Friedberg EC

(1993) Inhibition of Rad3 DNA helicase activity by DNA adducts and abasic sites: implications for the role of a DNA helicase in damage-specific incision of DNA. Biochemistry 1993; 32:613-621.

131. Naegeli H, Bardwell L, Harosh I et al. Substrate specificity of the Rad3 ATPase/DNA helicase of Saccharomyces cerevisiae and binding of Rad3 protein to nucleic acids. J Biol Chem 1992; 267:7839-7844.

132. Naegeli H, Modrich P, Friedberg EC. The DNA helicase activity of Rad3 protein of *Saccharomyces cerevisiae* and helicase II of *Escherichia coli* are differentially inhibited by covalent and non-covalent DNA modifications. J Biol Chem 1993; 268:10386-10392.

133. Friedberg EC. Yeast genes involved in DNA repair processes: new looks on old faces. Mol Microbiol 1991; 5:2303-2310.

134. Lesser DR, Kurpiewski MR, Waters T et al. Facilitated distortion of the DNA site enhances *Eco*RI endonuclease-DNA recognition. Proc Natl Acad Sci USA 1993; 90:7548-7552.

135. Kim Y, Grable JC, Love R et al. Refinement of *Eco*RI endonuclease crystal structure: a revised protein chain tracing. Science 1990; 249:1307-1309.

136. Draper DE. Protein-DNA complexes: the cost of recognition. Proc Natl Acad Sci USA 1993; 90:7429-7430.

137. Kim Y, Geiger JH, Hahn S. Crystal structure of a yeast TBP/TATA-box complex. Nature 1993; 365:512-520.

138. Erie DA, Yang G, Schultz HC et al. DNA bending by Cro protein in specific and nonspecific complexes: implications for protein site recognition and specificity. Science 1994; 266:1562-1566.

139. Sun D, Hurley LH. Cooperative binding of the 21-base-pair repeats of the SV40 viral early promoter by Sp1. Biochemistry 1994; 33:9578-9587.

140. Shi Q, Thresher R, Sancar A et al. An electron microscopic study of (A)BC excinuclease. J Mol Biol 1992; 226:425-432.

141. Burns JL, Guzder SN, Sung P et al. An affinity of human replication protein A for ultraviolet-damaged DNA. J Biol Chem 1996; 271:11607-11610.

142. Sung P, Watkins JF, Prakash L et al. Negative superhelicity promotes ATP-dependent binding of yeast Rad3 protein to ultraviolet-damaged DNA. J Biol Chem 1994; 269:8303-8308.

143. Kim J-K, Choi B-S. The solution structure of DNA duplex-decamer containing the (6-4) photoproduct of thymidylyl(3'→5')thymidine by NMR and relaxation matrix refinement. Eur J Biochem 1995; 228:849-854.

CHAPTER 7

MOLECULAR DETERMINANTS OF DAMAGE RECOGNITION BY MAMMALIAN NUCLEOTIDE EXCISION REPAIR

The mechanism by which XPA, RPA, XPC-HHR23B, TFIIH and possibly other factors discriminate a wide range of chemically dissimilar DNA lesions as substrates of mammalian nucleotide excision repair is poorly understood. The striking versatility of nucleotide excision repair led to the assumption that its recognition subunits detect conformational changes imposed on DNA at sites of damage rather than specific base modifications.[1-5] As indicated in Figure 7.1, this hypothesis was prompted by the observation that many base lesions alter the helical parameters of DNA by inducing kinks,[6,7] bends[8] or localized unwinding,[9] suggesting that damage-induced conformational distortion may constitute an important determinant of recognition by the nucleotide excision repair system.

One problem associated with recognition models based on damage-induced kinks, bends, unwinding or other conformational changes is that mammalian nucleotide excision repair factors should be sensitive to a broad range of qualitatively and quantitatively different abnormalities in the secondary structure of DNA. Another shortcoming of simple models involving damage-induced distortion as the sole determinant for recognition is that not all DNA lesions that are substrates of nucleotide excision repair cause site-specific distortion (Fig. 7.1).[5,10] On the other hand, helical deformations are also normally found in native DNA, particularly in critical regulatory sequences,[11,12] but such intrinsically kinked, bent or unwound DNA segments fail to elicit substantial nucleotide excision repair reactions (Fig. 7.1).[5]

The main purpose of this chapter is to provide experimental evidence in support of the idea that mammalian nucleotide excision repair operates by a more complex bipartite mechanism of substrate discrimination. In fact, recent work in our laboratory was aimed at identifying molecular determinants that trigger initiation of mammalian nucleotide excision repair. For that purpose, we compared recognition of several bulky adducts of which detailed structural information was available in the literature, and found that mammalian nucleotide excision repair is primarily targeted to sites at which the thermodynamic stability of the DNA double helix is reduced by disruption of Watson-Crick base pairing.[13] However, dual DNA incision and subsequent oligonucleotide excision occurs only if such sites suffering from thermodynamic instability contain chemically modified DNA constituents

Mechanisms of DNA Damage Recognition in Mammalian Cells, by Hanspeter Naegeli.
© 1997 R.G. Landes Company.

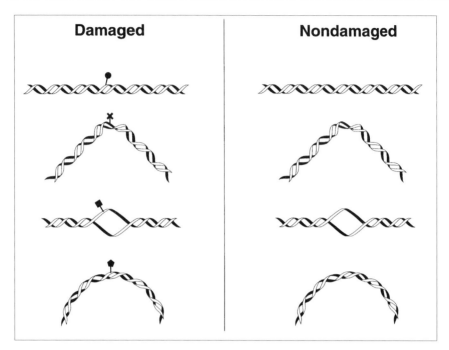

Fig. 7.1. Scheme illustrating different conformational changes that may serve as molecular determinants of recognition by nucleotide excision repair. This drawing emphasizes the fact that, although most types of damage perturb native DNA at least to some extent, certain DNA lesions fail to induce distortions of the double helix. On the other hand, irregular DNA conformations such as kinks, bends or localized unwinding are also found in nondamaged sequences.

(M.T. Hess, U. Schwitter, M. Petretta, B. Giese and H. Naegeli, manuscript in preparation). This requirement for two separate recognition determinants involving a base pair destabilizing change in the secondary structure (or conformation) of DNA in combination with a modification of its primary structure (or chemistry) led us to propose that mammalian nucleotide excision repair exploits a bipartite mechanism of recognition. In a parallel line of experimentation, the differential excision activity on stereoselective *anti*-BPDE-N^2-guanine adducts confirms the prominent role of defective Watson-Crick base pairing in this bipartite mode of substrate discrimination by mammalian nucleotide excision repair (M.T. Hess, D. Gunz, N.E. Geacintov, H. Naegeli, manuscript in preparation).

DNA CONFORMATION AT BULKY BASE ADDUCTS

Prototypical examples of bulky base adducts formed by genotoxic agents are illustrated in Figure 7.2. Calorimetric studies indicate that acetylaminofluorene (AAF), benzo[a]pyrene diol-epoxide (BPDE), 8-methoxypsoralen (8-MOP), anthramycin, and CC-1065 base adducts alter the conformational characteristics of DNA in very different ways. In thermodynamic terms, the DNA molecule is in equilibrium between the double-stranded helical form and its single-stranded counterpart. AAF adducts[14,15] and BPDE adducts[9,16] influence the conformational state of DNA by destabilizing the double helix relative to nonmodified DNA. UV radiation products also destabilize the double helix but in a heterogeneous manner.

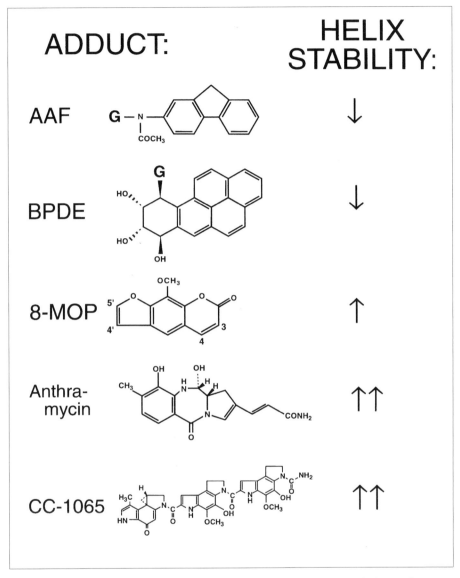

Fig. 7.2. Chemical structure of bulky carcinogen-DNA adducts and their effects on thermodynamic stability of the double helix.

Significant duplex destabilization is observed at sites of pyrimidine(6-4)pyrimidone photoproduct formation,[6] whereas cyclobutane pyrimidine dimers display minimal effects on the ability of DNA to assemble into the helical conformation.[17] In contrast, 8-MOP,[18,19] anthramycin,[20] and CC-1065 adducts[21] enhance the thermodynamic stability of DNA duplexes (Fig. 7.2).

The structural changes associated with these different thermodynamic effects have been elucidated by various techniques including nuclear magnetic resonance (NMR) conformation studies as well as chemical or enzymatic probing. For example, the major adduct formed by N-acetoxy-2-acetyl-aminofluorene (AAF-C^8-guanine; Fig. 7.2) completely abolishes base pairing between the adducted guanine and its complementary cytosine.[14] In this carcinogen-DNA adduct, the fluorene moiety stacks on an adjacent base pair within the double helix, thereby rotating the modified guanine out of the helix axis. Analysis with chloro-acetaldehyde or osmium tetroxide showed that the resulting helix destabilizing effect is not restricted to the site of AAF modification but rather extends to the neighboring base pairs.[22]

Reaction of benzo[a]pyrene diol-epoxide (BPDE) with double-stranded DNA generates four stereochemically distinct guanine modifications, of which the (+)-trans-*anti*-BPDE-N^2-guanine adduct shown in Figure 7.2 constitutes the major fraction. Quantitatively minor BPDE lesions derive from (-)-trans, (+)-cis and (-)-cis additions to the same position N^2 of guanine (see also Fig. 3.8 in chapter 3).[23-25] Depending on their precise stereochemistry, these BPDE-N^2-guanine adducts are either accommodated in the minor groove[24] or assume a base-displacement configuration with localized base pair disruption.[25] All four stereoisomers of BPDE-N^2-guanine produce unfavorable thermodynamic changes by reducing the stability of duplex DNA, although to slightly different degrees.[16]

The molecular basis of helix stabilization by 8-MOP, anthramycin or CC-1065 has also been solved. Treatment of DNA with 8-MOP and long wavelength UV light (> 320 nm) yields psoralen monoadducts and a small proportion of psoralen diadducts.[26] This photoaddition reaction occurs between the 5,6 double bond of pyrimidine bases and either the 3,4 (pyrone) or the 4',5' (furan) double bond of the psoralen moiety (Fig. 7.2). Covalent modification with psoralen induces considerable helical distortion by unwinding the duplex and enhancing backbone flexibility,[27] but fails to destabilize the secondary structure of DNA. On the contrary, thermostability measurements showed that both pyroneside and furanside monoadducts slightly increase double helix stability, presumably by mediating stacking interactions between the psoralen moiety and the surrounding base pairs.[18,19,27]

Anthramycin and CC-1065 are representative examples of two classes of compounds forming base adducts that cause substantial increments in helical stability. Anthramycin is a pyrrolo[1,4]benzo-diazepine antibiotic (Fig. 7.2) that binds selectively to N^2 of guanine and forms covalent adducts with essentially no distortion of the DNA helix.[10,20] CC-1065 consists of three pyrroloindole subunits joined by amide linkages (Fig. 7.2). One subunit of this composite molecule displays a reactive cyclopropyl ring which alkylates DNA at the position N^3 of adenine to generate covalent adducts that cause bending and winding of the double helix.[8,21,28,29] Both anthramycin and CC-1065 adducts enhance duplex stability through noncovalent interactions derived from hydrogen bonds (anthramycin) or van der Waal's and hydrophobic forces (CC-1065) within the minor groove of DNA.[10,20,28]

A REPAIR COMPETITION ASSAY TO ASSESS RECOGNITION BY HUMAN EXCISION REPAIR

We recently developed a novel repair competition assay to compare the capacity of mammalian nucleotide excision repair to recognize structurally different forms of DNA damage.[30] This assay is based on a site-specifically modified nucleotide excision repair substrate and measures the ability of damaged plasmids to compete for excision repair of the site-directed substrate (Fig. 7.3). This repair competition assay is performed by coincubating in the same reaction bacteriophage M13 double-stranded DNA with a single and uniquely located AAF-guanine adduct (M13-AAF) and, as a competitor, plasmid pUC19 that is multiply damaged with the lesion of interest (Fig. 7.3).[13,30] A standard soluble extract from human cells[31,32] is exploited as a source of mammalian nucleotide excision repair factors.

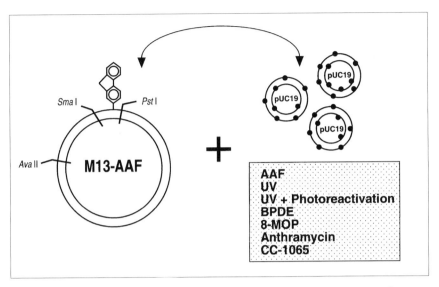

Fig. 7.3. Competition for nucleotide excision repair recognition factors. HeLa cell extract is incubated with M13 DNA substrate containing a site-directed AAF-guanine adduct and various amounts of multiply damaged plasmid pUC19 as competitor DNA. The assay measures the capacity of DNA adducts located on pUC19 to competitively inhibit nucleotide excision repair of the site-specific AAF-guanine modification. Nucleotide excision repair is quantified by monitoring the accumulation of radiolabeled deoxyribonucleotides within the 37-base pair long SmaI-PstI region of the M13 substrate.

To validate the repair competition assay, we first tested the situation in which both substrate and competitor DNA contain the same lesion, i.e., AAF-DNA adducts. For that purpose, we coincubated M13 DNA substrate with increasing amounts of pUC19 DNA containing 10.2 ± 0.9 AAF adducts/molecule. In these experiments, nucleotide excision repair operating on M13 DNA substrate was progressively reduced by competitive inhibition (Fig. 7.4). This inhibitory effect was strictly damage-specific, as no suppression of nucleotide excision repair on the substrate was observed when the reactions were supplemented with non-damaged pUC19 even in a large molar excess over M13 DNA.

Quantitative analysis showed that excision repair operating on the substrate was inhibited to 50% at a M13 substrate to pUC19 competitor molar ratio of 10.8:1

(Fig. 7.4). Since M13 DNA contains only one AAF adduct/molecule whereas pUC19 contains an average of 10.2 AAF adducts/molecule, these calculations yielded 50% inhibition of nucleotide excision repair at a 1:1 stoichiometry of these lesions. Thus, AAF modifications located on pUC19 were recognized, on the average, as efficiently as the single site-directed AAF adduct placed on M13 DNA. These results led us to conclude that the repair competition assay is a highly quantitative method to assess recognition of a particular form of DNA damage by human nucleotide excision repair.[13,30]

EVIDENCE FOR THERMODYNAMIC PROBING

The repair competition assay was employed to test the capacity of human nucleotide excision repair to recognize UV radiation products or bulky base adducts with

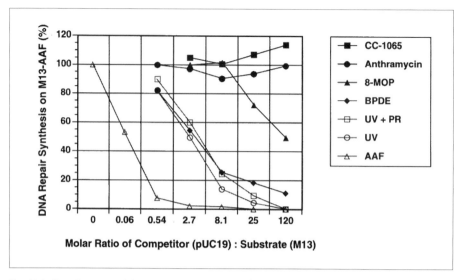

Fig. 7.4. Repair competition assay: comparison between bulky DNA adducts. HeLa cell extract was incubated with M13 DNA substrate and various amounts of competitor pUC19. The graph shows the level of nucleotide excision repair on M13 substrate as a function of competitor pUC19 added to the reactions. The differential capacity of the tested adducts to sequester recognition factors and, hence, compete with nucleotide excision repair of the substrate suggests a thermodynamic probing mechanism. The number of modifications per pUC19 molecule was as follows: AAF, 10.2; BPDE, 16.7; 8-MOP, 14.5; anthramycin, 11.9; CC-1065, 9.6. UV: UV irradiation at a fluence of 450 J/m^2, resulting in an average of 11-12 UV radiation products per pUC19 molecule; UV + PR: UV radiation at a fluence of 450 J/m^2 followed by treatment with DNA photolyase, yielding approximately 3 pyrimidine(6-4)pyrimidone photoproducts per pUC19 molecule.

AAF, BPDE, 8-MOP, anthramycin or CC-1065. To that end, we constructed pUC19 DNA containing a defined number of each of these adducts and determined their potential to sequester nucleotide excision repair recognition factors and, hence, compete with excision repair of the site-directed M13-AAF substrate.[13] This novel approach based on factor sequestration revealed >1,000-fold differences in the recognition of helix stabilizing and helix destabilizing DNA modifications (Fig. 7.4). The capacity of the tested lesions to sequester human nucleotide excision repair factors decreased with the following order: AAF > UV radiation products > BPDE > 8-MOP > anthramycin, CC-1065 (Fig. 7.4). Interestingly, photoreactivation of UV-irradiated plasmids with DNA photolyase and visible light to completely eliminate cyclobutane dimers reduced only

marginally their ability to compete with the site-directed AAF adduct (Fig. 7.4). This finding is consistent with many previous reports indicating that pyrimidine(6-4)pyrimidone photoproducts are recognized and processed in mammalian cells more efficiently than cyclobutane dimers (see chapter 7).

The quantitative competition values (summarized in Fig. 7.4) were used to calculate the stoichiometric excess of each base adduct required to inhibit excision repair of the site-directed AAF-guanine to 50% (Table 7.1).[13] The obtained numbers show that AAF, UV and BPDE adducts were 2-3 orders of magnitude stronger competitors than 8-MOP, anthramycin or CC-1065 adducts. For example, AAF modifications were able to compete with the nucleotide excision repair substrate ~1,740 times more efficiently than 8-MOP adducts (Table 7.1).

The resulting hierarchy of sequestration efficiency in the repair competition assay, combined with the known thermodynamic characteristics of the tested lesions, indicates that mammalian nucleotide excision repair is primarily targeted to structural defects that destabilize the double-helical conformation of DNA. Those lesions that destabilize the DNA helix (AAF or BPDE adducts, UV radiation products) were effective competitors. In contrast, those adducts that stabilize the helix (8-MOP, anthramycin, and CC-1065 adducts) displayed minimal or no competing effects. Thus, an early subset of nucleotide excision repair recognition factors is endowed with the capacity to sense the thermodynamic parameters of double-stranded DNA and preferentially interact with those sites that exhibit unfavorable changes in free energy. The XPA-RPA complex is a possible protein candidate for executing this recognition function involving thermodynamic probing of duplex stability. This hypothesis is supported by previous reports demonstrating that both XPA and RPA are single-stranded DNA binding proteins,[33,34] and that these two factors display strong cooperativity in binding to AAF- or UV-damaged DNA.[35,36]

COMPARISON WITH PROKARYOTIC (A)BC EXCINUCLEASE

It appears that the thermodynamic characteristics of damaged DNA are less important for recognition by (A)BC excinuclease, the multisubunit enzyme that recognizes DNA adducts and initiates nucleotide excision repair in prokaryotes. For example, helix-stabilizing psoralen monoadducts are efficient substrates for (A)BC excinuclease[37] but constitute a modest substrate for human nucleotide excision repair (Fig. 7.4). Also, the poor competition exerted by anthramycin-DNA adducts in the mammalian system (Fig. 7.4) contrasts with a previous

Table 7.1. Repair competition assay: stoichiometric excess of covalent base adducts required to inhibit excision repair of the site-directed AAF-guanine modification to 50%

Adduct	Stoichiometric excess over AAF-C⁸-guanine
AAF	1
UV radiation products[a]	42
Pyrimidine(6-4)pyrimidone photoproducts[b]	11
BPDE	45
8-MOP	1,740
Anthramycin	> 2,000
CC-1065	> 2,000

[a] containing a mixture of cyclobutane pyrimidine dimers and pyrimidine(6-4)pyrimidone photoproducts in a ratio of about 3 : 1
[b] prepared by photoreactivation of UV-irradiated pUC19 DNA (D. Gunz, L.H.F. Mullenders, H. Naegeli, unpublished results).

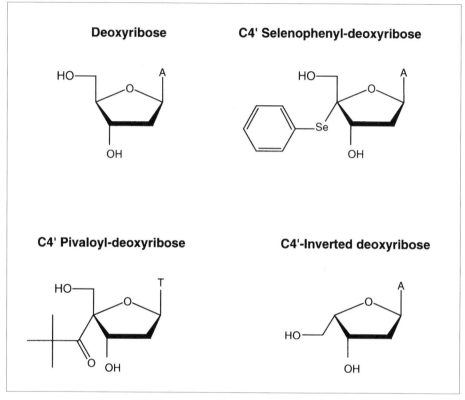

Fig. 7.5. C4' deoxyribose variants used as molecular tools to study DNA damage recognition in mammalian nucleotide excision repair.

report indicating that prokaryotic (A)BC excinuclease incises anthramycin-damaged DNA with a 4-5 times higher rate than UV-irradiated DNA.[38] Relatively efficient incision of CC-1065-damaged DNA by (A)BC excinuclease has also been reported,[39,40] again contrasting with the poor recognition of CC-1065 adducts in the mammalian system (Fig. 7.4). The diverging substrate preference between prokaryotic and mammalian nucleotide excision repair apparently reflects the lack of substantial sequence and structure homology between the two systems.

BIPARTITE RECOGNITION IN MAMMALIAN NUCLEOTIDE EXCISION REPAIR

A common feature of essentially all lesions that have been identified as substrates of nucleotide excision repair is the covalent modification of DNA bases[1-5] (see chapter 5 for further explanation of nucleotide excision repair). Other lesions that are processed by mammalian nucleotide excision repair include sites at which a base is completely lost (apurinic/apyrimidinic sites)[37] or replaced by bulky organic derivatives, such as a cholesterol moiety.[41] These previous studies on base modifications, base losses or base replacements led us to test whether DNA adducts located on the deoxyribose-phosphate backbone of DNA may also stimulate nucleotide excision repair. This line of research yielded a series of novel DNA modifications that escape processing by mammalian nucleotide excision repair and provide suitable substrates to test the hypothesis of thermodynamic probing.

We recently observed that C4' deoxyribose adducts are refractory to the human

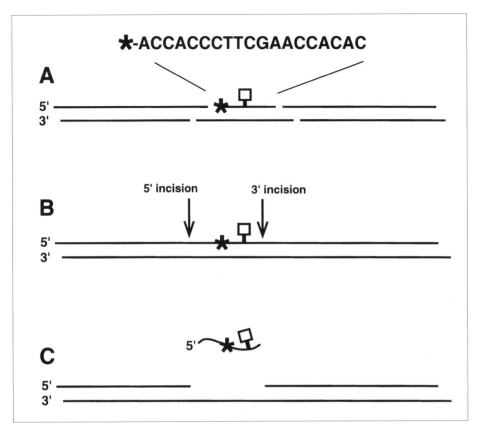

Fig. 7.6. Oligonucleotide excision assay. A, preparation of internally labeled substrates; B, double DNA incision by mammalian nucleotide excision repair; C, radiolabeled oligonucleotide excision products of 26-32 residues in length containing the damaged nucleotide. The asterisk denotes the site of radioactive labeling located approximately 10 nucleotides 5' to the modified residue.

nucleotide excision repair system in vitro.[42] These backbone variants were constructed by linking a bulky selenophenyl or pivaloyl group to the C4' position of the deoxyribose moiety (Fig. 7.5). Cyanoethyl phosphoramidite technology was then applied to introduce these altered deoxyribose residues into short oligonucleotides which, in turn, were used to obtain double-stranded DNA containing site-directed backbone modifications. The resulting selenophenyl- or pivaloyl-modified substrates failed to elicit detectable excision responses in a repair-proficient human cell extract.[42] In contrast, the same human cell extract was able to excise base adducts of comparable size. Hu-

man nucleotide excision repair was also completely inactive on DNA substrates containing a more subtle deoxyribose derivative generated by inverting the geometry of chemical bonds at position C4' (Fig. 7.5; M.T. Hess, U. Schwitter, M. Petretta, B. Giese and H. Naegeli, unpublished results). These findings suggest that covalent modification of DNA bases constitutes an important prerequisite for recognition by mammalian nucleotide excision repair. Backbone alterations, on the other hand, do not affect the structure of the bases and may therefore escape this repair process.

In these experiments performed with deoxyribose modifications, repair activity was measured in vitro using the oligonucle-otide excision assay originally devised by Huang et al.[43] These authors were the first to exploit the characteristic dual DNA incision pattern of human nucleotide excision repair for analytical purposes. As illustrated in Figure 7.6, site-specifically modified, duplex DNA substrates of 146 base pairs were constructed by ligating a 19-base oligonucle-otide (19-mer) carrying the modification with five other oligonucleotides. Prior to ligation, the central 19-mer was labeled with ^{32}P-ATP at its 5' end, such that the resulting 146-mer substrate contained an internal radiolabel in the vicinity of the site-directed modification (Fig. 7.6). After purification, the double-stranded fragments were incubated with a repair-competent HeLa cell extract in the presence of ATP, dATP, dCTP, dGTP and TTP.[42,43] Using these internally labeled substrates, damage-specific dual DNA incision by nucleotide excision repair generates radioactive products of 24-32 nucleotides in length (Fig. 7.6), which are subsequently resolved by denaturing gel electrophoresis and visualized by autoradiography.

The representative polyacrylamide gel of Figure 7.7 shows the complete lack of detectable oligonucleotide excision products after incubation of DNA substrates containing either an inverted or a selenophenyl-adducted deoxyribose residue with human cell extract. However, human nucleotide excision repair was able to catalyze highly efficient oligonucleotide excision when single C4'-modified deoxyribose residues were incorporated into 3-nucleotide long segments of unpaired bases (Fig. 7.7). The mismatches alone failed to induce oligo-nucleotide excision. A markedly weaker but nevertheless detectable excision reaction was also observed when C4'-modified deoxyribose residues were incorporated into 1-nucleotide mismatches (Fig. 7.7).

In summary, the experiment of Figure 7.7 illustrates that neither disruption of base pair complementarity nor the tested C4' backbone lesions were capable of eliciting substantial nucleotide excision repair activity, but the combination of these two sub-strate alterations constituted a potent signal for double DNA incision. Thus, DNA backbone modifications provide a useful molecular tool to dissect the recognition problem during nucleotide excision repair into discrete components. We propose the term "bipartite recognition" in mammalian nucleotide excision repair to denote the strict requirement for two separate determinants of recognition, i.e., a base pair destabilizing defect in the secondary structure (or conformation) of DNA accompanied by a modification of its primary structure (or chemistry).

CONFORMATION-DEPENDENT EXCISION OF BPDE-DNA ADDUCTS

Base adducts generated by the ultimate carcinogen benzo[a]pyrene diol-epoxide offer an excellent system to test the relationship between DNA double helix conformation and damage recognition by nucleotide excision repair. As indicated in chapter 3, enzymatic activation of the polycyclic aromatic hydrocarbon benzo[a]pyrene results in the formation of numerous diol-epoxide derivatives that differ in their stereochemistry, but share the common ability to covalently modify DNA.[44,45] Reaction of the epoxide moiety of either the (+)- or (-)-anti-BPDE with the position N^2 of guanine generates two pairs of enantiomeric adducts: (+)-trans-, (+)-cis-, (-)-trans- and (-)-cis-anti-BPDE-N^2-guanine (see Fig. 3.8).

The solution structures of short DNA duplexes (5'-CCATCGCTACC-3') each containing one of these stereoisomeric distinct guanine adducts have been characterized by high resolution NMR spectroscopy in combination with distance-constrained energy minimization.[23-25,46] According to these reports, the (+)-trans- and (-)-trans-anti-BPDE-N^2-guanine adducts adopt an external conformation in which the pyrene moiety is located in the minor groove with little perturbation of the B-DNA helix (Fig. 7.8). This external minor groove conformation weakens, but does not break the hydrogen bonds responsible for base pair-

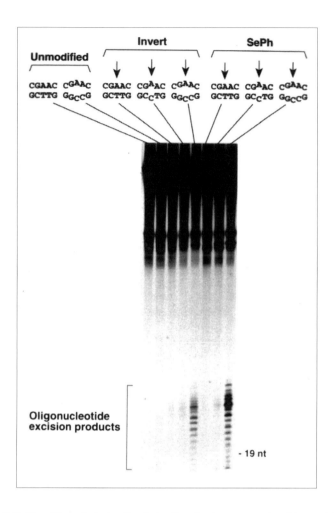

Fig. 7.7. Bipartite substrate discrimination by human nucleotide excision repair. Excision repair activity was determined by the excision assay described in Figure 7.6. In the cell-free extract, oligonucleotide excision products are partially degraded yielding an oligonucleotide ladder upon polyacrylamide gel electrophoresis. The arrows indicate the site of backbone modification with either C4'-inverted deoxyribose ("Invert") or selenophenyl deoxyribose ("SePh"). The sequence context of each substrate is indicated.

ing between the modified guanine and its complementary cytosine. The two enantiomeric forms [(+)-trans and (-)-trans] differ only in the orientation of the benzo[*a*]pyrene moiety within the minor groove, in that the pyrene points to the 5' end of the modified strand in the (+)-trans conformation, but to the 3' end in the (-)-trans conformation.[46,23]

In contrast to these trans-*anti*-BPDE adducts, both cis-*anti*-BPDE-N^2-guanine modifications adopt a helix-inserted, intercalative structure in which the aromatic pyrene moieties stack with the neighboring bases (Fig. 7.8). In this base-stacked conformation, the adducted guanine and the complementary cytosine are displaced with

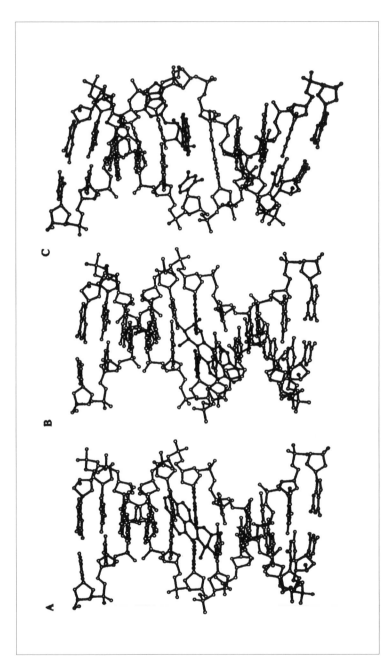

Fig 7.8. Solution structures of DNA containing stereoisomeric distinct BPDE-N^2guanine adducts. A, (+)-trans-BPDE-N^2guanine. B, (-)-trans-BPDE-N^2guanine. C, (+)-cis-BPDE-N^2guanine. The pyrenyl moiety of BPDE is shown in black, while the modified guanine and its complementary cytosine are gray. Reprinted with permission from: Zou Y, Liu T-M, Geacintov NE et al. Interaction of the UvrABC nuclease system with a DNA duplex containing a single stereoisomer of dG-(+)- or dG-(-)-anti-BPDE. Biochemistry 1995; 34:13582-13593. © 1995 American Chemical Society.

severe disruption of the hydrogen bonds between the two bases.[24,25] Thus, (+)-cis- and (-)-cis-BPDE-N^2-guanine show structural features that are reminiscent of the predominant AAF-C^8-guanine adduct, i.e., insertion of the covalently bound ligand into the helix accompanied by displacement of the modified base.

The capacity of human nucleotide excision repair to process these stereochemically distinct carcinogen-DNA adducts in the sequence 5'-CCATCGCTACC-3' has been determined in our laboratory using the in vitro oligonucleotide excision assay (M. Hess, D. Gunz, N.E. Geacintov, H. Naegeli, manuscript in preparation). A direct comparison yielded the following hierarchy of dual DNA incision: AAF-C^8-guanine > (-)-cis-BPDE-N^2-guanine > (+)-cis-BPDE-N^2-guanine > (-)-trans-BPDE-N^2guanine = (+)-trans-BPDE-N^2-guanine (Fig. 7.9). These results demonstrate that the rate of recognition and, hence, repair of BPDE-N^2-guanine adducts by human nucleotide excision repair is strictly dependent on their conformational properties. In particular, the 10-fold higher incision activity observed on cis-BPDE adducts compared to their counterparts in the trans conformation indicates that recognition by human nucleotide excision repair is facilitated by disruption of hydrogen bonding interactions between complementary bases.

Not surprisingly, differences in conformation also modulate the biological endpoints of BPDE adducts. For example, the (+)-*anti*-BPDE isomer is more tumorigenic than the (-)-*anti*-BPDE isomer on mouse skin[47] or in the lungs of newborn mice.[45] Also, (+)-*anti*-BPDE is more mutagenic than (-)-*anti*-BPDE in mammalian cell lines maintained in culture.[48,49] Interestingly, exactly the opposite is observed in cells of certain bacterial strains.[50] Shibutani et al[51] reported that BPDE adduct conformation is crucial to the type of mutation that may arise upon DNA replication, but the differential rate of excision repair is also likely to play a primary role in determining the mutagenic and carcinogenic potency of these compounds.

REPAIR OF BPDE ADDUCTS IN MAMMALIAN CELLS: AN IN VIVO STUDY

The relative persistence of BPDE-N^2-guanine adducts was recently investigated in mouse skin treated topically with benzo[*a*]pyrene.[52] The frequency of each stereoisomeric BPDE adduct was measured by ^{32}P-postlabeling, high performance liquid chromatography and low-temperature fluorescence spectroscopy. In agreement with many previous reports (+)-trans-BPDE-N^2-guanine was the quantitatively major DNA adduct.

When analyzed by fluorescence spectroscopy, this (+)-trans adduct exhibited a broad distribution of base-stacked (intercalative), partially base-stacked and helix-external conformations. Mouse skin DNA obtained at early time points after benzo[*a*]pyrene treatment contained mainly the external type of adduct, while samples from later time points after treatment (24-48 h) contained relatively more adducts of the internal, base-stacked conformation.[52] Quantitative evaluation of these results indicated that the latter adducts are repaired approximately three times more slowly than the former. Similarly, it was found in the mouse skin system that the quantitatively minor (+)-cis-adducts are repaired substantially less efficiently than (+)-trans-adducts. These results apparently contradict the hierarchy of excision repair observed in vitro when a set of defined BPDE-N^2guanine stereoisomers were incubated in a human cell extract. As shown in Figure 7.9, human nucleotide excision repair incised DNA substrates containing the intercalative cis isomers markedly more efficiently than substrates modified with the external trans isomers (see previous section). Further studies are in progress to solve these intriguing discrepancies with respect to BPDE-N^2-guanine recognition and repair.

CONCLUSIONS

Using a novel repair competition assay, we found experimental evidence indicating that mammalian damage recognition complexes are assembled on DNA preferentially at sites of thermodynamic instability. In fact,

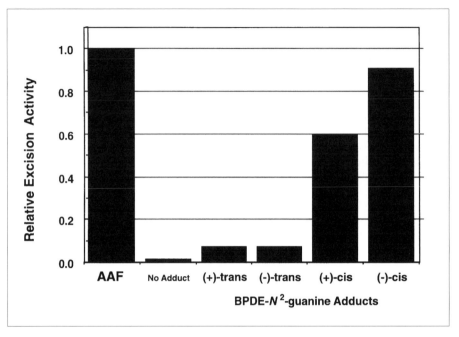

Fig. 7.9. Relative excision rate of BPDE-N^2guanine adducts compared to AAF-C^8-guanine. The graph shows the quantification of representative oligonucleotide excision assays performed by the method outlined in Figure 7.6.

bulky base adducts that destabilize the double helix are able to sequester damage recognition subunits of human nucleotide excision repair up to several orders of magnitude more efficiently than those adducts that induce opposite thermodynamic effects by stabilizing the double helix. Of course, DNA is a dynamic structure undergoing rapid conformational changes. Within the biological range of temperature and ionic strength, DNA duplexes are subject to spontaneous "breathing" involving localized melting of complementary base pairs. This normal "breathing" activity raises the question of how mammalian nucleotide excision repair may discriminate between damage-specific conformational distortion and transient destabilization events that also occur in nondamaged DNA.

The concept of bipartite substrate discrimination was introduced to indicate that dual endonucleolytic cleavage occurs at sites of helical instability only in the presence of concomitant alterations that affect the pri-

mary chemical structure of DNA. We used a series of novel C4' backbone derivatives as molecular tools to study DNA damage recognition, and found that neither the tested C4' variants nor short segments of unpaired bases are processed by human nucleotide excision repair in vitro. In combination, however, C4' backbone lesions situated within short segments of unpaired bases were able to trigger strong nucleotide excision repair responses in vitro. This requirement for two separate determinants may account for the failure of normal DNA "breathing" to induce substantial excision responses in intrinsically thermolabile sequences.

The bipartite mode of substrate discrimination in mammalian nucleotide excision repair prompted us to postulate the involvement of two separate molecular "calipers" during the initial damage recognition step. A molecular "caliper" for outside measurements senses the secondary structure (or conformation) of DNA. In addition, there is a molecular "caliper" for inside measurements that senses the primary structure (or chemistry) of DNA. Two known nucleotide excision repair factors (XPA, RPA) are potential candidates for probing the secondary structure of the double helix. These proteins may exploit their single-stranded DNA binding capacity to assemble on DNA molecules preferentially at sites of helical instability. TFIIH, on the other hand, may act as a sensor of the primary structure of DNA. As outlined in the hypothetical model of Figure 6.5 (chapter 6), the DNA helicase components of TFIIH may have been adopted by mammalian nucleotide excision repair to probe DNA chemistry in an ATP-dependent fashion. Future studies will be directed to testing this model of substrate discrimination by in vitro reconstitution of a damage recognition complex using pure components.

Finally, a structural basis for the striking preference of mammalian nucleotide excision repair for thermodynamically labile sites is provided by the NMR analysis of DNA fragments containing AAF-C^8-guanine, (+)-cis-*anti*-BPDE-N^2-guanine or (-)-cis-*anti*-BPDE-N^2-guanine adducts. In fact, highly efficient excision of these carcinogen lesions was observed and this preferential activity seems to correlate with their deleterious effects on complementary base pairing, i.e., disruption of hydrogen bonding interactions allowing displacement of the adducted guanine and its cytosine partner. In comparison, (+)-trans-*anti*-BPDE-N^2-guanine and (-)-trans-*anti*-BPDE-N^2-guanine display minimal effects on base pairing capability between the modified guanine and its complementary cytosine and, in fact, are excised with markedly lower efficiency.

REFERENCES

1. Sancar A. Excision repair in mammalian cells. J Biol Chem 1995; 270:15915-15918.
2. Sancar A. DNA excision repair. Annu Rev Biochem 1996; 65:43-81.
3. Hoeijmakers JHJ. Nucleotide excision repair II: from yeast to mammals. Trends Genet 1993; 9:211-217.
4. Grossman L, Thiangalingam S. Nucleotide excision repair, a tracking mecha-nism in search of damage. J Biol Chem 1993; 268:16871-16874.
5. Van Houten B. Nucleotide excision repair in *Escherichia coli*. Microbiol Rev 1990; 54:18-51.
6. Kim J-K, Choi B-S. The solution structure of DNA duplex-decamer containing the (6-4) photoproduct of thymidylyl(3'→5')thymidine by NMR and relaxation matrix refinement. Eur J Biochem 1995; 228:849-854.
7. Takahara PM, Rosenzweig AC, Frederick CA et al. Crystal structure of double-stranded DNA containing the major ad-duct of the anticancer drug cisplatin. Nature 1995; 377:649-652.
8. Lee C-S, Sun D, Kizu R et al. Determi-nation of the structural features of (+)-CC-1065 that are responsible for bend-ing and winding of DNA. Chem Res Toxicol 1991; 4:203-213.
9. Xu R, Birke S, Carberry SE et al. Differ-ences in unwinding of supercoiled DNA induced by the two enantiomers of *anti*-benzo[*a*]pyrene diol epoxide. Nucleic Acids Res 1992; 20:6167-6176.
10. Krugh TR, Graves DE, Stone MP. Two-dimensional NMR studies on the anthra-mycin-d(ATGCAT)$_2$ adduct. Biochemis-try 1989; 28:9988-9994.
11. Crothers DM, Haran TE, Nadeau JG. Intrinsically bent DNA. J Biol Chem 1990; 265:7093-7096.
12. Hagerman PJ. Sequence-directed curva-ture of DNA. Annu Rev Biochem 1990; 59:755-781.
13. Gunz D, Hess MT, Naegeli H. Recogni-tion of DNA adducts by human nucle-otide excision repair: evidence for a ther-modynamic probing mechanism. J Biol Chem 1996; 271:25089-25098.
14. O'Handley SF, Sanford SG, Xu R et al. Structural characterization of an *N*-acetyl-2-aminofluorene (AAF) modified DNA oligomer by NMR, energy minimi-zation, and molecular dynamics. Bio-chemistry 1993; 32:2481-2497.
15. Garcia A, Lambert IB, Fuchs RP. DNA adduct-induced stabilization of slipped frameshift intermediates within repeti-tive sequences: implications for mu-tagenesis. Proc Natl Acad Sci USA 1993; 90:5989-5993.

16. Zou Y, Liu T-M, Geacintov NE et al. Interaction of the UvrABC nuclease system with a DNA duplex containing a single stereoisomer of dG-(+)- or dG(-)-anti-BPDE. Biochemistry 1995; 34:13582-13593.

17. Taylor J-S, Garrett DS, Brockie IR et al. 1H NMR assignment and melting temperature study of cis-syn and trans-syn thymine dimer containing duplexes of d(CGTATTATGC)·(GCATAATACG). Biochemistry 1990; 29:8858-8866.

18. Shi YB, Hearst JE. Thermostability of double-stranded deoxyribonucleic acids: effects of covalent additions of a psoralen. Biochemistry 1986; 25:5895-5902.

19. Shi Y-B, Griffith J, Hearst JE. Evidence for structural deformation of the DNA helix by a psoralen diadduct but not by a monoadduct. Nucleic Acids Res 1988; 16:8945-8952.

20. Hurley LH, Petrusek R. Proposed structure of the anthramycin-DNA adduct. Nature 1979; 282:529-531.

21. Swenson DH, Li LH, Hurley, LH. Mechanism of interaction of CC-1065 (NSC 298223) with DNA. Cancer Res 1982; 42:2821-2828.

22. Schwartz A, Marrot L, Leng M. The DNA bending by acetylaminofluorene residues and by apurinic sites. J Mol Biol 1989; 207:445-450.

23. de los Santos C, Cosman M, Hingerty BE et al. Influence of benzo[a]pyrene diol epoxide chirality on solution conformations of DNA covalent adducts: the (-)-trans-anti-[BP]G·C adduct structure and comparison with the (+)-trans-anti-[BP]G·C enantiomer. Biochemistry 1992; 31:5245-5252.

24. Cosman M, de los Santos C, Fiala R et al. Solution conformation of the (+)-cis-anti-[BP]dG adduct in a DNA duplex: intercalation of the covalently attached benzo[a]pyrenyl ring into the helix and displacement of the modified deoxyguanosine. Biochemistry 1993; 32:4146-4155.

25. Cosman M, Hingerty BE, Luneva N et al. Solution conformation of the (-)-cis-anti-benzo[a]pyrenyl-dG adduct opposite dC in a DNA duplex: intercalation of the covalently attached BP ring into the he-

lix with base displacement of the modified deoxyguanosine into the major groove. Biochemistry 1996; 35:9850-9863.

26. Hearst JE, Isaacs ST, Kanne D et al. The reaction of psoralens with deoxyribonucleic acid. Quart Rev Biophys 1984; 17:1-44.

27. Spielmann HP, Dwyer TJ, Hearst JE et al. Solution structures of psoralen monoadducted and cross-linked DNA oligomers by NMR spectroscopy and restrained molecular dynamics. Biochemistry 1995; 34:12937-12953.

28. Reynolds VL, McGovren JP, Hurley LH. The chemistry, mechanism of action and biological properties of CC-1065, a potent antitumor antibiotic. J Antibiotics 1986; 39:319-334.

29. Gunz D, Naegeli H. A noncovalent binding-translocation mechanism for site-specific CC-1065-DNA recognition. Biochem Pharmacol 1996; 52:447-453.

30. Hess MT, Gunz D, Naegeli H. A repair competition assay to assess recognition by human nucleotide excision repair. Nucleic Acids Res 1996; 24:824-828.

31. Manley JL, Fire A, Samuels M et al. In vitro transcription: whole cell extract. Meth Enzymol 1983; 101:568-582.

32. Wood RD, Robins P, Lindahl T. Complementation of the xeroderma pigmentosum DNA repair defect in cell-free extracts. Cell 1988; 53:97-106.

33. Challberg MD, Kelly TJ. Animal virus DNA replication. Annu Rev Biochem 1989; 58:671-717.

34. Jones CJ, Wood RD. Preferential binding of the xeroderma pigmentosum group A complementing protein to damaged DNA. Biochemistry 1993; 32:12096-12104.

35. He Z, Henricksen LA, Wold MS et al. RPA involvement in the damage-recognition and incision steps of nucleotide excision repair. Nature 1995; 374:566-568.

36. Li L, Lu X, Peterson CA et al. An interaction between the DNA repair factor XPA and replication protein A appears essential for nucleotide excision repair. Mol Cell Biol 1995; 15:5396-5402.

37. Huang J-C, Hsu DS, Kazantsev A et al. Substrate spectrum of human excinuclease: repair of abasic sites, methylated bases, mismatches, and bulky ad-

ducts. Proc Natl Acad Sci USA 1994; 91:12213-12217.

38. Nazimiec M, Grossman L, Tang M-S. A comparison of the rates of reaction and function of UVRB in UVRAB-mediated anthramycin-N^2-guanine-DNA repair. J Biol Chem 1992; 267:24716-24724.

39. Tang M-S, Lee C-S, Doisy R et al. Recognition and repair of the CC-1065-(N^3-adenine)-DNA adduct by the UVRABC nucleases. Biochemistry 1988; 27:893-901.

40. Selby CP, Sancar A. ABC Excinuclease incises both 5' and 3' to the CC-1065-DNA adduct and its incision activity is stimulated by DNA helicase II and DNA polymerase I. Biochemistry 1988; 27:7184-7188.

41. Mu D, Park C-H, Matsunaga T et al. Reconstitution of human DNA repair excision nuclease in a highly defined system. J Biol Chem 1995; 270:2415-2418.

42. Hess MT, Schwitter U, Petretta M et al. Site-specific DNA substrates for human excision repair: comparison between deoxyribose and base adducts. Chem Biol 1996; 3:121-128.

43. Huang J-C, Svoboda DL, Reardon JT et al. Human nucleotide excision nuclease removes thymine dimers from DNA by incising the 22nd phosphodiester bond 5' and the 6th phosphodiester bond 3' to the photodimer. Proc Natl Acad Sci USA 1992; 89:3664-3668.

44. Newbold RF, Brookes P. Exceptional mutagenicity of a benzo[a]pyrene diol epoxide in cultured mammalian cells. Nature 1976; 261:52-54.

45. Buening MK, Wislocki PG, Levin W et al. Tumorigenicity of the optical enantiomers of the diastereomeric benzo[a]pyrene 7,8-diol-9,10-epoxides in newborn mice: exceptional activity of (+)-7β,8α-dihydroxy-9α,10α-epoxy-7,8,9,10-tetrahydrobenzo-[a]pyrene. Proc Natl Acad Sci USA 1978; 75:5358-5361.

46. Cosman M, de los Santos C, Fiala R et al. Solution conformation of the major adduct between the carcinogen (+)-*anti*-benzo[a]pyrene diol epoxide and DNA. Proc Natl Acad Sci USA 1992; 89:1914-1918.

47. Slaga TJ, Bracken WJ, Gleason G et al. Marked differences in the skin tumor-initiating activities of the optical enantiomers of the diastereomeric benzo[a]pyrene 7,8-diol-9,10-epoxides. Cancer Res 1979; 394:67-71.

48. Wood AW, Chang RL, Levin W et al. Differences in mutagenicity of the optical enantiomers of the diastereomeric benzo[a]pyrene 7,8-diol-9,10-epoxides. Biochem. Biophys Res Commun 1977; 77:1389-1396.

49. Wei SJC, Chang RL, Wong C-Q et al. Dose-dependent differences in the profile of mutations induced by an ultimate carcinogen from benzo[a]pyrene. Proc Natl Acad Sci USA 1991; 88:11227-11230.

50. Stevens CW, Bouck N, Burgess JA et al. Benzo[a]pyrene diol-epoxides: different mutagenic efficiency in human and bacterial cells. Mutat Res 1985; 152:5-14.

51. Shibutani S, Margulis LA, Geacintov NE et al. Translesion synthesis on a DNA template containing a single stereoisomer of dG-(+)- or dG(-)-*anti*-BPDE (7,8-dihydroxy-*anti*-9,10-epoxy-7.8,9,10-tetrahydrobenzo[a]pyrene. Biochemistry 1993; 32:7531-7541.

52. Suh M, Ariese F, Small GJ et al. Formation and persistence of benzo[a]pyrene-DNA adducts in mouse epidermis in vivo: importance of adduct conformation. Carcinogenesis 1995; 16:2561-2569.

CHAPTER 8

MOLECULAR CROSSTALKS AT CARCINOGEN-DNA ADDUCTS

To avoid damage-induced mutagenesis and minimize cytotoxicity, carcinogen-DNA adducts should be immediately channeled into appropriate repair pathways. Problems may arise, however, when a damaged sequence serves as a substrate not only for DNA repair but, simultaneously, for other nuclear functions such as transcription, replication or homologous recombination. These different processes may interfere with each other by competing for the same DNA substrate. For example, it is well established that many base lesions constitute effective blocks to transcription[1] and replication.[2] Analysis of UV-induced mutations in *SUP4-o* (a yeast transfer RNA gene transcribed by RNA polymerase III) demonstrated a strong strand bias with approximately 90% of the nucleotide sequence changes in the transcribed template strand.[3] This observation indicates that excision repair in transfer RNA genes is inhibited by concurrent transcription. Exactly the opposite is observed in genes transcribed by eukaryotic RNA polymerase II, where excision repair of the transcribed template strand is stimulated by transcription.[4] Thus, RNA (and DNA) polymerases are important modulators of the biological consequences of DNA damage. The role of transcription and replication in the cellular response to genotoxic insults will be reviewed in chapters 9 and 10.

The main purpose of this chapter is to elaborate on the hypothesis that many carcinogen-DNA adducts may exert cytotoxic effects by novel, unexpected mechanisms that do not involve the simple inhibition of DNA tracking or DNA metabolizing enzymes. For example, it has been repeatedly observed that certain nuclear proteins are able to bind specific classes of bulky base adducts with considerable affinity and avidity.[5,6] Such nuclear proteins were originally thought to participate in the initial stages of DNA repair mechanisms. These observations became more intriguing when the respective DNA binding proteins were isolated and characterized, and it was concluded that they do not participate in DNA repair processes directly.[5-9] Instead, the ability to bind damaged DNA is a fortuitous event unrelated to the normal activity of the proteins implicated in this phenomenon. It is now believed that certain bulky DNA adducts cause structural distortions in the double helix that mimic normal conformational intermediates occurring during activation of nuclear functions such as transcription, replication or recombination. As a consequence, such bulky DNA adducts provide artificial binding substrates for cellular factors involved in these important nuclear

Mechanisms of DNA Damage Recognition in Mammalian Cells, by Hanspeter Naegeli.
© 1997 R.G. Landes Company.

activities. The resulting protein-DNA complexes at sites of bulky base lesions may potentially modulate the mutagenic or cytotoxic effects of DNA damaging agents by inhibiting or stimulating DNA repair, by diverting essential regulatory factors away from their physiological sites of action or, alternatively, by generating nucleation centers for the assembly of transcription initiation complexes at improper sites in the genome. Sequestration of transcription or recombination factors has been proposed as a possible mechanism of cytotoxicity of DNA adducts generated by the antitumor drug cis-diamminedichloroplatinum(II).[5,7] In most cases, however, the biological significance of such carcinogen-mediated crosstalks between DNA repair, transcription or recombination remains poorly understood.

TRANSCRIPTION FACTOR HIJACKING

The concept of transcription factor hijacking was proposed by Treiber et al[5] to indicate a novel mechanism by which DNA damaging agents may disrupt nuclear functions. Central to their model is the conception that specific DNA lesions are potentially able to titrate essential regulatory proteins away from their natural binding sites in the genome.

At least two different bulky DNA adducts have been shown to provide high affinity binding sites for transcription factors. The first example is provided by the binding of high mobility group (HMG) domain proteins to intrastrand crosslinks formed by cisplatin [cis-diamminedichloroplatinum(II)].[5,8,9] Another report showed that base adducts formed by the polycyclic aromatic hydrocarbon benzo[*a*]pyrene diol-epoxide constitute potential binding sites for the transcriptional activator Sp1.[6] In both cases, a well-known genotoxic agent was found to produce illegitimate binding sites for positively acting transcription factors.

Although not formally proven, these carcinogen-DNA adducts may disrupt transcriptional control mechanisms. In fact, assuming that sequestration of transcription

factors on DNA adducts occurs in a quantitative manner one would expect reduced levels of transcription from relevant genes. In addition, sequestration of regulatory factors at artificial sites on the chromosomes may lead to the assembly of transcription initiation complexes in the promoter region of genes, including perhaps oncogenes, that are normally tightly regulated or not expressed at all. This alternative mechanism may result in abnormal oncogene activation by carcinogen-DNA adducts. A third possibly deleterious effect is increased persistence of base adducts because of protein-induced protection, or shielding, from DNA repair processes.[5,6]

THE INTERACTION BETWEEN CISPLATIN ADDUCTS AND HMG-DOMAIN PROTEINS

Cisplatin is one of the most commonly used anticancer drugs, and is particularly effective against testicular tumors. Before cisplatin was available, only about 5% of patients were cured. With cisplatin treatment, up to 90% of patients with testicular cancer can expect long-term survival.[10] Cisplatin is also useful in the treatment of ovarian, bladder, head, neck and small-cell lung cancer, although the basis for its selectivity against specific types of tumors is unknown.[11]

The intracellular target of cisplatin is DNA and chemotherapeutic efficacy is conferred by the formation of covalent cisplatin-DNA crosslinks. Exposure of DNA to this drug generates mainly 1,2-intrastrand d(GpG) cisplatin crosslinks (65%), 1,2-intrastrand d(ApG) cisplatin crosslinks (25%) and a smaller fraction (6%) of 1,3-intrastrand d(GpNpG) cisplatin crosslinks (Fig. 8.1).[12,13] To a minor extent, cisplatin also forms DNA-DNA interstrand crosslinks and protein-DNA adducts. These bulky cisplatin lesions are known inhibitors of RNA and DNA synthesis,[14,15] and it has been suggested that cisplatin crosslinks may trigger pathways of apoptotic cell death.[16-18]

A novel hypothetical mechanism of cytotoxicity was prompted by the finding that

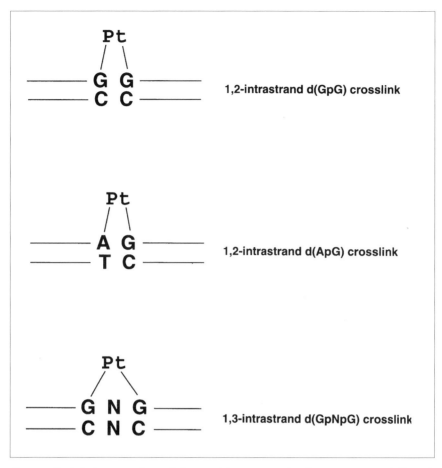

Fig. 8.1. Cisplatin intrastrand crosslinks. HMG box proteins bind to 1,2-intrastrand d(GpG) crosslinks and 1,2-intrastrand d(ApG) cisplatin crosslinks, but not to 1,3-intrastrand d(GpNpG) cisplatin crosslinks.

proteins containing high mobility group (HMG) domains bind specifically to the major cisplatin-DNA adducts, i.e., 1,2-intrastrand d(GpG) and d(ApG) cross-links. These HMG box proteins do not bind, however, to other quantitatively minor cisplatin adducts such as the 1,3-intrastrand d(GpNpG) crosslink.[8,9,19-21] Similarly, HMG box proteins fail to inter-act with DNA adducts formed by the clini-cally ineffective trans isomer of cisplatin [trans-diamminedichloroplatinum(II)], a compound that is unable to produce 1,2-crosslinks because of steric constraints. Moreover, HMG box proteins do not bind

monodentate base adducts formed by monofunctional platinum derivatives.

The HMG box is an evolutionary con-served 80-amino acid region that displays a characteristic pattern of basic and aromatic residues which normally serve to mediate sequence-specific DNA binding interac-tions.[9,22-24] Proteins containing this motif include HMG-1 and HMG-2, human up-stream binding factor (hUBF), the testis determining factor SRY (product of the sex determining region on the Y chromo-some), mitochondrial transcription fac-tor (h-mtTFA), lymphoid enhancer bind-ing factor 1 (LEF-1), a T cell-specific transcription factor (TCF-1α) and the yeast autonomously replicating sequence factor ABF2 (reviewed in ref. 9).

Human upstream binding factor (hUBF) is an activator of ribosomal RNA transcription and, as stated before, also belongs to the HMG box family. A quantitative DNase I footprint analysis showed that hUBF binds to 1,2-intrastrand d(GpG) adducts with striking affinity ($K_d = 6 \times 10^{-11}$ M).[5] Interestingly, the affinity of hUBF for its specific promoter ($K_d = 1.8 \times 10^{-11}$ M) is similar to that measured for the 1,2-intrastrand d(GpG) lesion, suggesting that cisplatin adducts should be effective competitive inhibitors of hUBF-promoter complex formation. Support for this hypothesis came from experiments in which binding of hUBF to the natural ribosomal RNA promoter was efficiently antagonized by the addition of DNA fragments containing 1,2-intrastrand d(GpG) cisplatin adducts but not by the addition of nondamaged fragments.[5] Unfortunately, in vivo experiments aimed at confirming that platinum adducts may indeed divert HMG box proteins from their normal cellular locus have failed.[25]

Shielding from Excision Repair

Studies on the yeast *Saccharomyces cerevisiae* have identified a gene product that confers enhanced sensitivity to cisplatin.[21] Ixr1 (for *i*ntrastrand *c*rosslink *r*ecognition) was isolated from an expression library for its ability to bind to cisplatin-damaged DNA. Cloning and sequence analysis of the *IXR1* gene showed that it contains two tandemly repeated HMG boxes. When the *IXR1* gene was deleted, the resulting yeast mutant grew normally but displayed increased resistance to cisplatin.[21] The possibility that the lower cytotoxicity of cisplatin in the ixr1 strain is the result of enhanced excision repair was investigated in a follow-up study. When the *IXR1* deletion was introduced in nucleotide excision repair-deficient rad2, rad4 or rad14 strains, the differential sensitivity caused by removing Ixr1 protein was nearly abolished.[26] These results indicate that inhibition of excision repair is the most plausible mechanism by which Ixr1 sensitizes yeast towards cisplatin.[26]

Shielding of platinum adducts from DNA repair processes was confirmed in vitro using human cell extracts proficient in nucleotide excision repair.[27] In this system, repair of 1,2-intrastrand d(GpG) cisplatin adducts was detected by monitoring the excision of oligonucleotide segments of 24-32 residues in length from substrates that carry a site-specific lesion (see Fig. 7.6 in chapter 7 for details on the oligonucleotide excision assay). As expected from the in vivo yeast data, excision of d(GpG) platinum adducts was progressively reduced when increasing amounts of HMG-1 or h-mtTFA (two different HMG box proteins) were added to the repair reactions. In contrast, excision of 1,3-intrastrand d(GpTpG) crosslinks was not affected by the presence of HMG proteins.[27] Recently, these findings were confirmed using a defined system in which human nucleotide excision repair was reconstituted in vitro from purified components.[28]

These results confirmed that HMG domain proteins specifically protect d(GpG) platinum adducts from nucleotide excision repair. It was noted that relatively high concentrations (0.5-5 μM) of HMG-1 or h-mtTFA proteins were required to suppress excision repair in vitro, and these concentrations are certainly higher than those found in vivo. On the other hand, it was argued that there are multiple HMG box proteins that are likely to exert similar inhibitory effects on the repair of cisplatin lesions. The total cellular concentration of all HMG box proteins may match or even exceed the concentrations used in the in vitro repair assay.[27,28]

The attractive idea of shielding from recognition by the excision repair system has at least three important clinical implications.[27,28] First, tissue-specific overexpression of proteins that recognize cisplatin-DNA adducts should enhance the cytotoxicity of cisplatin and may explain the extraordinary sensitivity of certain tumor types to this particular drug. Second, cancer cells may avoid this mechanism of cytotoxicity and acquire cisplatin resistance by suppressing expression of HMG box proteins. Third, future pharmacological strategies may exploit the shielding mechanism to increase therapeutic effectiveness of platinum-based anticancer drugs and generate new derivatives that form bulky base adducts with even higher affinity for HMG domain proteins. Alternatively, enhanced effectiveness may also be achieved by the overexpression of HMG box proteins in target tumor tissues using appropriate recombinant vectors. Unfortunately, cisplatin resistance is a common clinical occurrence and effective strategies are urgently needed to circumvent this problem.

THE STRUCTURAL BASIS FOR PLATINATED DNA-HMG PROTEIN INTERACTIONS

The crystal structure of a short double-stranded DNA fragment containing a single 1,2-intrastrand d(GpG) cisplatin adduct revealed the following conformational features.[29]

(1) The platinated DNA duplex is bent towards the major groove (Fig. 8.2). The angle could not be determined exactly, but ranges between 39° and 55°. The extent of cisplatin-induced bending detected in the crystal is consistent with previous reports based on nuclear magnetic resonance (NMR)[30] or gel electrophoretic analysis.[31,32]

(2) The cisplatin 1,2-intrastrand d(GpG) adduct does not disrupt Watson-Crick hydrogen bonding in the crystal. However, the four bases at and around the site of modification are propeller twisted (unwound) by a value that ranges from -8° to -36°.

(3) The minor groove width of the modified duplex is in the range of 0.92-1.12 nm, which is closer to the 1.1 nm width of A-DNA than the 0.6 nm width of normal B-DNA (see chapter 3). This conformational dis-

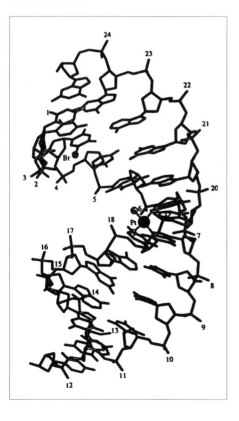

Fig. 8.2. Crystal structure of a DNA duplex containing a cisplatin 1,2-intrastrand d(GpG) crosslink. After platination, the duplex is bent, unwound and contains an abnormally widened and hydrophobic minor groove. The platinum adduct is indicated with the filled circle in the center. Reprinted with permission from: Takahara PM, Rosenzweig AC, Frederick CA et al. Crystal structure of double-stranded DNA containing the major adduct of the anticancer drug cisplatin. Nature 1995; 377: 649-652. © 1995 Macmillan Magazines Limited.

tortion exposes the bases to the minor groove and generates an unusual hydrophobic surface.[29]

It was concluded from this crystallographic analysis that the geometric changes at 1,2-intrastrand d(GpG) cisplatin adducts share structural features with the DNA conformation found at physiological binding sites of HMG box proteins. For example, the interaction of HMG domain proteins with platinated sites is consistent with the preference of these proteins for bent and unwound DNA structures.[33,34] Also, it had been shown that the HMG domain of SRY intercalates an isoleucine side chain into the minor groove.[35] Thus, the intrinsic capacity of cisplatin 1,2-intrastrand adducts to stabilize a widened and hydrophobic minor groove may facilitate, in energetic terms, conformational switches required for HMG box protein binding and, hence, specifically attract these proteins to platinated sites.

THE INTERACTION BETWEEN BPDE-DNA ADDUCTS AND Sp1

Sp1 is a transcriptional activator that binds reversibly to GC-rich consensus sequences (for example 5'-GGGGCGGGGC-3') called GC boxes.[36,37] These GC boxes are commonly found in the promoter and upstream regulatory regions of many housekeeping genes, oncogenes and genes involved in cell cycle control. The binding of Sp1 to GC boxes is mediated by three zinc-finger domains located near the carboxy terminus of the protein and involves specific contacts with the major groove of double-stranded DNA. Activation of transcription by Sp1 depends on its glutamine-rich domains, which are required to mediate numerous protein-protein interactions.

The interaction of purified human Sp1 with GC-rich sequences in the promoter of the hamster adenosine phosphoribosyl transferase gene was investigated using electrophoretic mobility shift assays, competition experiments and DNase I footprinting.[38] In the course of these studies, it was found that Sp1 binds with 5-10 times greater affinity to substrates containing GC boxes when

DNA was previously exposed to the ultimate carcinogen benzo[*a*]pyrene diol epoxide (BPDE). Subsequent experiments showed that even BPDE-modified DNA fragments that do not contain canonical GC boxes provide strong binding sites for Sp1.[38] These results suggest that BPDE adducts generate artificial Sp1 target sites on DNA, presumably as a consequence of specific conformational changes imposed on the double helix upon covalent modification with BPDE.

Like many other transcription factors, Sp1 induces DNA bending and unwinding upon binding to its target sequence. Sp1 interaction with GC boxes in the simian virus 40 genome produces a 60° bend.[39] As was postulated for cisplatin 1,2-intrastrand crosslinks, structural alterations at sites of BPDE modification may reduce the free energy of Sp1 protein-DNA interactions, thereby facilitating the formation of Sp1 protein-DNA complexes. The high frequency of BPDE modification required to observe these effects (2.6-8.0% of nucleotides modified) suggests that Sp1 recognizes one of the quantitatively minor BPDE adducts rather than the predominant (+)-trans-*anti*-BPDE-N^2-guanine lesion. Additional work using site-directed stereoisomers of BPDE-N^2-guanine or BPDE-N^6-adenine adducts will be necessary to clarify this issue.

RECOMBINATION FACTOR HIJACKING

DNA-PK is a large heterotrimeric complex that binds to DNA breaks, gaps or hairpins, and depends on these structural discontinuities for activation of its kinase activity.[40] The complex is composed of a catalytic subunit (the ~350 kDa DNA-PK$_{CS}$ polypeptide) and two regulatory DNA binding subunits (Ku70 and Ku86) with molecular masses of 70 and 86 kDa, respectively.[41] Several lines of investigation led to the conclusion that DNA-PK is a major player in the recombinational repair of DNA strand breaks.

The Ku heterodimer was originally discovered as a protein in normal human cells that reacts with sera obtained from some patients afflicted by autoimmune diseases.[42]

The designation "Ku" derives from the first two letters of a patient's name. The same Ku heterodimer was reisolated from human cells as a DNA-dependent ATPase with associated DNA helicase activity.[43] A breakthrough in understanding the function of Ku and its protein kinase (DNA-PK$_{CS}$) partner came with the discovery that several radiosensitive Chinese hamster cell lines impaired in DNA double strand break repair are deficient in Ku DNA binding activity.[44-47] Thompson and Jeggo[48] proposed the designation *XRCC* (*X-ray Repair Cross Complementing*) for human genes that complement the radiosensitivity of these rodent cell lines. The genes encoding Ku86 and Ku70 are identical to *XRCC5* and *XRCC6*, respectively.[44,48-50] The gene encoding DNA-PK$_{CS}$ is the same as *XRCC7*.[48,51] *DNA-PK$_{CS}$/XRCC7* is also identical to the gene mutated in mice affected by severe combined immunodeficiency (SCID; see chapter 2).[52-55] These spontaneously mutated animals suffer from hypersensitivity to ionizing radiation and lack mature B and T cells.[56-58] At the molecular level, SCID mice are unable to repair double strand DNA breaks and are deficient in V(D)J recombination.[56-59] Collectively, this striking association between radiation sensitivity and immunodeficiency in the SCID phenotype indicates that DNA-PK is an essential factor for recognition and repair of strand breaks generated by genotoxic reactions as well as for ligation of DNA breaks introduced by the V(D)J rearrangement machinery.

SEQUESTRATION OF DNA-PK ON CISPLATIN-DAMAGED DNA

Turchi and Henkels[7] isolated two protein complexes based on their affinity for DNA treated with cisplatin. The first complex has two subunits with apparent molecular masses of 83 and 68 kDa. The second complex is a heterotrimer of 350, 83 and 68 kDa subunits in a 1:1:1 stoichiometry. The 83 kDa subunit in these cisplatin binding complexes was identified as Ku86 by N-terminal protein sequence analysis and its reactivity with monoclonal antibody raised against human Ku86. The 350 kDa subunit was identified as DNA-PK$_{CS}$ because the trimeric complex displayed DNA-stimulated protein kinase activity.

Binding of the two different complexes to DNA containing cisplatin crosslinks was confirmed using electrophoretic mobility shift assays and competition binding assays.[7] Interestingly, kinase activity of DNA-PK was stimulated only by nondamaged DNA, whereas cisplatin-damaged DNA failed to activate DNA-PK activity despite the ability of Ku to bind to cisplatin-modified substrates. This lack of DNA-PK stimulation correlated with the frequency of cisplatin crosslinks.

The observation that Ku can bind cisplatin-damaged DNA but is unable to activate the catalytic DNA-PK$_{CS}$ subunit has multiple functional implications. This mechanism may, for example, block the normal DNA-PK$_{CS}$ response to damage inflicted on DNA, i.e., phosphorylation of downstream protein targets such as p53, RPA, various transcription factors (cJun, cFos, Oct1, Sp1, cMyc, CTF/NF-I and TFIID), enzymes (topoisomerase I and II), histones, heat shock proteins or microtubule-associated proteins to name a few possible examples (reviewed in ref. 60). Turchi and Henkels[7] proposed that this capacity to damage DNA without activating a significant cellular DNA damage response pathway may contribute to the exceptional cytotoxicity of cisplatin. Alternatively, cisplatin cytotoxicity may be enhanced by sequestration of DNA-PK on cisplatin-damaged substrates, thereby diverting an essential recombination factor from its normal physiological function. Conceivably, a fraction of active DNA-PK may be constantly required to repair DNA strand breaks that arise spontaneously in rapidly replicating tumors and, hence, malignant cells are particularly vulnerable to this sequestration mechanism. Of course, binding of Ku proteins to cisplatin intrastrand crosslinks may also result in inhibition of excision repair by shielding.

RECOGNITION OF DNA ADDUCTS BY THE MISMATCH CORRECTION SYSTEM

In mammalian cells, MutSα initiates mismatch correction by binding to mispaired bases.[61,62] MutSα is a heterodimer consisting of two polypeptides (MSH2 and GTBP/p160). Using electrophoretic mobility shift assays, Duckett et al were able to show that MutSα binds to short DNA fragments containing one of the following base adducts: O^6-methylguanine, O^4-methylthymine or a 1,2-intrastrand d(GpG) cisplatin crosslink.[63] O^6-methylguanine was recognized by MutSα regardless of whether the methylated base was paired with cytosine or with thymine. O^4-methylthymine was recognized when the methylated base was paired with adenine. In all cases, the binding affinity for alkylated bases was estimated to be about one order of magnitude less than that measured in the presence of a G·T mismatch.[63]

Mismatch repair defects have frequently been implicated in cellular resistance to alkylating agents.[64,65] However, the precise mechanism by which interaction of mismatch repair proteins with alkylated bases may lead to cell death is still a matter of controversy. One favored model involves translesion synthesis when the replication machinery encounters an alkylated base (for example O^6-methylguanine) in the DNA template. The resulting base pair abnormality (for example O^6-methylguanine paired to thymine) induces DNA incision and processing by the mismatch correction system. Because mismatch repair is restricted to the newly replicated DNA strand, the alkylated base persists in the parental strand and, as a consequence, induces repeated rounds of DNA incision and resynthesis. Thus, by executing futile repair cycles that eventually lead to chromosome degradation mismatch correction is thought to confer cytotoxicity to alkylating agents. Alternatively, or additionally, binding of MutSα to alkylated sites may cause nonproductive sequestration of the mismatch recognition factor and divert the mismatch correction system from its normal cellular functions. Finally, it is again possible that MutSα may enhance the cytotoxicity of alkylated bases by shielding them from repair processes.

THE DNA PHOTOLYASE PARADIGM

Studies on DNA photolyase show that no a priori prediction can be made on whether a protein bound to a lesion will stimulate or inhibit excision repair. DNA photolyase is a light-dependent DNA repair enzyme implicated in DNA damage reversal (see chapter 4).[66,67] The enzyme binds to cyclobutane pyrimidine dimers and, after excitation with blue light, restores the adducted pyrimidines to their monomeric form. When DNA photolyase binds to cyclobutane dimers in the dark it cannot split the cyclobutane ring because of lack of the activating photon. Under these conditions, DNA photolyase does not inhibit but actually stimulates DNA excision repair in *Escherichia coli* and the yeast *Saccharomyces cerevisiae*.[68-70] It has been proposed that photolyase bound to a cyclobutane dimer looks like a bulkier adduct to the recognition subunits compared to the naked dimer.[71] Alternatively, DNA photolyase may recruit the nucleotide excision repair system through specific protein-protein interactions.

More recently, DNA photolyase was shown to bind to cisplatin 1,2-d(GpG) intrastrand crosslinks as well.[18,72,73] DNA photolyase is unable to reverse cisplatin crosslinks with or without light.[73] However, one report concludes that, as was previously found with cyclobutane pyrimidine dimers, the interaction with cisplatin intrastrand crosslinks stimulates nucleotide excision repair and confers increased resistance to killing by cisplatin.[73] These results obtained with the prokaryotic DNA photolyase from *Escherichia coli* sharply contrast with similar experiments performed using its eukaryotic counterpart.[18,72] In fact, the loss of photolyase function in *Saccharomyces cerevisiae* strains containing a mutated *PHR1* gene led to increased resistance to cisplatin and other alkylating agents such as *N*-methyl-*N'*-nitro-*N*-nitrosoguanidine and 4-nitroso-quinoline-*N*-oxide. Conversely, transformation of yeast photolyase mutants with wild-type *PHR1* sensitized cells to killing by these genotoxic compounds.[72] Thus, the prokaryotic photolyase stimulates repair of certain chemical adducts, whereas the available evidence indicates that its yeast counterpart may rather have inhibitory effects.

CONCLUSIONS

This chapter summarizes some of the most provocative and controversial, but nevertheless important themes at the interface between different DNA damage pathways. In most cases, the proposed mechanisms and their putative biological, toxicological or clinical endpoints remain poorly understood. It is not clear, for example, how and to what extent binding of Sp1 to base adducts formed by BPDE influences the mutagenicity and cytotoxicity exerted by benzo[*a*]pyrene and other polycyclic aromatic hydrocarbons.

Intrastrand 1,2-(GpG) or 1,2(ApG) crosslinks generated by the chemotherapeutic agent cisplatin appear capable of disrupting multiple processes of nuclear metabolism. First, like many other bulky adducts cisplatin crosslinks inhibit DNA and RNA polymerases. Second, triggering of apoptosis is discussed as a possible pathway for its anticancer activity. In addition, there is strong evidence indicating that cisplatin intrastrand crosslinks are protected from excision repair mechanisms through binding of certain transcription factors. Whether transcription factor hijacking augments the cytotoxic effects of cisplatin is currently unknown. Similarly, there is circumstantial evidence suggesting that cell killing by cisplatin may involve the recognition of intrastrand crosslinks by recombination and mismatch correction subunits.

The question of how cisplatin kills cancer cells in a selective manner is of extraordinary importance to facilitate future approaches in rational drug design. In fact, chemotherapeutic strategies could be considerably improved by dissecting those interactions that are necessary for antitumor efficacy from those that are responsible for unwanted cytotoxic or mutagenic side effects. For example, hijacking of recombination (Ku) or mismatch correction (MutSα) proteins may constitute a basic principle for inducing lethality in highly proliferating tumor cells. It may also be possible, however, that recombination or mismatch correction factor hijacking increases the probability of secondary tumor formation in nontarget tissues. Thus, detailed knowledge of the mode of action of cisplatin would provide a rational basis for devising new drugs that either enhance or reduce these molecular crosstalks with components of the transcription, recombination or mismatch correction machinery. In summary, it is expected that research on the mechanism of action of cisplatin will facilitate the development of new clinically effective antitumor drugs.

REFERENCES

1. Michalke H, Bremer H. RNA synthesis in Escherichia coli after irradiation with ultraviolet light. J Mol Biol 1969; 41:1-23.
2. Moore PD, Bose KK, Rabkin SD et al. Sites of termination of in vitro DNA synthesis on ultraviolet- and N-acetylaminofluorene-treated φX174 templates by prokaryotic and eukaryotic DNA polymerases. Proc Natl Acad Sci USA 1981; 78:110-114.
3. Armstrong JD, Kunz BA. Site and strand specificity of UVB mutagenesis in the SUP4-o gene of yeast. Proc Natl Acad Sci USA 1990; 87:9005-9009.
4. Bohr VA, Smith CA, Okumoto DS et al. DNA repair in an active gene: removal of pyrimidine dimers from the *DHFR* gene of CHO cells is much more efficient than in the genome overall. Cell 1985; 40:359-369.
5. Treiber DK, Zhai X, Jantzen H-M et al. Cisplatin-DNA adducts are molecular decoys for the ribosomal RNA transcription factor hUBF (human upstream binding factor). Proc Natl Acad Sci USA 1994; 91:5672-5676.
6. MacLeod MC, Powell KL, Tran N. Binding of the transcription factor, Sp1, to non-target sites in DNA modified by benzo[*a*]pyrene diol epoxide. Carcinogenesis 1995; 16:975-983.

7. Turchi JJ, Henkels K. Human Ku auto-antigen binds cisplatin-damaged DNA but fails to stimulate human DNA-activated protein kinase. J Biol Chem 1996; 271:13861-13867.

8. Toney J, Donahue B, Kellett P et al. Isolation of cDNAs encoding a human protein that binds selectively to DNA modified by the anticancer drug cis-diamminedichloroplatinum. Proc Natl Acad Sci USA 1989; 86:8328-8332.

9. Bruhn SL, Pil PM, Essigmann JM et al. Isolation and characterization of human cDNA clones encoding a high mobility group box protein that recognizes structural distortions to DNA caused by binding of the anticancer agent cisplatin. Proc Natl Acad Sci USA 1992; 89:2307-2311.

10. Abrams MJ, Murrer BA. Metal compounds in therapy and diagnosis. Science 1993; 261:725-730.

11. Andrews PA, Howell SB. Cellular pharmacology of cisplatin: perspectives on mechanisms of acquired resistance. Cancer Cells 1990; 2:35-43.

12. Pinto AL, Lippard SJ. Binding of the antitumor drug cis-diamminedichloroplatinum(II) (cisplatin) to DNA. Biochim Biophys Acta 1985; 780:167-180.

13. Eastman A. The formation, isolation and characterization of DNA adducts produced by anticancer platinum complexes. Pharmacol Ther 1987; 34:155-166.

14. Howle JA, Gale GR. Cis-dichlorodiammineplatinum (II). Persistent and selective inhibition of deoxyribonucleic acid synthesis in vivo. Biochem Pharmacol 1970; 19:2757-2762.

15. Ciccarelli RB, Solomon MJ, Varshavsky A et al. In vivo effects of cis- and trans-diamminedichloroplatinum(II) on SV40 chromosomes: differential repair, DNA-protein cross-linking, and inhibition of replication. Biochemistry 1985; 24:7533-7540.

16. Barry MA, Behnke CA, Eastman A. Activation of programmed cell death (apoptosis) by cisplatin, other anticancer drugs, toxins and hyperthermia. Biochem Pharmacol 1990; 40:2353-2362.

17. Eastman A. Activation of programmed cell death by anticancer agents: cisplatin as a model system. Cancer Cells 1990; 2:275-280.

18. Chu G. Cellular responses to cisplatin. J Biol Chem 1994; 269:787-790.

19. Pil PM, Lippard SJ. Specific binding of chromosomal protein HMG1 to DNA damaged by the anticancer drug cisplatin. Science 1992; 256:234-237.

20. Hughes EN, Engelsberg BN, Billings PC. Purification of nuclear proteins that bind to cisplatin-damaged DNA. Identity with high mobility group proteins 1 and 2. J Biol Chem 1992; 267:13520-13527.

21. Brown SJ, Kellett PJ, Lippard SJ. Ixr1, a yeast protein that binds to platinated DNA and confers sensitivity to cisplatin. Science 1993; 261:603-605.

22. Jantzen HM, Admon A, Bell SP et al. Nuclear transcription factor hUBF contains a DNA-binding motif with homology to HMG proteins. Nature 1990; 344:830-836.

23. Lilley DM. DNA-protein interactions. HMG has DNA wrapped up. Nature 1992; 357:282-283.

24. Bianchi ME, Falciola L, Ferrari S et al. The DNA binding site of HMG1 protein is composed of two similar segments (HMG boxes), both of which have counterparts in other eukaryotic regulatory sequences. EMBO J 1992; 11:1055-1063.

25. McA'Nulty MM, Whitehead JP, Lippard SJ. Binding of Ixr1, a yeast HMG-domain protein, to cisplatin-DNA adducts in vitro and in vivo. Biochemistry 1996; 362:75-86.

26. McA'Nulty MM, Lippard SJ. The HMG-domain protein Ixr1 blocks excision repair of cisplatin-DNA adducts in yeast. Mutat Res 1996; 362:75-86.

27. Huang J-C, Zamble DB, Reardon JT et al. HMG-domain proteins specifically inhibit the repair of the major DNA adduct of the anticancer drug cisplatin by human excision nuclease. Proc Natl Acad Sci USA 1994; 91:10394-10398.

28. Zamble DB, Mu D, Reardon JT et al. Repair of cisplatin-DNA adducts by the mammalian excision nuclease. Biochemistry 1996; 35:10004-10013.

29. Takahara PM, Rosenzweig AC, Frederick CA et al. Crystal structure of double-

stranded DNA containing the major adduct of the anticancer drug cisplatin. Nature 1995; 377:649-652.

30. den Hartog JH, Altona C, van Boom JH et al. *Cis*-diamminedichloroplatinum(II) induced distortion of a single and double stranded deoxydecanucleosideenonaphosphate studied by nuclear magnetic resonance. J Biomol Struct Dyn 1985; 2:1137-1155.

31. Rice JA, Crothers DM, Pinto AL et al. The major adduct of the antitumor drug *cis*-diamminedichloroplatinum(II) with DNA bends the duplex by ≈40° toward the major groove. Proc Natl Acad Sci USA 1988; 85:4158-4161.

32. Bellon SF, Lippard SJ. Bending studies of DNA site-specifically modified by cisplatin, *trans*-diamminedichloroplatinum(II) and *cis*-[Pt(NH₃)₂(N³-cytosine)Cl]⁺. Biophys Chem 1990; 35:179-188.

33. Bianchi ME, Beltrame M, Paonessa G. Specific recognition of cruciform DNA by nuclear protein HMG1. Science 1989; 243:1056-1059.

34. Shirakata M, Hüppi K, Usuda S et al. HMG1-related DNA-binding protein isolated with V-(D)-J recombination signal probes. Mol Cell Biol 1991; 11:4528-4536.

35. King C-Y, Weiss MA. The SRY high-mobility-group box recognizes DNA by partial intercalation in the minor groove: a topological mechanism of sequence specificity. Proc Natl Acad Sci USA 1993; 90:11990-11994.

36. Gidoni D, Dynan WS, Tjian R. Multiple specific contacts between a mammalian transcription factor and its cognate promoter. Nature 1984; 312:409-413.

37. Kadonaga JT, Jones KA, Tjian R. Promoter-specific activation of RNA polymerase II transcription by Sp1. Trends Biochem Sci 1986; 11:20-23.

38. MacLeod MC, Powell KL, Tran N. Binding of the transcription factor, Sp1, to non-target sites in DNA modified by benzo[a]pyrene. Carcinogenesis 1995; 16:975-983.

39. Sun D, Hurley LH. Cooperative bending of the 21-base-pair repeats of the SV40 viral early promoter by human Sp1. Bio-

chemistry 1994; 33:9578-9587.

40. Gottlieb TM, Jackson SP. The DNA-dependent protein kinase: requirement for DNA ends and association with Ku antigen. Cell 1993; 72:131-142.

41. Gottlieb TM, Jackson SP. Protein kinases and DNA damage. Trends Biochem Sci 1994; 19:500-503.

42. Mimori T, Hardin JA. Mechanism of interaction between Ku protein and DNA. J Biol Chem 1986; 261:10375-10379.

43. Tuteja N, Tuteja R, Ochem A et al. Human DNA helicase II: a novel DNA unwinding enzyme identified as the Ku autoantigen. EMBO J 1994; 13:4991-5001.

44. Taccioli GE, Gottlieb TM, Blunt T et al. Ku80: product of the XRCC5 gene and its role in DNA repair and V(D)J recombination. Science 1994; 265:1442-1445.

45. Smider V, Rathmell WK, Lieber MR et al. Restoration of X-ray resistance and V(D)J recombination in mutant cells by Ku cDNA. Science 1994; 266:289-291.

46. Boubnov NV, Hall KT, Wills Z et al. Complementation of the ionizing radiation sensitivity, DNA end binding, and V(D)J recombination defects of double-strand break repair mutants by the p86 Ku autoantigen. Proc Natl Acad Sci USA 1995; 92:890-894.

47. Getts RC, Stamato TD. Absence of a Ku-like DNA end binding activity in the xrs double-strand DNA repair-deficient mutant. J Biol Chem 1995; 269:15981-15984.

48. Thompson LH, Jeggo PA. Nomenclature of human genes involved in ionizing radiation sensitivity. Mutat Res 1995; 337:131-134.

49. Mimori TY, Ohosone N, Hama A et al. Isolation and characterization of cDNA encoding the 80 kDa subunit protein of the human autoantigen Ku (p70/p80) recognized by autoantibodies from patients with skleroderma-polymyositis overlap syndrome. Proc Natl Acad Sci USA 1990; 87:1777-1781.

50. Reeves WH, Sthoeger ZM. Molecular cloning of cDNA encoding the p70 (Ku) lupus autoantigen. J Biol Chem 1989; 264:5047-5052.

51. Sipley JD, Menninger JC, Hartley KO et al. Gene for the catalytic subunit of the

human DNA-activated protein kinase maps to the site of the XRCC7 gene on chromosome 8. Proc Natl Acad Sci USA 1995; 92:7515-7519.

52. Blunt T, Finnie NJ, Taccioli GE et al. Defective DNA-dependent protein kinase activity is linked to V(D)J recombination and DNA repair defects associated with the murine scid mutation. Cell 1995; 80:813-823.

53. Kirchgessner CU, Patil CK, Evans JW et al. DNA dependent kinase (p350) as a candidate gene for the murine SCID defect. Science 1995; 267:1178-1182.

54. Miller RD, Hogg J, Ozaki JH et al. Gene for the catalytic subunit of mouse DNA-dependent protein kinase maps to the *scid* locus. Proc Natl Acad Sci USA 1995; 92:10792-10795.

55. Peterson SR, Kurimasa A, Oshimura M et al. Loss of the catalytic subunit of the DNA-dependent protein kinase in DNA double-strand-break-repair mutant mammalian cells. Proc Natl Acad Sci USA 1995; 92:3171-3174.

56. Schuler W, Weiler IJ, Schuler A et al. Rearrangement of antigen receptor genes is defective in mice with severe combined immune deficiency. Cell 1986; 46:963-972.

57. Biedermann KA, Sung J, Giaccia AJ et al. Scid mutation in mice confers hypersensitivity to ionizing radiation and a deficiency in DNA double-strand break repair. Proc Natl Acad Sci USA 1991; 88:1394-1397.

58. Hendrickson EA, Qin XQ, Bump EA et al. A link between double-strand break-related repair and V(D)J recombination: the scid mutation. Proc Natl Acad Sci USA 1991; 88:4061-4065.

59. Taccioli GE, Rathbun G, Oltz E et al. Impairment of V(D)J recombination in double-strand break repair mutants. Science 1993; 260:207-210.

60. Anderson CW. DNA damage and the DNA-activated protein kinase. Trends Biochem Sci 1993; 18:433-437.

61. Drummond JT, Li G-M, Longley MJ et al. Isolation of an hMSH2-p160 heterodimer that restores DNA mismatch repair to tumor cells. Science 1995; 268:1909-1912.

62. Palombo F, Gallinari P, Iaccarino I et al.

GTBP, a 160-kilodalton protein essential for mismatch binding activity in human cells. Science 1995; 268:1912-1914.

63. Duckett DR, Drummond JT, Murchie AIH et al. Human MutSα recognizes damaged DNA base pairs containing O^6-methylguanine, O^4-methylthymine, or the cisplatin-d(GpG) adduct. Proc Natl Acad Sci USA 1996; 93:6443-6447.

64. Kat A, Thilly WG, Fang WH et al. An alkylation-tolerant, mutator human cell line is deficient in strand-specific mismatch repair. Proc Natl Acad Sci USA 1993; 90:6424-6428.

65. Branch P, Aquilina G, Bignami M et al. Defective mismatch binding and a mutator phenotype in cells tolerant to DNA damage. Nature 1993; 362:652-654.

66. Hearst JE. The structure of photolyase: using photon energy for DNA repair. Science 1995; 268:1859-1859.

67. Kim ST, Sancar A. Photochemistry, photophysics, and mechanism of pyrimidine dimer repair by DNA photolyase. Photochem Photobiol 1993; 57:895-904.

68. Yamamoto K, Satake M, Shinigawa H et al. Amelioration of the ultraviolet sensitivity of an *Escherichia coli* RecA mutant in the dark by photoreactivating enzyme. Mol Gen Genet 1983; 190:511-515.

69. Sancar A, Franklin KA, Sancar GB. *Escherichia coli* DNA photolyase stimulates UvrABC excision nuclease in vitro. Proc Natl Acad Sci USA 1984; 81:7397-7401.

70. Sancar GB, Smith FW. Interactions between yeast photolyase and nucleotide excision repair proteins in Saccharomyces cerevisiae and Escherichia coli. Mol Cell Biol 1989; 9:4767-4776.

71. Sancar A, Tang M-S. Nucleotide excision repair. Photochem Photobiol 1993; 57:905-921.

72. Fox ME, Feldman BJ, Chu G. A novel role for DNA photolyase: binding to DNA damaged by drugs is associated with enhanced cytotoxicity in *Saccharomyces cerevisiae*. Mol Cell Biol 1994; 14:8071-8077.

73. Özer Z, Reardon JT, Hsu DS et al. The other function of DNA photolyase: stimulation of excision repair of chemical damage to DNA. Biochemistry 1995; 34:15886-15889.

CHAPTER 9

INTRAGENOMIC HIERARCHIES OF DNA DAMAGE RECOGNITION

Nuclear DNA in mammalian cells is a highly nonuniform substrate with respect to its susceptibility to genotoxic reactions. For example, cyclobutane pyrimidine dimer formation is modulated at the nucleosomal level with a 10.3 base average periodicity.[1,2] The frequency of pyrimidine(6-4)pyrimidone photoproducts is greater in nuclease-sensitive DNA than in nuclease-resistant fractions of chromatin.[3] Binding of transcription factors may either enhance or suppress photoproduct formation within active genes.[4] Similarly, the distribution of DNA adducts formed by chemical reagents is dependent on the protein environment and the nucleotide sequence context.[5-7]

Nuclear DNA is also a heterogeneous substrate with respect to repair capabilities. Studies in which repair rates following DNA damage were compared in different genomic regions revealed a hierarchy of efficiencies. The level of DNA repair in a particular chromosomal location appears to be determined by multiple parameters including the transcriptional status, interactions with sequence-specific proteins, local chromatin compaction, associations with the nuclear matrix and the nucleotide sequence surrounding the lesion.[8-10] The present chapter elaborates mainly on the role of transcription in DNA damage recognition, as this aspect has been most extensively investigated during recent years.

RECOGNITION OF DNA DAMAGE BY RNA POLYMERASES

Several studies have examined the capacity of RNA polymerases to recognize damaged templates in vitro.[11-19] Depending on the type of damage and the nucleotide sequence context, at least four distinct types of damage-induced interference with transcription have been recognized.

(1) Certain DNA lesions are efficiently bypassed during transcription. For example, aminofluorene-C^8-guanine adducts located in the transcribed (template) strand were shown to constitute weak pausing sites for mammalian RNA polymerase II.[11] Upon prolonged incubation, aminofluorene lesions were bypassed and RNA synthesis was completed.

(2) Other types of lesions constitute strong blocks to transcription. When acetyl-aminofluorene-C^8-guanine adducts (the acetylated form of aminofluorene adducts) were located in the transcribed strand, transcript elongation by RNA polymerase II

was completely arrested. Interestingly, acetylaminofluorene-C^8-guanine adducts promoted dissociation of RNA polymerase II from the damaged DNA template with concomitant release of the nascent transcript.

(3) In contrast, aminofluorene- or acetylaminofluorene-C^8-guanine adducts located in the nontranscribed (coding) strand failed to impede transcription directly. In this situation, the lesions exerted a transient effect on in vitro transcription by prolonging the swell time of RNA polymerase II at a natural sequence-dependent pause site about 15 base pairs downstream of the site of modification.[11]

(4) Yet another mechanism is observed in the presence of cyclobutane pyrimidine dimers. When these major UV lesions were placed in the transcribed strand, in vitro transcription was blocked but, in this case, RNA polymerase II formed a stable ternary complex and protected the damaged site from recognition by a model DNA repair enzyme (DNA photolyase).[13] Conversely, a cyclobutane dimer located in the nontranscribed strand had no effect on transcription. Previous studies have demonstrated that a pyrimidine dimer is also a strong block to *Escherichia coli* RNA polymerase and that the stalled enzyme inhibits recognition of the lesion by the (A)BC subunits of bacterial nucleotide excision repair.[20]

PREFERENTIAL GENE-SPECIFIC REPAIR

These reports demonstrating inhibition of UV radiation damage repair by RNA polymerases in reconstituted in vitro systems contrast sharply with the preferential gene-specific repair of cyclobutane pyrimidine dimers observed in both prokaryotic[21] and eukaryotic organisms.[22-29] In fact, cyclobutane dimers and some other forms of DNA damage are excised more rapidly from transcriptionally active genes than from transcriptionally silent chromosomal regions or from the genome overall. This

phenomenon, frequently referred to as transcription-repair coupling, was discovered when Hanawalt and collaborators devised an appropriate technique to monitor the formation and kinetics of repair of cyclobutane dimers in the dihydrofolate reductase (*DHFR*) gene of UV-irradiated Chinese hamster ovary cells is considerably faster than the average repair rate determined in bulk genomic DNA. The examined rodent cells typically repair 10-20% of pyrimidine dimers from the genome overall in a 24 h period. In the same time frame, about 70% of the dimers were removed from a genomic fragment containing the *DHFR* gene.[22]

The same laboratory also discovered that the preferential gene-specific removal of cyclobutane pyrimidine dimers in Chinese hamster ovary cells is almost entirely attributable to strand-selective repair in active genes.[26] Cyclobutane pyrimidine dimers were repaired in the transcribed (template) strand of *DHFR* about 10 times faster than in the nontranscribed (coding) strand, which is processed at a similar rate as the genome overall. Subsequent studies confirmed the preferential repair of cyclobutane pyrimidine dimers in the transcribed strand of active genes in various eukaryotic systems, ranging from yeast to men.[23-25,27-29] Taking into account all these experimental findings, it can be concluded that the removal of cyclobutane pyrimidine dimers from the transcription template strand is enhanced approximately 10-fold in rodents, 5-fold in yeast and 2-fold in humans.

In all eukaryotes, the mechanism of preferential strand-selective repair is limited to genes transcribed by RNA polymerase II. Ribosomal RNA genes (transcribed by RNA polymerase I) and, presumably, transfer RNA genes (transcribed by RNA polymerase III) are not repaired preferentially relative to the genome overall.[30-32] On the contrary, mutagenesis studies suggest that repair of the template strand of genes transcribed by RNA polymerase I or III is inhibited by transcription.[33] These observations imply that coupling of repair to transcription in eukaryotes (and mammals) depends on specific signals associated with RNA polymerase II. It has been noted that only some bulky base lesions that are typically substrates of nucleotide excision repair are processed in a strand-biased manner. A prominent exception to this rule is the preferential repair of ionizing radiation-induced damage observed in human cells.[34]

THE RODENT PARADOX

Transcription-repair coupling results in the almost exclusive removal of cyclobutane pyrimidine dimers from the transcribed strand of active genes in rodent cells, and provides an explanation for the apparent paradox that rodent cells repair their genome overall with poor efficiency, yet exhibit the same degree of UV resistance as human cells that are highly proficient in overall genomic repair.[35-37]

When removal of cyclobutane pyrimidine dimers and clonal survival were studied in Chinese hamster and human cells in culture, the removal of cyclobutane dimers was only 15% in hamster cells but 80% of dimers were removed in human cells during a 24 h period.[38] Despite this striking difference in overall repair efficiency, both types of cells survived UV exposure equally well. It appears that rodent cells mainly rely on the preferential repair of DNA sequences that are immediately essential for viability, primarily the transcribed strand of active genes. In UV-irradiated rodent cell lines, major silent genomic regions remain essentially unrepaired whereas the smaller fraction of transcribed genes is repaired quickly. Therefore, rodent cells are considered largely deficient in the nucleotide excision repair subpathway that is responsible for the slower removal of cyclobutane pyrimidine dimers from the genome overall.[9,26]

Where they are not excised from nontranscribed sequences, cyclobutane dimers must cause problems to DNA replication and display cytotoxic or mutagenic effects. It is, in fact, tempting to associate inefficient repair of UV damage in rodents with their higher susceptibility to mutagenesis and spontaneous tumor formation, and with the characteristic short rodent lifetime.[39] The molecular basis for these interspecies differ-

ences in the repair capacity of bulk DNA is unknown. Also, repair rates in other mammalian species have not been systematically compared.

THE ROLE OF RNA POLYMERASES IN TRANSCRIPTION-REPAIR COUPLING

Although the mechanism of transcription-coupled repair in mammalian cells is not yet understood, it is clear that this process requires an active RNA polymerase II. The experimental evidence supporting this conclusion is as follows.

(1) Ribosomal RNA and transfer RNA genes are not repaired preferentially,[30-32] indicating that RNA polymerase I and III are unable to stimulate transcription-coupled repair.

(2) Strand selectivity in the repair of genes transcribed by RNA polymerase II strictly depends on transcriptional activity. For example, inducing transcription of the human metallothionein genes with either a glucocorticoid or a heavy metal was shown to increase the rate of cyclobutane dimer repair on the transcribed strand.[40] However, glucocorticoid or heavy metal treatment had no effect on repair rates in the nontranscribed strand of metallothionein genes or in the genome overall. These observations were confirmed using the tightly regulated *GAL7* gene of yeast *Saccharomyces cerevisiae*.[41]

(3) Selective repair of transcribed strands could be abolished by maintaining cells in the presence of a specific inhibitor of RNA polymerase II. α-Amanitin eliminated the difference in repair rate between template and coding strands in human metallothionein[40] and Chinese hamster *DHFR* genes.[42]

(4) The direct involvement of RNA polymerase II in transcription-repair coupling was unequivocally demonstrated using a yeast strain containing a temperature sensitive mutation in one of the subunits of RNA polymerase II.[41] Under permissive

temperature, repair of cyclobutane dimers within the yeast *GAL7* gene was faster in the transcribed strand than in the nontranscribed strand. When RNA polymerase II was inactivated by shifting the temperature to a nonpermissive level, this strand bias disappeared completely.

In summary, these reports indicate that the transcriptional complex, primarily RNA polymerase II, plays a critical role in directing excision repair to the transcribed strand of active genes. A favored model involves formation of a transcription-repair coupling intermediate in which RNA polymerase II is arrested at DNA lesions in the template strand. This model limits the range of substrates that are subject to transcription-repair coupling to those which physically block transcript elongation by RNA polymerase II, a prediction that is consistent with the observation that certain types of DNA damage are not preferentially processed. For example, repair of aminofluorene-C[8]-guanine adducts does not appear to be coupled to transcription[43] and, in fact, these lesions were unable to arrest RNA polymerase II in transcription reactions performed in vitro.[11] Other examples of lesions that are not preferentially removed from active genes, presumably because they fail to block transcription by RNA polymerase II, are those inflicted by monofunctional alkylating agents.[44-45]

Transcription-repair coupling may require, in addition to inhibition of transcript elongation, the formation of a stable ternary complex involving RNA polymerase II, the template DNA strand and the nascent RNA transcript. Presumably, such a ternary complex at the damaged site should persist long enough to induce subsequent steps in the coupling mechanism. In fact, the presence of cyclobutane pyrimidine dimers yields stable enzyme-DNA-RNA complexes after arresting RNA polymerase. In at least one case, however, a bulky base adduct (acetyl-aminofluorene-C[8]-guanine) was found to promote rapid dissociation of RNA polymerase II from the DNA substrate.[11] The differential dissociation rate of RNA poly-

merase II from various types of transcription blocking lesions may determine the efficiency of subsequent transcription-repair coupling. Rapid dissociation of ternary complexes may, for example, explain the lack of correlation between the degree by which various DNA adducts interfere with transcription in vitro and the rate of strand-specific repair of the same lesions detected in a human gene in vivo.[12]

MECHANISMS OF TRANSCRIPTION-REPAIR COUPLING

The molecular mechanisms that govern coupling of repair to transcription are the subject of intense scrutiny in many laboratories, but the issue of mammalian transcription-repair coupling remains highly controversial. In contrast, the biochemical details of gene-specific repair in the prokaryote *Escherichia coli* have been elucidated to a considerable extent. It seems appropriate to briefly review the prokaryotic model system before summarizing recent concepts and hypotheses that are pertinent to transcription-repair coupling in mammals.

TRANSCRIPTION-REPAIR COUPLING IN *ESCHERICHIA COLI*

In *Escherichia coli,* transcription-repair coupling requires only one extra polypeptide of 130 kDa.[46-48] The search for such a coupling factor was initiated by the observation that prokaryotic RNA polymerase is arrested by bulky DNA lesions in the template strand, and that the stalled RNA polymerase makes the lesion inaccessible to the nucleotide excision repair system.[20] Using *Escherichia coli* cell-free extracts, a protein factor was identified that is able to displace the stalled RNA polymerase and couple transcription to nucleotide excision repair.[47,49] This protein, designated transcription-repair coupling factor (TRCF), was found to be encoded by a previously known gene, *mfd* (for *m*utation *f*requency *d*ecline).[50,51] TRCF binds to RNA polymerase stalled at a lesion, displaces both the RNA polymerase and the nascent transcript, and simultaneously recruits the UvrA subunit of

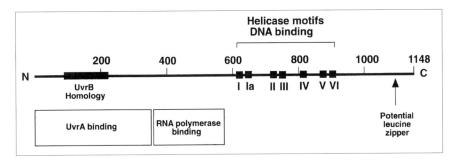

Fig. 9.1. Domain structure of the prokaryotic transcription-repair coupling factor (TRCF). This factor interacts with UvrA, prokaryotic RNA polymerase and DNA to couple nucleotide excision repair to transcription. Despite the presence of helicase motifs I to VI, TRCF has no detectable DNA helicase activity. See text for references.

(A)BC excinuclease to the site of damage.[46-48] This transcription-repair coupling activity exerted by a single polypeptide results in an approximately 10-fold increased repair of the template strand.

Structure-function analysis of TRCF showed that residues 379-571 are involved in binding to a stalled RNA polymerase (Fig. 9.1).[52] The TRCF sequence also contains classical signatures of a DNA and RNA helicase (motifs I through VI), but the purified TRCF protein has no detectable strand separating activity.[53] Perhaps, these helicase motifs reflect the ability to induce local melting and kinking of the DNA substrate. In any case, the helicase motif domain of TRCF is indispensable for its interactions with DNA (Fig. 9.1). Consistent with the presence of an ATP binding box in the protein sequence (helicase motifs I and II), TRCF displays an ATP hydrolyzing (ATPase) activity, suggesting that release of the stalled RNA polymerase and subsequent coupling to excision repair is an energy consuming process. Finally, TRCF shares with UvrB a 140-residue long region of high sequence homology that was considered a probable UvrA-binding domain for the formation of transient TRCF-UvrA complexes. This prediction was confirmed with appropriate deletion mutants that were used to demonstrate that residues 1-378 of TRCF protein are required for the interaction with UvrA (Fig. 9.1).[52,53] On the other side of TRCF, near its carboxy terminus, the protein con-

tains a potential leucine zipper with four leucine residues at intervals of 7 amino acids. This region provides another potential site for protein-protein interactions.

COCKAYNE SYNDROME: A MANIFESTATION OF DEFICIENT TRANSCRIPTION-REPAIR COUPLING IN HUMAN CELLS

Cockayne syndrome is a human genetic disorder characterized by cachectic dwarfism, mental retardation, premature aging and UV sensitivity (summarized in chapter 1). At the cellular level, this syndrome is mainly characterized by hypersensitivity to UV light and the inability to resume RNA synthesis after exposure to DNA damaging agents.[54] The molecular pathology of Cockayne syndrome involves a specific defect in the removal of cyclobutane pyrimidine dimers from transcriptionally active sequences, despite a normal excision repair capacity at the genome overall level.[55] As a consequence, prolonged depression of RNA synthesis in Cockayne cells is generally attributed to a defect in transcription-repair coupling, resulting in abnormally slow removal of transcription blocking lesions from the transcribed strand of active genes.

The pattern of cyclobutane pyrimidine dimer excision was investigated in several different genomic sites of both normal and Cockayne syndrome fibroblasts. The most frequently examined loci range from actively transcribed genes such as adenosine deaminase (*ADA*) or dihydrofolate reductase (*DHFR*) to X-chromosomal nontranscribed segments (for example locus 754 and the gene encoding coagulation factor IX). These locus-specific studies revealed a hierarchy of excision repair with at least three different levels of efficiency in normal cells:[56,57]

(1) transcribed strands of active genes are repaired faster than nontranscribed strands,

(2) nontranscribed, but potentially active genes are repaired slower than transcribed genes,

(3) nontranscribed repressed loci are processed with lowest priority.

The defect in Cockayne syndrome affects two distinct levels in this hierarchy of excision repair. Cockayne syndrome cells are primarily deficient in the preferential repair of the transcribed strand of active genes relative to the nontranscribed strand. In addition, van Hoffen et al[56] observed that both strands of active genes in Cockayne fibroblasts are repaired slower than the nontranscribed strand of the same genes in wild type cells. This finding prompted the hypothesis that Cockayne syndrome cells may be defective in factors required to provide general access to transcriptionally active chromatin (see section on the accessibility hypothesis below).

The lack of strand-selective repair in Cockayne syndrome fibroblasts was confirmed when Kantor and Bastin[57] reexamined the rates of cyclobutane dimer excision in several genomic regions. These authors selected for their study the active β-actin and *ADA* genes, as well as the inactive insulin gene and locus 754. The β-actin gene was of particular interest because its product is the most abundant protein in human fibroblasts and the gene is one of the most transcriptionally active.[58] In Cockayne cells, strand selectivity in the expressed genes (β-actin and *ADA*) was completely abolished, as their template and coding strands were repaired at identical rates.[57] These expressed genes (β-actin and *ADA*) were nevertheless repaired considerably faster than the transcriptionally inactive insulin gene, in which cyclobutane dimers were processed at the same slow rate as the genome overall. In this hierarchy of repair efficiency, the repressed X-chromosomal locus 754 was found to be highly refractory to excision repair.

The biochemical basis of the gene-specific repair defects identified in Cockayne cells is not understood, but various models have been proposed to accommodate the ability of RNA polymerase II to couple excision repair to transcription. These models propose different, in part contradicting functions for the protein factors that are missing in Cockayne syndrome cells.

THE COUPLING FACTOR HYPOTHESIS

Several genetic complementation groups have been defined for Cockayne, as this syndrome can be caused by mutations in at least five different genes: *CSA*, *CSB*, *XPB*, *XPD* and *XPG* (see chapter 1 for details on the genetic background of Cockayne syndrome). Examination of the amino acid sequence of the translated *CSB* gene revealed some similarity to TRCF. In particular, CSB is a protein of 1,493 amino acids with a predicted molecular mass of 168 kDa containing seven consecutive domains conserved between DNA and RNA helicases.[59] A consensus ATPase motif is contained in helicase domains I and II. These sequence homologies with TRCF have led to the suggestion that the *CSB* gene product may function as a transcription-repair coupling factor in human cells.[29,47,48] Specifically, this model predicts that Cockayne syndrome arises from a deficiency in a factor that normally couples repair to transcription by RNA polymerase II. At the time of writing, no convincing biochemical evidence has been provided to either support or confute this coupling hypothesis, mainly because of the lack of an appropriate mammalian in vitro system to study the biochemistry of preferential repair of transcribed genes.

THE UNCOUPLING FACTOR HYPOTHESIS

Recently, Van Oosterwijk et al[60] reported that Cockayne fibroblasts are hypersensitive to treatment with *N*-acetoxy-2-acetyl-aminofluorene and are unable to resume RNA synthesis after treatment with this carcinogen. This observation is intriguing because aminofluorene-C^8-guanine adducts (the prevalent intracellular lesion formed by *N*-acetoxy-2-acetylaminofluorene) are not preferentially removed from the transcribed strand of active genes in normal cells.[43,61] Thus, the sensitivity of Cockayne syndrome cells to a particular type of DNA damage does not correlate with the extent of transcription-coupled repair observed for that type of damage in wild type cells. Based on these findings, Van Oosterwijk et al[60] concluded that the *CSA* and *CSB* gene products are not coupling factors. Instead, they postulated that these proteins may be required to convert a multifunctional factor (presumably TFIIH) from its excision repair function to the transcription function. If this converting activity is defective, TFIIH remains sequestered in genomic DNA as an excision repair factor and is not available for transcription of essential genes. In this model, Cockayne syndrome cells are deficient in specialized factors that uncouple TFIIH from its role in DNA excision repair and allow transcription initiation despite the presence of DNA damage. The uncoupling hypothesis also implies that the CSA and CSB proteins are not functional homologs of TRCF in *Escherichia coli*, but rather have a novel function with no apparent prokaryotic precedent.

The Transcription Hypothesis

As outlined in chapter 1, the clinical manifestations of Cockayne syndrome, primarily postnatal growth retardation, suggest that this disease may define a transcription syndrome. Recently, Friedberg and collaborators observed that extracts of cells obtained from patients afflicted by Cockayne syndrome display abnormally low levels of RNA polymerase II-dependent transcription in vitro.[62,63] This finding supports the idea that *CSA* and *CSB* may encode transcription factors and prompted a hypothetical model in which these two proteins induce preferential repair of template strands in an indirect manner by promoting transcription. This model predicts that the primary defect in Cockayne syndrome resides in transcription by RNA polymerase II. Impaired transcription results in a secondary phenotype of deficient repair of transcribed sequences when cells are exposed to DNA damaging agents.[64] A possible role of CSA and CSB proteins in transcription is suggested by a recent report that implicates these two factors in a novel posttranslational modification reaction that generates ubiquinated RNA polymerase II.[65] For example, CSA and CSB may facilitate transcription by reorganizing protein-protein interactions whenever the elongation complex encounters either natural pausing sites or transcription blocking lesions.

The TFIIH Hypothesis

TFIIH is a multisubunit and multifunctional factor involved in transcription and nucleotide excision repair (see section on TFIIH in chapter 6). TFIIH is required for both nucleotide excision repair subpathways, i.e., global repair throughout the genome overall and preferential repair of expressed genes. Additionally, the dual role of TFIIH in transcription initiation by RNA polymerase II and nucleotide excision repair suggests an immediate mechanism for coupling repair to transcription.[63,66]

In the absence of TFIIH, RNA polymerase II is unable to initiate transcript elongation and the enzyme complex remains arrested within the promoter region. Current studies suggest that TFIIH dissociates from the RNA polymerase II elongation complex when the nascent transcript becomes longer than approximately 30 residues.[67,68] Thus, the TFIIH hypothesis of transcription-repair coupling predicts that TFIIH is reloaded onto the damaged template strand following arrested transcription. Perhaps, RNA polymerase II stalled at damaged sites resembles RNA polymerase II within the initiation complex and TFIIH has a special affinity for such arrested RNA polymerase molecules.[63] Hence, an elongating RNA polymerase II enzyme that is physically blocked at damaged sites may attract TFIIH, which would then displace the RNA polymerase and recruit the nucleotide excision repair machinery. Alternatively, TFIIH may return to arrested transcription complexes as component of a larger "repairosome" that contains all subunits necessary for damage-specific DNA incision and excision.[69]

THE CHROMATIN HYPOTHESIS

DNA within mammalian interphase nuclei is associated with histones and other proteins to form a highly condensed structure known as chromatin. Roughly three levels of chromatin organization are recognized in mammalian nuclei. The first level involves the assembly of nucleosomes, in which DNA segments of 146 base pairs wind around an octamer of two each of the four histones H2A, H2B, H3 and H4. Alternating along the DNA with these nucleosomal cores are regions of linker DNA of variable lengths. Some regions of DNA lack nucleosomes and, as a consequence, are hypersensitive to enzymatic probing by nucleases. The second level of organization involves compaction of nucleosomes to form fibers with a diameter of about 30 nm. Finally, these fibers are organized in loop domains which on the average contain 20-100 kilobases of DNA.[70-72] As a consequence of chromatin organization, the substrate of repair processes is not naked DNA but a highly heterogeneous nucleic acid-protein complex in which DNA is nonuniformly packaged. Some sites are nuclease-hypersensitive whereas some other chromosomal regions are highly condensed to form heterochromatin.

It has been suggested in many instances that the preferential repair of transcriptionally active genes simply reflects a more accessible configuration of chromatin in such genes compared to transcriptionally inactive counterparts.[56,57,63,73,74] This view is supported by the finding that nucleosome assembly strongly suppresses excision repair of UV radiation products in human cell extracts in vitro.[75] Even the strand selectivity observed in the repair of transcriptionally active genes may be caused by differential chromatin compaction. Conceivably, arrest of transcript elongation may promote relaxation of histone-DNA interactions asymmetrically on the side of the transcribed strand. Thus, a subtle strand-specific bias

in chromatin condensation may provide a simple mechanism for the preferential repair of the transcription template strand.[63]

Chromatin compaction has been invoked to accommodate the observation that repair efficiency varies between different inactive loci in both normal and Cockayne syndrome cells. For example, the insulin gene, although not expressed in human fibroblasts, is repaired more rapidly than two different X-chromosomal loci confined to heterochromatin (locus 754 and coagulation factor XI).[56,57] Removal of cyclobutane dimers was also analyzed in a cell line derived from a patient suffering form severe combined immunodeficiency (SCID).[74] In humans, this disease is caused by a deletion in the *ADA* gene involving the promoter and the first exon. Transcription of this gene is completely abolished in the cells of affected individuals. Consistent with the complete absence of *ADA* transcripts in human SCID cells, the two strands of the *ADA* gene are processed at identical rates. Nevertheless, it was noted that the inactive *ADA* gene is repaired more efficiently than the repressed loci 754 and factor IX.[74] Conceivably, a more relaxed chromatin configuration may facilitate the repair of nontranscribed but potentially active sites, such as the *ADA* gene in human SCID, relative to the repair of highly repressed loci.

An alternative model related to the chromatin hypothesis proposes that the rate by which a particular gene is repaired correlates with its association with the nuclear matrix, an insoluble three-dimensional proteinaceous scaffold that organizes the higher order chromatin loop structure.[76,77] This view is supported by the finding that essential nuclear functions, primarily transcription, have been found to occur preferentially on the nuclear matrix.[78]

THE ACCESSIBILITY HYPOTHESIS

At least two separate observations suggest that the factors defective in Cockayne syndrome cells (CSA and CSB) are required to facilitate access of DNA excision repair to transcriptionally active genes. First, as mentioned before, the repair of cyclobutane dimers in active genes (*ADA, DHFR*) of Cockayne fibroblasts was detectably slower than that measured in the nontranscribed strand of active genes in wild type cells.[56] Similarly, both strands of the expressed *c-abl* protooncogene are repaired with abnormally poor efficiency in Cockayne cells.[79] Second, even the repair of ribosomal RNA genes (transcribed by RNA polymerase I) is deficient in Cockayne syndrome cells compared to normal human cells.[79] These results led to propose that the Cockayne syndrome defect may resides in components that facilitate access to all transcriptionally active genes, not only those transcribed by RNA polymerase II. In this model, CSA and CSB serve to deliver repair enzymes to active chromatin domains.

TRANSCRIPT ABORTION OR RESUMPTION

Hanawalt[29] suggested that it would seem extremely inefficient to abort large transcripts (up to 2.5 megabases) every time RNA polymerase encounters a lesion. Because of the presence of long and multiple introns, it takes nearly 24 h to transcribe large human genes such as the dystrophin gene. Such genes may never become expressed if truncated transcripts were released during transcription-repair coupling, as it is observed in the prokaryotic system. Biochemical characterization of transcription elongation factor SII indeed suggests a mechanism by which arrested transcripts may be reused after transcription-coupled repair.

Transcription elongation factor SII acts at natural pausing sites by binding to RNA polymerase II, thereby activating its intrinsic RNase activity. This enables the RNA polymerase to hydrolyze 2-10 nucleotides from the 3' end and shorten the nascent transcript. As a consequence, the polymerase backs off from its trapped conformation and resumes its efforts to bypass the critical transcription pausing site.[80] Interestingly, a site-directed cyclobutane pyrimidine dimer located in the template strand has been shown to induce essentially the same reaction mediated by elongation factor SII.[13] Thus, shortening of RNA transcripts and

backing off of RNA polymerase II may allow strand-specific repair without aborting transcription in mammalian cells.[13,29]

CONCLUSIONS

In mammalian cells, repair of cyclobutane pyrimidine dimers and other lesions takes place via two subpathways: global repair of the genome overall and transcription-coupled repair of active genes. The latter subpathway is believed to allow for rapid resumption of transcription after UV exposure. Most of this gene-specific repair is due to the fact that the transcribed (template) strand is repaired very efficiently during transcription compared to the nontranscribed (coding) strand or to transcriptionally silent sequences. An important consequence of strand-specific repair is that damage-induced mutations are often generated in a biased manner, i.e., at sites of longer persisting lesions located on the nontranscribed strand.[81]

The molecular mechanism by which excision repair of cyclobutane dimers is coupled to transcription in mammalian genes is highly disputed. It has been firmly established experimentally that transcription-coupled repair depends on some aspect of RNA polymerase II translocation along damaged DNA templates. Presumably, the preferential repair of transcribed sequences involves transient sequestration of RNA polymerase II at sites of damage in the template strand. The trapped enzyme is apparently capable of inducing a signal that triggers transcription-repair coupling. What exactly this signal might be and what factors are involved in this coupling mechanism remains unknown. Most studies performed to solve these problems have failed to provide conclusive answers.

The preferential repair of transcribed strands is abolished in cells obtained from individuals afflicted by Cockayne syndrome. This observation prompted many researchers to focus their investigations on the possible function of the gene products that are deficient in Cockayne syndrome. In analogy to the prokaryotic system, it has been postulated that CSA and CSB proteins are transcription-repair coupling factors. The uncoupling hypothesis proposes, on the other hand, that CSA and CSB may serve to convert TFIIH from its repair function to the transcription function. Another model predicts that CSA and CSB are neither coupling nor uncoupling factors, but rather transcription factors required for RNA polymerase II-dependent RNA synthesis. Yet another hypothesis predicts that CSA and CSB serve to facilitate access of repair enzymes specifically to genomic segments that are transcriptionally active. Alternative models propose that transcription-repair coupling operates through the action of the multisubunit and multifunctional transcription-repair factor TFIIH. An unexpected mechanism is suggested by the observation that mismatch correction-deficient cell lines have lost the capacity to perform transcription-coupled repair of UV radiation damage.[82] Thus, the mismatch correction system may be directly involved in transcription-repair coupling. A final hypothesis postulates that transcription-repair coupling simply reflects a more open chromatin configuration within transcribed sequences. Clearly, these diverging views on the phenomenon of transcription-repair coupling in mammalian cells do not allow to propose a unifying mechanistic model.

REFERENCES

1. Gale JM, Nissen KA, Smerdon MJ. UV-induced formation of pyrimidine dimers in nucleosome core DNA is strongly modulated with a period of 10.3 bases. Proc Natl Acad Sci USA 1987; 84:6644-6648.

2. Schieferstein U, Thoma F. Modulation of cyclobutane pyrimidine dimer formation in a positioned nucleosome containing poly(dA·dT) tracts. Biochemistry 1996; 35:7705-7714.

3. Mitchell DL, Nguyen TD, Cleaver JE. Nonrandom induction of pyrimidine-pyrimidone (6-4) photoproducts in ultraviolet-irradiated human chromatin. J Biol Chem 1990; 265:5353-5356.

4. Pfeifer GP, Drouin R, Riggs AD et al. Binding of transcription factors creates hot spots for UV photoproducts in vivo.

Mol Cell Biol 1992; 12:1798-1804.

5. Cimino GD, Gamper HB, Isaacs ST et al. Psoralens as photoactive probes of nucleic acid structure and function: organic chemistry, photochemistry, and biochemistry. Annu Rev Biochem 1985; 54:1151-1193.

6. Pierce JR, Nazimiec M, Tang MS. Comparison of sequence preference of tomamycin- and anthramycin-DNA bonding by exonuclease III and lambda exonuclease digestion and UvrABC nuclease incision analysis. Biochemistry 1993; 32:7069-7078.

7. Smith BL, MacLeod MC. Covalent binding of the carcinogen benzo[a]pyrene diol epoxide to Xenopus laevis 5 S DNA reconstituted into nucleosomes. J Biol Chem 1993; 268:20620-20629.

8. Bohr VA, Wassermann K. DNA repair at the level of the gene. Trends Biochem Sci 1988; 13:429-433.

9. Vreeswijk MPG, van Hoffen A, Westland BE et al. Analysis of repair of cyclobutane pyrimidine dimers and pyrimidine 6-4 pyrimidone photoproducts in transcriptionally active and inactive genes in Chinese hamster cells. J Biol Chem 1994; 269:31858-31863.

10. Gao S, Drouin R, Holmquist GP. DNA repair rates mapped along the human *PGK1* gene at nucleotide resolution. Science 1994; 263:1438-1440.

11. Donahue BA, Fuchs RPP, Reines D et al. Effects of aminofluorene and acetylaminofluorene DNA adducts on transcriptional elongation by RNA polymerase II. J Biol Chem 1996; 271:10588-10594.

12. McGregor WG, Mah MC-M, Chen R-H et al. Lack of correlation between degree of interference with transcription and rate of strand specific repair in the HPRT gene of diploid human fibroblasts. J Biol Chem 1995; 270:27222-27227.

13. Donahue BA, Yin S, Taylor J-S et al. Transcript cleavage by RNA polymerase II arrested by a cyclobutane pyrimidine dimer in the DNA template. Proc Natl Acad Sci USA 1994; 91:8502-8506.

14. Choi D-J, Marino-Alessandri DJ, Geacintov NE et al. Site-specific benzo[a]-pyrene diol epoxide-DNA adducts inhibit

transcription elongation by bacteriophage T7 RNA polymerase. Biochemistry 1994; 33:780-787.

15. Chen Y-H, Bogenhagen DF. Effects of DNA lesions on transcription elongation by T7 RNA polymerase. J Biol Chem 1993; 268:5849-5855.

16. Zhou W, Doetsch PW. Effects of abasic sites and DNA single-strand breaks on prokaryotic RNA polymerases. Proc Natl Acad Sci USA 1993; 90:6601-6605.

17. Chen YH, Matsumoto Y, Shibutani S et al. Acetylaminofluorene and aminofluorene adducts inhibit in vitro transcription of a *Xenopus* 5S RNA gene only when located on the coding strand. Proc Natl Acad Sci USA 1991; 88:9583-9587.

18. Selby CP, Witkin E, Sancar A. *Escherichia coli mfd* mutant deficient in "mutation frequency decline" lacks strand-specific repair: in vitro complementation with purified coupling factor. Proc Natl Acad Sci 1991; 88:11574-11578.

19. Shi Y-b, Gamper H, Hearst JE. Interaction of T7 RNA polymerase with DNA in an elongation complex arrested at a specific psoralen adduct site. J Biol Chem 1988; 263:527-534.

20. Selby CP, Sancar A. Transcription preferentially inhibits nucleotide excision repair of the template DNA strand in vitro. J Biol Chem 1990; 265:21330-21336.

21. Mellon I, Hanawalt PC. Induction of the Escherichia coli lactose operon selectively increases repair of its transcribed DNA strand. Nature 1989; 342:95-98.

22. Bohr VA, Smith CA, Okumoto DS et al. DNA repair in an active gene: removal of pyrimidine dimers from the *DHFR* gene of CHO cells is much more efficient than in the genome overall. Cell 1985; 40:359-369.

23. Mellon I, Bohr VA, Smith CA et al. Preferential DNA repair of an active gene in human cells. Proc Natl Acad Sci USA 1986; 83:8878-8882.

24. Madhani HD, Bohr VA, Hanawalt PC. Differential DNA repair in transcriptionally active and inactive proto-oncogenes: c-*abl* and c-*mos*. Cell 1986; 45:417-423.

25. Ruven HJT, Berg RJW, Seelen CM et al. Ultraviolet-induced cyclobutane pyrimidine dimers are selectively removed from transcriptionally active genes in the epidermis of the hairless mouse. Cancer Res 1993; 53:1642-1645.

26. Mellon I, Spivak G, Hanawalt PC. Selective removal of transcription-blocking DNA damage from the transcribed strand of the mammalian DHFR gene. Cell 1987; 51:241-249.

27. Smerdon MJ, Thoma F. Site-specific DNA repair at the nucleosome level in a yeast minichromosome. Cell 1990; 61:675-684.

28. Venema J, Bartosova, Natarajan AT et al. Transcription affects the rate but not the extent of repair of cyclobutane pyrimidine dimers in the human adenosine deaminase gene. J Biol Chem 1992; 267:8852-8856.

29. Hanawalt PC. Transcription-coupled repair and human diseases. Science 1994; 266:1957-1958.

30. Christians FC, Hanawalt PC. Lack of transcription-coupled repair in mammalian ribosomal RNA genes. Biochemistry 1993; 32:10512-10518.

31. Fritz LK, Smerdon MJ. Repair of UV damage in actively transcribed ribosomal genes. Biochemistry 1995; 34:13117-13124.

32. Vos J-M, Wauthier EL. Differential introduction of DNA damage and repair in mammalian genes transcribed by RNA polymerase I and II. Mol Cell Biol 1991; 11:2245-2252.

33. Armstrong JD, Kunz BA. Site and strand specificity of UVB mutagenesis in SUP4-0 gene of yeast. Proc Natl Acad Sci USA 1990; 87:9005-9009.

34. Leadon SA, Cooper PK. Preferential repair of ionizing radiation-induced damage in the transcribed strand of an active human gene is defective in Cockayne syndrome. Proc Natl Acad Sci USA 1993; 90:10499-10503.

35. Zelle B, Reynolds RJ, Kottenhagen MJ et al. The influence of the wavelength radiation on survival, mutation induction and DNA repair in irradiated Chinese hamster cells. Mutat Res 1980; 72:491-509.

36. van Zeeland AA, Smith CA, Hanawalt PC. Sensitive determination of pyrimidine dimers in DNA of UV-irradiated mammalian cells: introduction of T4 endonuclease V into frozen and thawed cells. Mutat Res 1981; 82:173-189.

37. Smith, C.A. DNA repair in specific sequences in mammalian cells. J Cell Sci 1987; Suppl 6:225-241.

38. Bohr VA, Okumoto DS, Hanawalt. Survival of UV-irradiated mammalian cells correlates with efficient DNA repair in an essential gene. Proc Natl Acad Sci USA 1986; 83:3830-3833.

39. Downes CS, Ryan AJ, Johnson RT. Fine tuning of DNA repair in transcribed genes: mechanisms, prevalence and consequences. BioEssays 1993; 15:209-216.

40. Leadon SA, Lawrence DA. Preferential repair of DNA damage on the transcribed strand of the human metallothionein genes requires RNA polymerase II. Mutat Res 1991; 255:67-78.

41. Leadon SA, Lawrence DA. Strand-selective repair of DNA damage in the yeast GAL7 gene requires RNA polymerase II. J Biol Chem 1992; 267:23175-23182.

42. Christians FC, Hanawalt PC. Inhibition of transcription and strand-selective DNA repair by α-amanitin in Chinese hamster ovary cells. Mutat Res 1992; 274:93-101.

43. Tang MS, Bohr VA, Zhang X et al. Quantification of aminofluorene adduct formation and repair in defined sequences in mammalian cells using the UVRABC nuclease. J Biol Chem 1989; 264:14455-14462.

44. Nose K, Nikaido O. Transcriptionally active and inactive genes are similarly modified by chemical carcinogens and X-ray in normal human fibroblasts. Biochim Biophys Acta 1984; 781:273-278.

45. Bartlett JD, Scicchitano DA, Robison SH. Two expressed human genes sustain slightly more DNA damage after alkylation agent treatment than an inactive gene. Mutat Res 1991; 255:247-256.

46. Selby CP, Witkin EM, Sancar A. Escherichia coli mfd mutant deficient in "Mutation Frequency Decline" lacks strand-specific repair: in vitro complementation with purified coupling factor. Proc Natl Acad Sci USA 1991; 88:11574-11578.

47. Selby CP, Sancar A. Molecular mechanism of transcription-repair coupling. Science 1993; 260:53-58.

48. Selby CP, Sancar A. Mechanisms of transcription-repair coupling and mutation frequency decline. Microbiol Rev 1994; 58:317-329.

49. Selby CP, Sancar A. Gene- and strand-specific repair in vitro: partial purification of a transcription-repair coupling factor. Proc Natl Acad Sci USA 1991; 88:8232-8236.

50. Witkin EM. Radiation-induced mutations and their repair. Science 1966; 152:1345-1353.

51. Witkin EM. Mutation frequency decline revisited. BioEssays 1994; 16:437-444.

52. Selby CP, Sancar A. Structure and function of transcription-repair coupling factor. I. Structural domains and binding properties. J Biol Chem 1995; 270:4882-4889.

53. Selby CP, Sancar A. Structure and function of transcription-repair coupling factor. II. Catalytic properties. J Biol Chem 1995; 270:4890-4895.

54. Mayne LV, Lehmann AR. Failure of RNA synthesis to recover after UV irradiation: an early defect in cells from individuals with Cockayne's syndrome and xeroderma pigmentosum. Cancer Res 1982; 42:1473-1478.

55. Venema J, Mullenders LHF, Natarajan AT et al. The genetic defect in Cockayne syndrome is associated with a defect in repair of UV-induced DNA damage in transcriptionally active DNA. Proc Natl Acad Sci USA 1990; 87:4707-4711.

56. van Hoffen A, Natarajan AT, Mayne LV et al. Deficient repair of the transcribed strand of active genes in Cockayne's syndrome cells. Nucleic Acids Res 1993; 21:5890-5895.

57. Kantor GJ, Bastin SA. Repair of some active genes in Cockayne syndrome cells is at the genome overall rate. Mutat Res 1995; 336:223-233.

58. Leavitt J, Gunning P, Porreca S-Y et al. Molecular cloning and characterization of mutant and wild-type human β-actin genes. Mol Cell Biol 1984; 4:1961-1969.

59. Troelstra C, van Gool A, de Wit J et al. ERCC6, a member of a subfamily of putative helicases, is involved in Cockayne's syndrome and preferential repair of active genes. Cell 1992; 71:939-953.

60. Van Oosterwijk MF, Versteeg A, Filon R

et al. The sensitivity of Cockayne's syndrome cells to DNA-damaging agents is not due to defective transcription-coupled repair of active genes. Mol Cell Biol 1996; 16:4436-4444.

61. Wade MH, Chu EHY. Effects of DNA damaging agents on cultured fibroblasts derived from patients with Cockayne syndrome. Mutat Res 1979; 59:49-60.

62. Henning KA, Li L, Iyer N et al. The Cockayne syndrome group A gene encodes a WD repeat protein that interacts with CSB protein and a subunit of RNA polymerase II TFIIH. Cell 1995; 82:555-564.

63. Friedberg EC. Relationships between DNA repair and transcription. Annu Rev Biochem 1996; 65:15-42.

64. Friedberg EC. Cockayne syndrome-a primary defect in DNA repair, transcription, both or neither? BioEssays 1996; 18:731-738.

65. Bregman DB, Halaban R, van Gool AJ et al. UV-induced ubiquination of RNA polymerase II: a novel modification deficient in Cockayne syndrome. Proc Natl Acad Sci USA 1996; 93:11586-11590.

66. Buratowski S. DNA repair and transcription: the helicase connection. Science 1993; 260:37-38.

67. Goodrich JA, Tjian R. Transcription factor IIE and IIH and ATP hydrolysis direct promoter clearance by RNA polymerase II. Cell 1994; 77:145-156.

68. Zawel L, Kumar KP, Reinberg D. Recycling of the general transcription factors during RNA polymerase II transcription. Genes Dev 1995; 9:1479-1490.

69. Svejstrup JQ, Wang Z, Feaver WJ et al. Different forms of TFIIH for transcription and DNA repair: holo-TFIIH and a nucleotide excision repairosome. Cell 1995; 80:21-28.

70. Laemmli UK. Levels of organization of the DNA in eukaryotic chromosomes. Pharmacol Rev 1979; 30:469-476.

71. Simpson RT, Thoma F, Brubaker JM. Chromatin reconstituted from tandemly repeated cloned DNA fragments and core histones: a model system for study of higher order structure. Cell 1985; 42:799-808.

72. Felsenfeld G, McGhee JD. Structure of the 30 nm chromatin fiber. Cell 1986; 44:375-377.

73. Tornaletti S, Pfeifer GP. UV damage and repair mechanisms in mammalian cells. BioEssays 1996; 18:221-228.

74. Venema J, Bartosova Z, Natarajan AT et al. Transcription affects the rate but not the extent of repair of cyclobutane pyrimidine dimers in the human adenosine deaminase gene. J Biol Chem 1992; 267:8852-8856.

75. Wang Z, Wu XH, Friedberg EC. Nucleotide excision repair of DNA by human cell extracts is suppressed in reconstituted nucleosomes. J Biol Chem 1991; 266:22472-22478.

76. Mullenders LHF, Van Kesteren-Van Leeuwen AC, Van Zeeland AA et al. Nuclear matrix associated DNA is preferentially repaired in normal human fibroblasts, exposed to a low dose of ultraviolet light but not in Cockayne's syndrome fibroblasts. Nucleic Acids Res 1988; 16:10607-10622.

77. Koehler DR, Hanawalt PC. Recruitment of damaged DNA to the nuclear matrix in hamster cells following ultraviolet irradiation. Nucleic Acids Res 1996; 24:2877-2884.

78. Cook PR. RNA polymerase: structural determinant of the chromatin loop and the chromosome. BioEssays 1994; 16:425-430.

79. Christians FC, Hanawalt PC. Repair in ribosomal RNA genes is deficient in xeroderma pigmentosum group C and Cockayne's syndrome cells. Mutat Res 1994; 323:179-187.

80. Izban MG, Luse AT. The RNA polymerase II ternary complex cleaves the nascent transcript in a $3' \rightarrow 5'$ direction in the presence of elongation factor S-II. Genes Dev 1992; 6:1342-1356.

81. Carothers AM, Zhen W, Mucha J et al. DNA strand-specific repair of (±)-3α,4β-dihydroxy-1α,2α-epoxy-1,2,3,4-tetrahydrobenzo[c]phenanthrene adducts in the hamster dihydrofolate reductase. Proc Natl Acad Sci USA 1992; 89:11925-11929.

82. Mellon I, Fajpal DK, Koi M. Transcription-coupled repair deficiency and mutations in human mismatch repair genes. Science 1996; 272:557-560.

CHAPTER 10

RECOGNITION OF DNA DAMAGE DURING REPLICATION

DNA lesions can be eliminated from metazoan organisms by DNA repair processes or by cell death. If DNA damage remains unrepaired and cells survive, the efficiency and fidelity of essential nuclear functions is seriously threatened. In particular, replication of damaged DNA represents a major mechanism of genetic instability.[1,2] Covalent modification of DNA bases may disrupt the hydrogen bonding information and generate noninstructional sites. During subsequent DNA replication, damaged templates are unable to mediate the recruitment of complementary deoxyribonucleotides using their hydrogen-bonding pattern and, in many cases, DNA polymerases have only a 25% probability of adding the correct base. As a general rule, DNA polymerases tend to select purines, preferentially adenine residues, across such noninstructional sites.[3]

DNA REPLICATION AND CARCINOGENESIS

The frequency by which unrepaired DNA lesions are converted to potentially carcinogenic mutations in the nucleotide sequence of DNA is primarily determined by the rate of DNA replication. Thus, proliferative cells are considerably more susceptible than quiescent cells to the mutagenic consequences of DNA damage.[4] The cell cycle dependence of tumor initiation was investigated using the regenerating rat liver system, in which a synchronous wave of cell division was induced by a two-thirds partial hepatic resection.[5-7] In this model system, the frequency of initiation of hepatocarcinogenesis was highly dependent on the cell cycle stage at which a single carcinogen dose was administered. The highest yield of tumors was observed in the animals treated when hepatocytes were crossing the G_1/S border. Animals treated during G_1 or during G_2/M were considerably less susceptible to tumor formation (see Fig. 1.4 for the sequence of cell cycle events).

The cell cycle-dependent variation in susceptibility to initiation of carcinogenesis was also studied in human fibroblasts maintained in culture.[8] Fibroblast proliferation was arrested at confluence and cells were subsequently released into growth by replating them at subconfluent density. After this procedure, fibroblasts started a synchronous wave of DNA synthesis and mitosis. These cells were UV-irradiated at various times after growth stimulation, and chromosomal aberrations were monitored by quantifying the frequency of chromatid breaks and exchanges in metaphase spreads. The highest frequency of chromatid le-

Mechanisms of DNA Damage Recognition in Mammalian Cells, by Hanspeter Naegeli.
© 1997 R.G. Landes Company.

sions was observed when cells were exposed to the genotoxic agent at the G_1/S border. These and numerous other studies (see for example refs. 9 and 10) demonstrate that dividing cells are particularly vulnerable to damage-induced mutagenesis when exposed to genotoxic agents during S phase of the cell cycle.

DNA SYNTHESIS ARREST AND REPLICON REGULATION

The induction of mutagenic replication errors is not the only problem encountered by mammalian DNA polymerases at sites of damage. In particular, cytotoxicity exerted by many DNA-reactive agents is due to inhibition of DNA replication.[11-14] When tested in vitro, purified DNA polymerases tend to stop DNA synthesis near sites of base damage in the template strand. Most experiments aimed at investigating the interaction of DNA polymerases with damaged templates involved the use of single-stranded DNA that was exposed to a variety of genotoxic agents including ultraviolet (UV) radiation or bulky carcinogens. The covalently modified template strands were annealed with specific primers and DNA synthesis was tested in the presence of various polymerases. Under these in vitro conditions, DNA elongation is arrested either 3' to or opposite damaged nucleotides.[15-20]

It is generally thought that DNA damage interferes with primer elongation by disrupting the pairing ability of template bases.[3] According to this model, progression of DNA polymerases is impeded by their inability to promote complementary base pairing downstream of defective template residues. However, we have been able to provide experimental evidence indicating that backbone distortion and, particularly, steric hindrance are important determinants of DNA synthesis arrest on defective templates.[21] This conclusion is prompted by the observation that bulky adducts linked to the position C4' of a single deoxyribose backbone residue are able to block DNA synthesis completely, although these backbone lesions fail to disturb correct hydro-

gen bonding interactions at the site of modification.[21]

The complete and faithful replication of mammalian genomes is achieved by the coordinated action of many different replication units (or replicons), each containing a functional origin where bidirectional DNA synthesis is initiated.[22,23] Estimates on the number of such replicons in mammalian cells range from approximately 10,000 up to 100,000. Interestingly, these multiple replicons are not simultaneously active: some undergo duplication early in S phase while others are replicated late in S phase.[24] This temporal organization of replicon activity is poorly understood, but recent studies in our laboratory suggest that DNA lesions situated in one replicon may effectively delay DNA synthesis in neighboring replicons (T. Morozova and H. Naegeli, manuscript in preparation). This hypothesis of replicon regulation is prompted by experiments in which we exploited complementary DNA synthesis in *Xenopus* egg[25] or human HeLa cell[26] lysates to study the mechanisms of genotoxicity following exposure to UV radiation or alkylating agents. In this in vitro model system of eukaryotic replication, nondamaged substrates (for example ϕX174 single-stranded DNA) are replicated in an efficient manner and result in the formation of covalently closed circular genomes (Fig. 10.1; reaction A). As expected, replication of damaged substrates (for example UV-irradiated M13 single-stranded DNA) is inhibited, yielding low levels of incomplete intermediates (Fig. 10.1; reaction B). More surprisingly, we found that damaged templates are able to inhibit the replication of nondamaged templates in trans (Fig. 10.1; reaction C). For example, complementary DNA synthesis on ϕX174 single-stranded DNA was inactivated when the reaction mixtures were supplemented with equal amounts of UV-irradiated or *N*-methyl-*N*-nitrosourea-treated M13 DNA substrate (Fig. 10.1; reaction C). However, replication of nondamaged ϕX174 DNA was rescued by the addition of T7 DNA polymerase, indicating that inactivation of DNA synthesis in

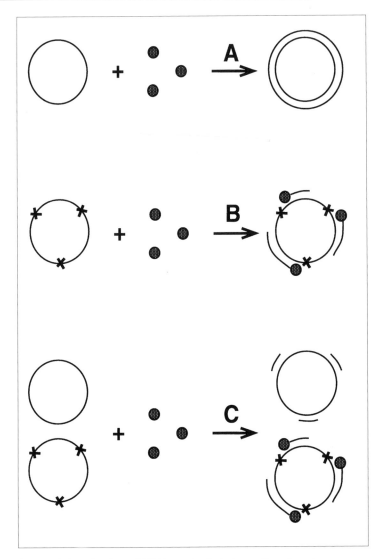

Fig. 10.1. Inactivation of DNA synthesis in trans. Reaction A, enzymes and replication factors contained in *Xenopus laevis* egg or HeLa cell lysates catalyze complementary DNA synthesis on single-stranded templates, yielding double-stranded products. Reaction B, UV radiation damage or carcinogen-DNA adducts inhibit complementary DNA synthesis. Incubation of damaged templates with cell lysates generates incomplete replication intermediates containing large single strand gaps. Reaction C, damaged templates inhibit complementary DNA synthesis on nondamaged substrates in trans. For example, DNA synthesis on ϕX174 single-stranded DNA (4 ng/μl) is essentially abolished in the presence of equivalent amounts of UV-C-irradiated M13 single-stranded DNA (100 J/m^2). Replication activity on ϕX174 DNA is rescued by the addition of T7 DNA polymerase, indicating that a polymerase complex that normally catalyzes primer elongation becomes trapped on damaged templates.

trans occurs at the level of primer elongation. We conclude from these experiments that a DNA polymerase complex that normally catalyzes the elongation phase of DNA synthesis during eukaryotic replication is trapped on UV-irradiated or alkylated template DNA. This mechanism involving inhibition of DNA synthesis in trans may be employed by eukaryotic cells to adapt the temporal organization of chromosome replication when the genome is subject to genotoxic insults.

THE CHECKPOINT FUNCTION OF DNA POLYMERASES

DNA polymerase ε is not only one of the four essential replicative polymerases in eukaryotes, but also serves as an early factor in a checkpoint regulatory system.[27] Proliferating cells are normally able to regulate cell cycle by delaying progression into S, G2 or M phase in response to DNA damage.[28,29] A transient block in these cell cycle transitions is important to provide the opportunity to remove DNA lesions and reestablish the structural integrity of chromosomes before initiating DNA replication or chromosome segregation. The significance of such checkpoint feedback controls in preventing irreversible genetic alterations is exemplified by the role of p53 tumor suppressor mutations in human cancer (see chapter 1) or by the phenotype of transgenic p53 knock-out mice (see chapter 2). Yeast DNA polymerase ε is required for an S phase-specific checkpoint function and displays properties that are consistent with a role as a sensor of incomplete intermediates of DNA synthesis arising, for example, when DNA polymerization is arrested by replication blocking lesions.[27]

Several *Saccharomyces cerevisiae* genes have been isolated on the basis of their strict requirement for induction of ribonucleotide reductase in response to DNA damage.[30] These genes are designated *DUN* (for *dam-age-un*inducible). For example, *DUN1* encodes a serine/threonine kinase that is a mediator in the signal transduction pathway that coordinates transcription and cell cycle regulation in response to replication blocks.[30] *DUN2*, on the other hand, is identical to *POL2*, the gene that encodes the large subunit of DNA polymerase ε. Genetic and molecular analysis of *DUN2/POL2* indicates that DNA polymerase ε contains two separable domains: the N-terminal half of DNA polymerase ε has the polymerase active site, whereas the C-terminal half contains an S phase checkpoint domain characterized by the presence of a zinc finger motif. Mutants defective in this C-terminal domain die rapidly in the presence of hydroxyurea, an inhibitor of ribonucleotide reductase, because they enter mitosis without prior DNA replication.[27] *DUN2/POL2* seems to function upstream of *DUN1*, perhaps by sensing the accumulation of strand termini or single strand gaps in response to blocks to DNA synthesis. Thus, it appears that DNA polymerase ε and perhaps other DNA polymerases are used in eukaryotic cells to monitor the functional and structural integrity of their own reaction products.

REPLICATION OF DAMAGED DNA IN MAMMALIAN CELLS

The many in vitro experiments indicating marked inhibition of DNA polymerases by template lesions[12-21] contradict studies in vivo indicating that mammalian cells possess a remarkable ability to circumvent blocks to DNA synthesis and complete replication of their genomes despite the presence of persisting sites of damage.[31-33]

The following sections of this chapter are primarily concerned with the molecular mechanisms that promote DNA replication despite the presence of DNA lesions. In the literature, such processes are frequently referred to as "postreplication repair", because it was discovered about 30 years ago that the major pathway for replication of damaged DNA in the prokaryote *Escherichia coli* is associated with the "repair" of gaps in daughter DNA strands, primarily by recombinational strand exchanges.[34] Thus, "postreplication repair" functions to convert the low molecular weight DNA initially synthesized on damaged templates into the high molecular weight form characteristic of control cells. The term "repair" is misleading, however, because such mechanisms serve to overcome DNA synthesis blocks without eliminating the offending lesion. On the other hand, these replicative bypass mechanisms allow another round of true DNA repair in the daughter cells. Also, such replicative bypass mechanisms result in a gradual dilution of DNA lesions as the cells continue to divide. In general, the molecular details of DNA damage bypass pathways are poorly understood and the interpretation of experiments aimed at investigating these processes has been highly controversial.[35,36]

Efficient replication of damaged human chromosomes was first suggested by studies in which bromodeoxyuridine density labeling was employed to detect the frequency of cyclobutane pyrimidine dimers before and after DNA replication. This technique involves the separation of replicated DNA, consisting of heavy-light chains, from nonreplicated DNA containing only light-light chains. In these studies, exactly the same number of UV radiation products was found in replicated and nonreplicated DNA. In addition, treatment of replicated DNA with cyclobutane pyrimidine dimer-DNA glycosylase failed to produce double strand breaks, suggesting continuous DNA synthesis across sites of damage.[31,32]

These studies were confirmed many years later, when the persistence of DNA lesions was analyzed in specific regions of mammalian genomes, i.e., in the human or hamster dihydrofolate reductase (*DHFR*) gene.[33,37] Using the same bromodeoxyuridine density labeling technique, the efficiency of DNA synthesis across DNA lesions was estimated by comparing the amount of DNA damage found in the replicated gene with the level of damage observed before DNA replication. As was previously demonstrated for bulk DNA, this approach showed that neither UV radiation products nor bulky DNA adducts induced by the carcinogen 4'-hydroxymethyl-4,5',8-trimethylpsoralen were able to suppress DNA replication within the *DHFR* gene.

An early experiment involving UV irradiation of simian virus 40 (SV40)-infected monkey cells suggested that mammalian cells are equipped with specific factors that circumvent DNA polymerase blocking lesions. In order to synchronize viral DNA replication, Sarasin and Hanawalt[38] used a temperature-sensitive SV40 mutant arrested at the initiation of DNA replication. They observed that cyclobutane pyrimidine dimers suppressed replication fork progression and viral DNA synthesis when located on the leading strand of replication. In contrast, pyrimidine dimers located on the lagging strand were unable to block replication fork progression. The sizes of viral DNA fragments synthesized after UV irradiation closely approximated the average interdimer distance in the template DNA, as if replication of SV40 genomes proceeded bidirectionally from the origin until the leading strand DNA polymerases encountered the first cyclobutane dimers. DNA synthesis did not readily resume beyond these sites. However, when pulse-labeled cultures were subsequently incubated in medium containing nonlabeled DNA precursors, viral DNA molecules that were initially small became progressively longer, indicating the presence of specialized mechanisms for lesion bypass.

MOLECULAR MECHANISMS IN PROKARYOTES

The paradigm of DNA replication in the prokaryote *Escherichia coli* indicates at least two distinctly different mechanisms by which damaged DNA templates may be bypassed. In bacteria, both an error-prone and an error-free pathway are active at DNA lesions encountered during replication.[1]

TRANSLESION REPLICATION

"Translesion replication" consists of an error-prone bypass of DNA damage in the template strand and is responsible for mutagenesis (Fig. 10.2). This pathway involves DNA synthesis past sites of damage by DNA polymerase III in an altered low-fidelity mode mediated by RecA, UmuC and UmuD'. These proteins form a complex that catalyzes the

Fig. 10.2. Scheme illustrating the error-prone (mutagenic) bypass of damaged bases in prokaryotic and eukaryotic organisms by "translesion replication". DNA damage is represented by the rectangle. Thin lines, parental strands; thick lines, daughter strands.

addition of nucleotides opposite and beyond the site of damage.[39] Since the inserted nucleotides are often not directed by the hydrogen bonding pattern of template bases, this pathways generates frequent mutations.

RECOMBINATION STRAND TRANSFER

The "recombinational strand transfer" model proposed by Rupp and Howard-Flanders[34] appears to be the major route of replication recovery at sites of DNA damage in *Escherichia coli* (Fig. 10.3). This pathway also requires RecA but is independent of UmuC and UmuD'. "Recombinational strand transfer" involves repriming of DNA synthesis some distance downstream from the damaged base, thus producing a daughter strand gap of 1-2 kilobases opposite the site of damage (Fig. 10.3). RecA protein presumably binds to this single-stranded segment and initiates repair of the gap by homologous recombination between the two sister duplexes at the replication fork. This mechanism is essentially error-free, but results in the transfer of DNA from the nondamaged duplex to the homologous strand of the damaged duplex.[1,34]

MOLECULAR MECHANISMS IN MAMMALS

Error-prone strategies similar to the bacterial "translesion replication" (Fig. 10.2) may also exist in mammalian cells. For example, the processivity factor PCNA (*Proliferating Cell Nuclear Antigen*) considerably enhances the ability of DNA polymerase δ to elongate DNA primers past cyclobutane pyrimidine dimers located in the template strand.[40] DNA polymerase β, the most distributive and least accurate of the four

nuclear DNA polymerases in mammals, is capable of primer extension past bulky DNA adducts generating highly mutagenic replication products.[41,42] In the model eukaryote *Saccharomyces cerevisiae*, a two-subunit DNA polymerase encoded by *REV3* and *REV7*, designated DNA polymerase ζ (DNA polymerase zeta), has been identified as a nonessential factor required for damage-induced mutagenesis.[43,44] These reports suggest several different mechanisms of translesion synthesis in mammalian cells. For example, polymerase δ may catalyze error-prone bypasses of DNA damage in the presence of PCNA or additional accessory factors specialized on mutagenesis. Alternatively, DNA polymerase β or a novel ζ-like DNA polymerase which remains to be discovered in mammals may be implicated in the error-prone bypass of damaged sites without being required for normal replicative processes, in a pathway similar to "translesion replication" in *Escherichia coli*.

THE TEMPLATE SWITCHING HYPOTHESIS

Although "translesion replication" appears conserved throughout evolution, bacteria and mammalian cells have diverged with respect to other aspects of DNA replication across sites of damage. In particular, single strand gaps opposite damaged nucleotides and subsequent recombinational strand exchanges are not detectable at all in UV-irradiated or carcinogen-treated mammalian cells, or are detectable at levels significantly lower than in prokaryotes.[45-48] These observations prompted Strauss and coworkers to propose a distinctly different scenario for the error-free bypass of DNA damage in mammalian cells.[49] The main fea-

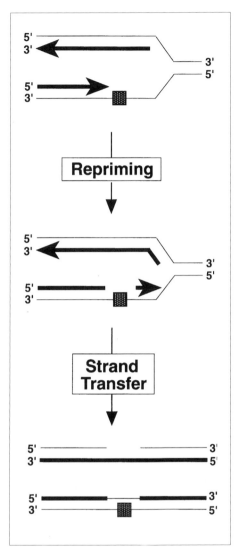

Fig. 10.3. Model of error-free (nonmutagenic) "recombinational strand transfer", involving repriming of DNA synthesis downstream from the damaged site followed by strand transfer. DNA damage is symbolized by the rectangle. Thin lines, parental strands; thick lines, daughter strands.

daughter strand instead of the damaged parental strand as a detour around the lesion, and then switches back to the parental strand after bypassing the damaged site (Fig. 10.4). One prediction of this "template switching" model is the transient formation of DNA duplexes in which both stands are newly synthesized.

In order to test the "template switching" model, Strauss and collaborators analyzed DNA fragments from human cells grown in the presence of bromodeoxyuridine.[49] After the first round of replication, untreated control cells yielded DNA fragments of hybrid density characteristic of heavy-light chains. In contrast, human cells treated with the chemical carcinogen methyl methanesulphonate produced three peaks of incorporation of bromodeoxyuridine on density gradients: a major peak characteristic of hybrid heavy-light DNA, a second peak of lower density which presumably included partially replicated molecules with single-stranded regions, and a third peak, banding at 1.80 g/cm^3, characteristic of duplex molecules substituted in both strands with bromodeoxyuridine (heavy-heavy DNA). A very small fraction of higher density was also found in the material obtained from control cells, but the relative amount of heavy-heavy DNA was consistently increased in response to carcinogen treatment. Pairing interactions between daughter strands, as predicted by the "template switching" model, were also identified in electron micrographs of DNA fragments obtained from damaged cells.[49] An important follow-up study showed that a large proportion of these heavy-heavy DNA intermediates, suggestive of pairing interactions between newly synthesized strands, were generated

ture of their model is to use a newly synthesized strand as template instead of the damaged parental strand. As depicted in Figure 10.4, when replication of one DNA strand is blocked by DNA damage, replication of the other strand can continue. "Template switching" is then initiated by pairing interactions between the two daughter DNA strands, allowing the nascent DNA initially blocked to bypass the lesion. Thus, the replicative polymerase uses the complementary

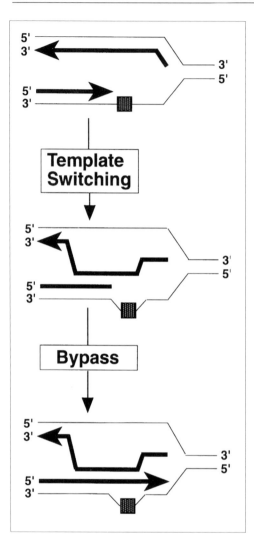

Fig. 10.4. Proposed model for the error-free bypass of damaged bases by "template switching". Rectangle, DNA damage; thin lines, parental strands; thick lines, daughter strands.

replication is bidirectional. According to the Jesuit model proposed by DePamphilis[23] ("many are called, but few are chosen"), effects of chromosome structure and nuclear organization normally suppress initiation of semiconservative DNA synthesis at most of these sites. However, when progression of an active replication fork is blocked by DNA damage, DNA replication may be completed by initiation of another adjacent site on the same chromosome (Fig. 10.5). This compensatory replication fork would then approach the site of damage from the other side, complete DNA synthesis, and eventually leave a single-strand gap in the daughter strand opposite the lesion (Fig. 10.5).[36,37,52] Assuming that such a mechanism produces newly synthesized DNA containing short gaps, the available evidence suggests that these gaps are rapidly closed by either error-prone or error-free mechanisms.[45-48]

Griffith and Ling[53] provided experimental data in favor of the idea that alternative origins of replication are activated in response to DNA damage. These authors were able to measure the distances between origins of replication by DNA fiber autoradiography. To that end, UV-irradiated or control hamster cells were sequentially incubated with [³H]thymidine of high and low specific activity. After DNA replication, the cells were spread across glass slides for autoradiographic analysis. This labeling protocol produced DNA molecules that exhibit centers of high grain density trailing off to lower grain density on either side. The site where DNA replication was initiated

in vitro during the cell lysis and DNA isolation procedure.[50] Nevertheless, "template switching" remains an attractive mechanism for error-free bypass of DNA lesions in mammalian cells.

MULTIPLE ORIGINS OF REPLICATION

A fundamental difference between prokaryotes and eukaryotes resides in the fact that, unlike bacteria, eukaryotic chromosomes contain multiple (10,000-100,000) replicons. Each of these eukaryotic replication units has its own initiation site from which

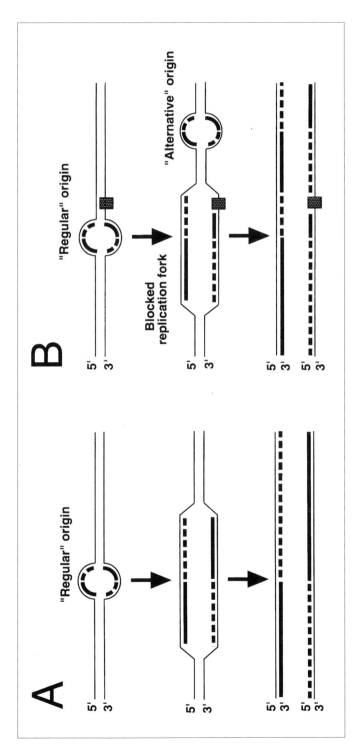

Fig. 10.5. Schematic diagram illustrating the activation of alternative origins of replication in damaged mammalian chromosomes. DNA damage is represented by the rectangles. A, nondamaged chromosome; B, damaged chromosome with one additional origin of replication. Thin lines, parental strands; thick solid lines, leading strands; thick dashed lines, lagging strands.

is located in the center of each of the high grain density regions. When cells were damaged by UV irradiation, the distances between adjacent sites of initiation (inter-origin distances) were significantly shorter than in control cells. After exposure to 10 J/m², for example, these inter-origin distances were decreased by approximately 40%, indicating that DNA damage results in the activation of many additional origins of replication which are not employed under normal circumstances, i.e., in the absence of exogenous DNA damage. This activation occurred in both repair-proficient and repair-deficient cell lines but, consistent with the slower kinetics of repair, persisted for a longer time in the repair-deficient line.[53]

A GENETIC AND BIOCHEMICAL FRAMEWORK IN LOWER EUKARYOTES

A large number of genes that are putatively involved in processes that catalyze the replication of damaged DNA have been identified in the model eukaryote *Saccharomyces cerevisiae*. One group of genes has been isolated by searching for mutations that are abnormally sensitive to killing by UV or ionizing radiation, and these yeast strains were designated *rad* mutants (for *rad*iation sensitive).[54] Several *rad* mutants are perfectly proficient with respect to the classical responses to DNA damage, including damage reversal, excision repair, recombinational repair or checkpoint functions, but are reduced in their capacity to replicate damaged DNA templates.[55]

Another line of investigations has focused on yeast mutants that exhibit abnormally low levels of mutagenesis when exposed to DNA damaging agents. Such hypomutable yeast strains are designated *rev* mutants (for defective mutation *rev*ersion)[56] or *umr* mutants (for *U*V *m*utation resistance).[57] Mutations in the *REV* and *UMR* series of genes confer moderate sensitivity to genotoxic agents, but their absolute requirement for DNA damage-induced mutagenesis during mitotic growth indicates that these genes encode proteins that par-

ticipate in error-prone modes of bypass during DNA replication.[56,57]

MUTAGENIC DNA SYNTHESIS

Overexpression, purification and biochemical characterization of different *REV* genes led to the discovery of 2 distinct enzymatic functions that are responsible for enhanced mutagenesis. *REV1* encodes a 112 kDa protein with deoxycytidyl transferase activity.[58,59] This unique enzyme is able to transfer a dCMP residue from dCTP to the 3' end of a DNA primer in a template-dependent reaction. Efficient transfer occurs opposite a guanine, an adenine and even opposite an abasic site in the template strand, but not opposite a cytosine or a thymine in the template. Similarly, Rev1 protein does not insert dCMP residues opposite a cyclobutane pyrimidine dimer.

As mentioned earlier, Rev3 and Rev7 form a tight complex with DNA polymerase activity.[43,44,60] This heterodimeric enzyme of 173 and 29 kDa subunits was designated DNA polymerase ζ (zeta). The activity of this novel DNA polymerase is rather distributive, it lacks 3'-5' proofreading and it is insensitive to the classical inhibitors aphidicolin and dideoxynucleotide triphosphates. In contrast to DNA polymerase α, DNA polymerase ζ is capable of extending DNA primers past cyclobutane dimers with 10% efficiency, suggesting a role in the bypass of DNA damage by translesion synthesis.[44] This requirement in yeast of nonessential deoxycytidyl transferase and DNA polymerase activities for damage-induced mutagenesis supports a model in which the chromosomal replication complex is temporarily disassembled at blocked replication forks, and replaced by a machinery specialized on error-prone translesion DNA synthesis.[56]

POSTTRANSLATIONAL PROTEIN MODIFICATION AND PROCESSING

The *RAD6* gene encodes a protein that enzymatically transfers ubiquitin, a highly conserved polypeptide of 76 residues, to histones H2A and H2B.[56,61] *RAD6* is apparently required for both error-free and error-prone modes of bypass.[62] As a consequence, *Rad6* null mutants are extremely sensitive to killing by genotoxic agents and have lost the capacity for damage-induced mutagenesis. Also, the conversion of low molecular weight DNA after UV irradiation to the high molecular weight form characteristic of nondamaged cells is defective in replicating *rad6* mutants. The pleiotropic phenotype of *rad6* deletion mutants also involves a defect in sporulation.[56]

Additional proteins implicated in the replication of damaged DNA are the products of *PRB1*, a proteolytic enzyme,[63] and the product of *CDC7*, a protein kinase.[64] In light of these observations, it is tempting to speculate that posttranslational protein modification (mono- or polyubiquitination, phosphorylation) or protein processing by proteolytic enzymes plays a central role in the replication of damaged templates by either modulating chromatin structure or regulating the activity of enzymes participating in DNA synthesis. For example, the involvement of a protease in damage-induced mutagenesis suggests that certain factors involved in the replication of damaged DNA are activated by proteolytic cleavage.[63] The cleavage of UmuD protein to UmuD', which is considered the active form for mutagenesis in *Escherichia coli*, provides a biochemical precedent for this hypothesis.[1,39]

DNA Helicases

Gene products with DNA helicase activity participate in both error-prone and error-free modes of bypass of damaged DNA in *Saccharomyces cerevisiae*. These enzymes catalyze the separation of duplex DNA through disruption of the hydrogen bonds holding the two strands of the duplex together. As an energy source, these enzymes utilize the hydrolysis of nucleoside 5'-triphosphates (see Fig. 5.2 in chapter 5 for the basic biochemical reaction catalyzed by DNA helicases).

DNA helicase activity has been unequivocally demonstrated for the product of the *RADH* gene.[65] Genetic experiments

showed that *RADH* acts in an error-prone pathway for replication of damaged DNA, presumably by catalyzing an early, possibly the first step in this process.[66] In the absence of functional *RADH*, DNA damage is channeled into a recombinogenic pathway, while in the presence of functional *RADH* the damaged substrate is processed by an error-prone mechanism.[66] This finding led to the hypothesis that the *RADH* gene product may function to scan DNA for sites of damage, bind to the damaged site, and recruit a DNA polymerase to proceed with error-prone DNA synthesis.[65]

This particular role of a DNA helicase was inferred from biochemical studies on the yeast nucleotide excision repair protein Rad3, indicating that certain DNA helicases may have adopted a specialized function in reading the chemistry of DNA and discriminating between nondamaged and damaged templates. In fact, the DNA helicase activity of Rad3 protein is inhibited by covalent alterations in the primary structure of DNA but not by noncovalent changes in the secondary structure of DNA.[67-69] In addition, Rad3 protein was shown to form stable complexes with the covalently modified strand of damaged DNA duplexes. These biochemical properties of a DNA helicase provide a potential mechanism whereby sites of DNA damage are targeted for processing by the nucleotide excision repair pathway (Rad3) or by error-prone replication (*RADH* gene product). Interestingly, the nucleotide sequence of the cloned *RAD5* gene, which together with *RAD18* is implicated in an error-free pathway for replication of damaged DNA templates, suggests the involvement of another DNA helicase in the initiation of these processes.[70,71]

Molecular Flags

Although only fragmentary, the genetic and biochemical evidence from yeast indicates that several factors that normally regulate enzyme function or modulate macromolecular conformation (ubiquitin conjugating enzymes, kinases, proteases, ATPases/DNA helicases) participate in the replication of damaged DNA. These factors may act to prime chromatin proteins or damaged templates for error-prone and error-free pathways of replication.

Tightly regulated high precision events such as initiation of transcription or replication are effected in living cells by large multiprotein complexes (reviewed in chapter 6). Thus, it is possible that eukaryotes also assemble complex biochemical machineries to catalyze the replication of damaged DNA templates. Ubiquitin modification or partial proteolytic digestion of chromatin proteins, as well as sequestration of specific ATPases/DNA helicases at damaged sites, may signal the presence of replication blocking lesions and provide "molecular flags" necessary for the recruitment of multiple factors involved in either error-prone translesion synthesis or error-free bypasses of DNA damage. Subsequently, ubiquitination, proteolytic degradation or ATPase/DNA helicase activities may facilitate protein-protein interactions and conformational changes required for the assembly of multisubunit complexes.

GENES IMPLICATED IN THE REPLICATION OF DAMAGED DNA IN HIGHER EUKARYOTES

At the time of writing, it is not known whether the mechanisms of replicative bypass that have been identified in *Saccharomyces cerevisiae* also operate in mammalian cells. However, the existence of *RAD6* homologs in higher eukaryotes and mammals, the characterization of *Drosophila mus* mutants and a particular (variant) form of xeroderma pigmentosum suggest that at least some of these mechanism are evolutionary conserved.

RAD6 HOMOLOGS

Eukaryotes show a striking evolutionary conservation of RAD6 structure and function. A conserved homolog of *Saccharomyces cerevisiae RAD6* has been found in the fruitfly *Drosophila melanogaster*.[72] Unexpectedly, human cells possess at least two *RAD6* homologs.[73] These genes are desig-

nated *Dhr6* (for *Drosophila* homolog of *RAD6*), *HHR6A*, and *HHR6B* (for human homolog of *RAD6 A* and *B*, respectively). Thus, in contrast to yeast and *Drosophila*, where *RAD6* is a single gene, the *RAD6* homologous gene is duplicated in humans, and the proteins encoded by the two genes *HHR6A* and *HHR6B* have an amino acid sequence identity of 95%. The two human proteins also share approximately 70% sequence identity with the yeast homolog, strongly indicating that the human genes also encode ubiquitin-conjugating enzymes.[73]

The striking degree of amino acid sequence conservation between yeast and humans is also reflected at the functional level, as both human *RAD6* homologs (as well as the *Drosophila* homolog) restore normal levels of DNA damage-induced mutagenesis in *Saccharomyces cerevisiae rad6* null mutants. In addition, the human (and *Drosophila*) *RAD6* homologs also correct the UV hypersensitivity of the yeast *rad6* null mutants to near normal levels.[72,73] Therefore, the corresponding yeast and human proteins appear to act in an identical manner to reduce the cytotoxic effect of DNA damage. Further support for this contention is provided by the observation that the strict conservation of eukaryotic Rad6 proteins, particularly in their N-terminal region, does not extend to other ubiquitin-conjugating enzymes.[73] This uniqueness suggests that the N-terminal domains of Rad6 proteins are required for specific functions such as, for example, interaction with other components of the machinery that catalyzes DNA synthesis through sites of damage.

The murine version of one of the two human *RAD6* homologs (*mHR6B*) was recently inactivated by gene targeting.[74] The resulting knock-out mice were viable and phenotypically normal, perhaps because of a functional redundancy with *mHR6A*. Also, mHR6B-deficient females displayed normal fertility, but mHR6B-deficient male mice were infertile. Histologically, male infertility involved a developmental block at the spermatid stage of spermatogenesis that was associated with a defect in chromatin condensation. Thus, the requirement of *RAD6* for normal sporulation in yeast[56] is nicely reiterated by the requirement of *mHR6B* for normal spermatogenesis in mammals.

DROSOPHILA MUS MUTANTS

Dhr6 is not the only known *Drosophila* gene that has been implicated in the replication of damaged DNA. The search for *Drosophila* mutants which are hypersensitive to genotoxic agents led to the identification of over 30 genes whose mutants are defective in various DNA damage processing pathways.[75,76] One particular group of *Drosophila mus* mutants (for *mu*tagen sensitive) shows normal levels of DNA excision repair but is defective in the ability to replicate DNA after treatment with genotoxic agents.[76] Cells from these *Drosophila mus* strains respond to UV radiation or carcinogen treatment by synthesizing nascent DNA in segments shorter than those produced in equally damaged wild type cells, thus indicating a reduced capacity to circumvent DNA lesions during replication.[76,77] The formation of abnormally short replication products was particularly pronounced in UV-irradiated *mus310* cells, in which the size of nascent DNA fragments roughly corresponded to the expected average distance between UV radiation products, indicating a complete failure in bypassing these lesions.[77]

When the rate of DNA synthesis was measured by incorporation of radioactive thymidine, Brown and Boyd[77] were able to separate these *Drosophila* mutants into two distinct classes. Cells from most mutant strains (for example *mus101*, *mus108*) incorporate equivalent amounts of radioactive thymidine after treatment with DNA damaging agents as do wild type controls. However, two mutants (*mus302* and *mus310*) exhibit abnormally low levels of thymidine incorporation when DNA is damaged.[77] These observations indicate that the first class of mutants (*mus101*, *mus108*) is defective in the bypass of DNA damage located in the lagging strand of replication. This defect involves the formation of shorter DNA fragments than in wild type controls, but results in normal rates of DNA synthesis. The second class of mutants (*mus302* and *mus310*) appears to be defective in the bypass of DNA damage located in the leading strand of replication. This defect also produces shorter DNA segments but, additionally, results in abnormally low levels of overall DNA synthesis.[77]

A possible caveat in the interpretation of these results is that the same phenotype of mutagen sensitivity and abnormal replication may also be caused by deficient cell cycle checkpoint functions. For example, DNA replication in cells from the *Drosophila mei41* strain (for *mei*otic mutant) is also characterized by the formation of low molecular weight DNA following exposure to genotoxic agents.[77] Cloning and sequencing of the *mei41* gene revealed a striking homology with human *ATM* (for *a*taxia *t*elangiectasia *m*utated; see chapter 1) and, in fact, subsequent investigations showed that irradiated *mei41* cells are deficient in a checkpoint function that delays entry into mitosis to allow completion of DNA replication.[78]

XERODERMA PIGMENTOSUM VARIANT

The variant form of the human disease xeroderma pigmentosum may define a mammalian mutant that carries a deficiency in the replication of damage templates. As outlined in chapter 1, xeroderma pigmentosum is a rare genetic disorder with autosomal recessive pattern of inheritance and is characterized by skin hypersensitivity to the UV component of sunlight. Patients afflicted by xeroderma pigmentosum suffer from extreme dryness of the skin, alternating areas of hyper- and hypopigmentation and a very high incidence of malignant skin cancer. At the cellular level, patients having the classical form of the disease are defective in nucleotide excision repair of UV radiation damage and carcinogen-induced DNA adducts.[79]

A fraction of clinically typical xeroderma pigmentosum patients exhibit normal levels of DNA repair activity,[80] and the term "variant" was introduced to describe this form of the disease. This variant form represents from 20% (in western countries) to 45% (in Japan) of all xeroderma pigmentosum patients. Cells obtained from these individuals are thought to have a defect in the bypass of DNA damage during chromosomal replication.[35,36] As found in the *Drosophila mus* mutants discussed previously, the replication machinery is apparently blocked more frequently at sites of DNA damage in xeroderma pigmentosum variant cells than in normal controls. This damage-specific replication defect is indicated by the much smaller size of newly synthesized DNA fragments in UV irradiated xeroderma pigmentosum variant cells when compared to the high molecular weight fragments produced in normal cells.[35,36] However, during a subsequent incubation of up to six hours these fragments increased in size at the same rate as those in normal human cells.[36] This defect in the replication of damaged templates, originally discovered using UV radiation, was recently confirmed for bulky DNA adducts generated by treatment with 4'-hydroxymethyl-4,5',8-trimethyl-psoralen.[81] While psoralen-induced DNA adducts are bypassed in normal cells at an efficiency of nearly 100%, only approximately 50% of these adducts are bypassed in xeroderma pigmentosum variant cells.

Based on the phenotypic consequences of xeroderma pigmentosum variant (elevated levels of UV-induced mutagenesis),[82,83] the available data suggest that the disease is caused by a defect in the error-free pathway of replicative bypass. This view is supported by the studies of Maher, McCormick and colleagues, who compared the strand and sequence distribution of UV-induced lesions with the distribution of subsequent mutations using an appropriate reporter gene on a shuttle vector.[84,85] These authors observed that DNA replication in xeroderma pigmentosum variant cells is characterized by the failure to incorporate dGMP or dAMP opposite UV-cross-linked pyrimidines bases (either cytosines or thymines). The proposed defect in an error-free bypass mechanism accounts for the dramatically increased frequency of C·G→A·T transversions, a mutation that is rarely detected when normal cells are challenged with UV radiation.[85]

It may be possible that xeroderma pigmentosum variant cells are deficient in an error-free mechanism such as the "template switching" pathway outlined in Figure 10.4. Alternatively, the variant form of xeroderma pigmentosum may be caused by mutations in a DNA polymerase or accessory factor of DNA synthesis (PCNA, RFC, RPA) required for processivity and fidelity of DNA replication.[86] A replication complex containing such a mutated subunit may be particularly prone to arrest DNA synthesis at sites of template damage or may have a unique propensity to shift to an error-prone mode of bypass. In this alternative model, xeroderma pigmentosum variant patients may be affected by a general replication anomaly that is exacerbated in the presence of UV radiation damage or other DNA lesions. Yet another model predicts that xeroderma pigmentosum variant cells may be deficient in the activation of additional replicons when DNA synthesis is blocked on damaged templates.[36]

A MAMMALIAN SYSTEM FOR IN VITRO REPLICATION OF DAMAGED DNA

In the future, studies on the mechanisms of replication of damaged templates in mammalian cells will be facilitated by the availability of appropriate cell-free approaches.[87-89] The most promising in vitro system uses mammalian cell extracts to replicate UV-irradiated SV40-based shuttle vectors. As expected, replication was inhibited when the DNA substrate was UV-irradiated prior to incubation with cell extract. However, some plasmid molecules were completely replicated yielding covalently closed circular SV40 genomes despite the presence of UV radiation damage. The replicative bypass of cyclobutane pyrimidine dimers was confirmed by demonstrating that the DNA lesions were not removed during DNA synthesis but instead persisted in the newly replicated strands.[87,88] When mutagenesis of a particular target gene located on the plasmid was examined by transfection of replicated DNA into *Escherichia coli*, the bypass of DNA damage was associated with an up to 15-fold increase in the frequency of mutations.[87-90] Most of the mutations observed were single or tandem C·G→T·A transitions, indicating that this in vitro system reflects the process of UV-induced mutagenesis previously observed in mammalian cells or in skin cancers (see chapter 3).[89] SV40 replication in mammalian cell extracts has already been exploited to compare mutagenesis on the leading and lagging strand for replication, but no dramatic differences have been found.[89,90] The same in vitro system has also been used to test mutagenesis induced by psoralen or acetylaminofluorene adducts.[91,92] Moreover, this cell-free approach will be extremely useful to identify mammalian factors required for the mutagenic bypass of DNA lesions. In addition, this system provides the unique opportunity to test how the chemical and conformational nature of a particular lesion affects the efficiency and fidelity of mammalian DNA replication. It is possible, for example, that the pathway for replicative bypass varies depending on the chemistry of the lesion or the nucleotide sequence context in which it is situated.

CONCLUSIONS

DNA damage poses special problems to the fidelity of DNA replication because DNA polymerases are often unable to insert correct nucleotides opposite sites of base damage. As a consequence, the frequency by which DNA damage is converted to mutations increases dramatically when the cellular genome is replicated. A certain degree of genetic variability is essential in the evolutionary process to produce the phenotypic diversity from which to select the most advantageous functions. However, when normal regulatory cell cycle circuits (checkpoints) are disrupted, the ability to replicate despite the presence of damage constitutes a major cause of genetic instability. The base substitutions, deletions, translocation or amplification events cause birth defects, inherited diseases, cancer or lethality.

Despite the many inconveniences arising when replication forks encounter damaged templates, recognition of DNA lesions during replication is nevertheless important for the initiation of several damage response pathways. For example, DNA polymerase ε has been found to trigger a specific signal transduction system that delays entry into mitosis when cells are unable to complete chromosomal replication. Also, the newly discovered phenomenon of trapping of DNA polymerase complexes at sites of base damage is consistent with the idea that damaged replicons may directly inhibit DNA synthesis in adjacent replication units. This mechanism may serve to amplify the inhibitory effect of damaged templates, thereby potentiating a signal that results in the activation of extra origins of replication and transient cell cycle arrest.

Apparently, efficient mechanisms exist in mammalian cells to replicate the genome despite the presence of many damaged sites, raising the question of how, and to what extent, these putative mechanisms contribute to genetic instability. There are a number of hypothetical possibilities by which a growing fork may bypass replication blocking lesions. In eukaryotic (and mammalian) cells, the activation of alternative origins of replication may play a central role in facilitating the replication of damaged chromosomes. Genetic studies in yeast and other eukaryotes suggest that replication through DNA damage is highly regulated and catalyzed by complex biochemical machineries composed of many specialized gene products. Depending on which pathway is utilized, be it mutagenic or nonmutagenic, these replication machineries either increase or attenuate the carcinogenic consequences of DNA damage. Specialized machineries may also be used in mammalian cells to facilitate DNA synthesis across sites of damage. Knowledge of the molecular details by which such factors may operate during mammalian S phases should reveal important basic rules about how DNA damage induces mutagenesis. As a consequence, mechanistic studies on the action of these putative damage-dependent replication machineries are expected to shed new light on the carcinogenic responses to genotoxic agents.

REFERENCES

1. Echols H, Goodman M.F. Fidelity mechanisms in DNA replication. Annu Rev Biochem 1991; 60:477-511.

2. McBride TJ, Preston BD, Loeb LA. Mutagenic spectrum resulting from DNA damage by oxygen radicals. Biochemistry 1991; 30:207-213.

3. Strauss BS. The "A Rule" of mutagen specificity: a consequence of DNA polymerase bypass of non-instructional lesions? BioEssays 1991;13:79-84.

4. Kaufmann WK. Pathways of human cell post-replication repair. Carcinogenesis 1989; 10:1-11.

5. Kaufmann WK, Rice JM, Wenk ML et al. Reversible inhibition of rat hepatocyte proliferation by hydrocortisone and its effect on cell cycle-dependent hepatocarcinogenesis by N-methyl-N-nitrosourea. Cancer Res. 1981; 41:4653-4660.

6. Kaufmann WK, Rice JM, Wenk ML et al. Cell cycle-dependent initiation of hepatocarcinogenesis in rats by methyl-(acetoxymethyl)nitrosamine. Cancer Res 1987;47:1263-1266.

7. Kaufmann WK, Rahija RJ, MacKenzie SA et al. Cell cycle-dependent initiation of hepatocarcinogenesis in rats by (±)-7r,8t-dihydroxy-9t,10t-epoxy-7,8,9,10-tetrahydrobenzo[a]pyrene. Cancer Res 1987; 47:3771-3775.

8. Kaufmann WK, Wilson SJ. G1 arrest and cell cycle-dependent clastogenesis in UV-irradiated human fibroblasts. Mutat. Res 1994; 314:67-76.

9. Grisham JW, Greenberg DS, Kaufman DG et al. Cycle-related toxicity and transformation in 10T1/2 cells treated with N-methyl-N'-nitro-N-nitroso-guanidine. Proc Natl Acad Sci USA 1980; 77:4813-4817.

10. Bertram JS, Heidelberger C. Cell cycle dependency of oncogenic transformation induced by N-methyl-N'-nitro-N-nitroso-guanidine in culture. Cancer Res 1974; 34:526-537.

11. Strauss BS. Cellular aspects of DNA repair. Adv Cancer Res 1985; 45:45-105.

12. Hruszkewycz AM, Canella KA, Peltonen K et al. DNA polymerase action on benzo[a]pyrene-DNA adducts. Carcinogenesis 1992; 13:2347-2352.

13. Shibutani S, Margulis LA, Geacintov NE et al. Translesional synthesis on a DNA template containing a single stereoisomer of dG-(+) or dG-(-)-anti-BPDE (7,8-dihydroxy-anti-9,10-epoxy-7,8,9,10-tetrahydrobenzo[a]pyrene). Biochemistry 1993; 32:7531-7541.

14. Voigt JM, Topal MD. O^6-methylguanine-induced replication blocks. Carcinogenesis 1995; 16:1775-1782.

15. Moore PD, Bose KK, Rabkin SD et al. Sites of termination of in vitro DNA synthesis on ultraviolet- and N-acetylaminofluorene-treated ϕX174 templates by prokaryotic and eukaryotic DNA polymerases. Proc Natl Acad Sci USA 1981; 78:110-114.

16. Rabkin SD, Strauss BS. A role for DNA polymerase in the specificity of nucleotide incorporation opposite N-acetylaminofluorene adducts. J Mol Biol 1984; 178:569-594.

17. Woodgate R, Bridges BA, Herrera G et al. Mutagenic DNA repair in Escherichia coli. XIII. Proofreading exonuclease of DNA polymerase III holoenzyme is not operational during UV mutagenesis. Mutat Res 1987; 183:31-37.

18. Shwartz H, Shavitt O, Livneh Z. The role of exonucleolytic processing and polymerase-DNA association in bypass of lesions during replication in vitro. J Biol Chem 1988; 263:18277-18285.

19. Strauss BS, Wang J. Role of DNA polymerase 3'-5' exonuclease activity in the bypass of aminofluorene lesions in DNA. Carcinogenesis 1990; 11:2103-2109.

20. Hoffmann J-S, Moustacchi E, Villani G et al. In vitro synthesis by DNA polymerase I and DNA polymerase α on single-stranded DNA containing either purine or pyrimidine monoadducts. Biochem Pharmacol 1992; 44:1123-1129.

21. Hess MT, Schwitter U, Petretta M et al. DNA synthesis arrest at C4'-modified deoxyribose residues. Biochemistry 1997; in press.

22. Campbell J. Eukaryotic DNA replication. Annu Rev Biochem 1986; 55:733-772.
23. DePamphilis ML. Origins of DNA replication in metazoan chromosomes. J Biol Chem 1993; 268:1-4.
24. Diller JD, Raghuraman MK. Eukaryotic replication origins: control in space and time. Trends Biochem Sci 1994; 19:320-325.
25. Almouzni G, Méchali M. Assembly of spaced chromatin promoted by DNA synthesis in extracts from *Xenopus* eggs. EMBO J 1988; 7:665-672.
26. Krude T, Knippers R. Nucleosome assembly during complementary DNA strand synthesis in extracts from mammalian cells. J Biol Chem 1993; 268:14432-14442.
27. Navas TA, Zhou Z, Elledge SJ. DNA polymerase links the DNA replication machinery to the S phase checkpoint. Cell 1995; 80:29-39.
28. Hartwell LH, Weinert TA. Checkpoints: controls that ensure the order of cell cycle events. Science 1989; 246:629-634.
29. Kuerbitz SJ, Plunkett BS, Walsh WV et al. Wild-type p53 is a cell cycle checkpoint determinant following irradiation. Proc Natl Acad Sci USA 1992; 89:7491-7495.
30. Zhou Z and Elledge SJ. *DUN1* encodes a protein kinase that controls the DNA damage response in yeast. Cell 1993; 75:1119-1127.
31. Lehmann AR. The relationship between pyrimidine dimers and replicating DNA in UV-irradiated human fibroblasts. Nucleic Acids Res 1979; 7:1901-1912.
32. Waters R. Repair of DNA in replicating and unreplicating portions of the human genome. J Mol Biol 1979; 127:117-127.
33. Vos J-M, Hanawalt PC. Processing of psoralen adducts in an active human gene: repair and replication of DNA containing monoadducts and intrastrand cross-links. Cell 1987; 50:789-799.
34. Rupp WD, Howard-Flanders P. Discontinuities in the DNA synthesized in an excision-defective strain of *Escherichia coli* following ultraviolet irradiation. J Mol Biol 1968; 31:291-304.
35. Lehmann AR, Kirk-Bell S, Arlett CF et al. Xeroderma pigmentosum cells with normal levels of excision repair have a defect in DNA synthesis after UV-irradiation.
36. Park SD, Cleaver JE. Postreplication repair: questions of its definition and possible alteration in xeroderma pigmentosum cell strains. Proc Natl Acad Sci USA 1979; 76:3927-3931.
37. Spivak G, Hanawalt PC. Translesion DNA synthesis in the dihydrofolate reductase domain of UV-irradiated CHO cells. Biochemistry 1984; 31:6794-6800.
38. Sarasin AR, Hanawalt PC. Replication of ultraviolet-irradiated simian virus 40 in monkey kidney cells. J Mol Biol 1980; 138:299-319.
39. Rajagopalan M, Lu C, Woodgate R et al. (1992) Activity of the purified mutagenesis proteins UmuC, UmuD', and RecA in replicative bypass of an abasic DNA lesion by DNA polymerase III. Proc Natl Acad Sci USA 1992; 89:10777-10781.
40. O'Day CL, Burgers PMJ, Taylor JS. PCNA-induced DNA synthesis past cis-syn and trans-syn-I thymine dimers by calf thymus DNA polymerase δ in vitro. Nucleic Acids Res 1992; 20:5403-5406.
41. Hoffmann J-S, Pillaire M-J, Maga G et al. DNA polymerase β bypasses in vitro a single d(GpG)-cisplatin adduct placed on codon 13 of the *HRAS* gene. Proc Natl Acad Sci USA 1995; 92:5356-5360.
42. Hoffmann J-S, Pillaire M-J, Garcia-Estefania D et al. In vitro bypass replication of the cisplatin-d(GpG) lesion by calf thymus DNA polymerase β and human immunodeficiency virus type I reverse transcriptase is highly mutagenic. J Biol Chem 1996; 271:15386-15392.
43. Morrison A, Christensen RB, Alley J et al. *REV3*, a yeast gene whose function is required for induced mutagenesis, is predicted to encode a non-essential DNA polymerase. J Bacteriol 1989; 171:5659-5667.
44. Nelson JR, Lawrence CW, Hinkle DC. Thymine-thymine dimer bypass by yeast DNA polymerase ζ. Science 1996; 272:1646-1649.
45. Lehmann AR. Postreplication repair of DNA in ultraviolet-irradiated mammalian cells. No gaps in DNA synthesized late after ultraviolet irradiation. Eur J Biochemistry 1972; 31:438-445.
46. Clarkson JM, Hewitt RR. Significance of

dimers to the size of newly synthesized DNA in UV-irradiated Chinese hamster ovary cells. Biophys J 1976; 16:1155-1164.

47. Meneghini R, Hanawalt P. T4-endonuclease V-sensitive sites in DNA from ultraviolet-irradiated human cells. Biochim Biophys Acta 1976; 425:428-437.

48. Fujiwara Y, Tatsumi M. Low levels of DNA exchanges in normal human and xeroderma pigmentosum cells after UV irradiation. Mutat Res 1977; 43:279-290.

49. Higgins NP, Kato K, Strauss B. A model for replication repair in mammalian cells. J Mol Biol 1976; 101:417-425.

50. Tatsumi K, Strauss B. Production of DNA bifilarly substituted with bromodeoxyuridine in the first round of synthesis: branch migration during isolation of cellular DNA. Nucleic Acids Res 1978; 5:331-346.

51. Gasser SM. Replication origins, factors and attachment sites. Curr Opin Cell Biol 1991; 3:407-413.

52. Painter RB. Inhibition and recovery of DNA synthesis in human cells after exposure to ultraviolet light. Mutat Res 1985; 145:63-69.

53. Griffiths TD, Ling SY. Activation of alternative sites of replicon initiation in Chinese hamster cells exposed to ultraviolet light. Mutat Res 1987; 184:39-46.

54. Friedberg EC, Siede W, Cooper AJ. Cellular responses to DNA damage in yeast. In: Broach J, Jones E, Pringle J, ed. The Molecular and Cellular Biology of the Yeast *Saccharomyces*: Genome Dynamics, Protein Synthesis, and Energetics. Vol I. New York: Cold Spring Harbour Laboratory Press, 1991:147-192.

55. Di Caprio L, Cox BS. DNA synthesis in UV-irradiated yeast. Mutat Res 1981; 82:69-85.

56. Lawrence C. The RAD6 DNA repair pathway in *Saccharomyces cerevisiae*: what does it do, and how does it do it? BioEssays 1994; 16:253-257.

57. Lemontt JF. Pathways of ultraviolet mutability in *Saccharomyces cerevisiae*. II. Genetic analysis and properties of mutants resistant to ultraviolet-induced forward mutation. Mutat Res 1977; 43:179-204.

58. Larimer FW, Perry JR, Hardigree AA. The *REV1* gene of *Saccharomyces cerevisiae*: isolation, sequence, and functional analysis. J Bacteriol 1989; 171:230-237.

59. Nelson JR, Lawrence CW, Hinkle DC. Deoxycytidyl transferase activity of yeast REV1 protein. Nature 382; 382:729-731.

60. Torpey LE, Gibbs PE, Nelson J et al. Cloning and sequence of *REV7*, a gene whose function is required for DNA damage-induced mutagenesis in *Saccharomyces cerevisiae*. Yeast 1994; 10:1503-1509

61. Jentsch S, McGrath JP, Varshavsky A. The yeast DNA repair gene *RAD6* encodes a ubiquitin-conjugating enzyme. Nature 1987; 329:131-134.

62. Montelone BA, Prakash S, Prakash L. Recombination and mutagenesis in *rad6* mutants of *Saccharomyces cerevisiae*. Evidence for multiple functions of the *RAD6* gene. Mol Gen Genet 1981; 184:410-415.

63. Schwencke J, Moustacchi E. Proteolytic activities in yeast after UV irradiation. 2. Variation in proteinase levels in mutants blocked in DNA repair pathways. Mol Gen Genet 1982; 185:296-301.

64. Hollingsworth RE, Ostroff RM, Klein MB et al. Molecular genetic studies of the Cdc7 protein kinase and induced mutagenesis in yeast. Genetics 1992; 132:53-62.

65. Rong L, Klein HL. Purification and characterization of the SRS2 DNA helicase of the yeast *Saccharomyces cerevisiae*. J Biol Chem 1993; 268:1252-1259.

66. Palladino F, Klein HL. Analysis of mitotic and meiotic defects in *Saccharomyces cerevisiae* SRS2 DNA helicase mutants. Genetics 1992; 132:23-37.

67. Naegeli H, Bardwell L, Friedberg EC. The DNA helicase and adenosine triphosphatase activities of yeast Rad3 protein are inhibited by DNA damage. J Biol Chem 1992; 267:392-398.

68. Naegeli H, Bardwell L, Friedberg EC. Inhibition of Rad3 DNA helicase activity by DNA adducts and abasic sites: implications for the role of a DNA helicase in damage-specific incision of DNA. Biochemistry 1993; 32:613-621.

69. Naegeli H, Modrich P, Friedberg EC. The DNA helicase activity of Rad3 protein of *Saccharomyces cerevisiae* and helicase II of *Escherichia coli* are differentially inhibited by covalent and noncovalent DNA modifications. J Biol Chem 1993; 268:10386-10392.

70. Johnson RE, Henderson ST, Petes TD et al. *Saccharomyces cerevisiae RAD5*- encoded DNA repair protein contains DNA helicase and zinc-binding sequence motifs and affects the stability of simple repetitive sequences in the genome. Mol Cell Biol 1992; 12:3807-3818.

71. Jones JS, Weber S, Prakash L. The *Saccharomyces cerevisiae RAD18* gene encodes a protein that contains potential zinc finger domains for nucleic acid binding and a putative nucleotide binding sequence. Nucleic Acids Res 1988; 16:7119-7131.

72. Koken M, Reynolds P, Bootsma D et al. *Dhr6*, a *Drosophila* homolog of the yeast DNA-repair gene *RAD6*. Proc Natl Acad Sci USA 1991; 88:3832-3836.

73. Koken M, Reynolds P, Jaspers-Dekker I et al. Structural and functional conservation of two human homologs of the yeast DNA repair gene *RAD6*. Proc Natl Acad Sci USA 1991; 88:8865-8869.

74. Roest HP, van Klaveren J, de Wit J et al. Inactivation of the HR6B ubiquitin-conjugating DNA repair enzyme in mice causes male sterility associated with chromatin modification. Cell 1996; 86:799-810.

75. Boyd JB, Setlow RB. Characterisation of postreplication repair in mutagen-sensitive strains of *Drosophila melanogaster*. Genetics 1976; 84:507-526.

76. Boyd JB, Mason JM, Yamamoto AH et al. A genetic and molecular analysis of DNA repair in *Drosophila*. J Cell Sci 1987; Suppl 6:39-60.

77. Brown TC, Boyd JB. Postreplication repair-defective mutants of *Drosophila melanogaster* fall into two classes. Mol Gen Genet 1981; 183:356-362.

78. Hari KL, Santerre A, Sekelsky JJ et al. The *mei-41* gene of *D. melanogaster* is a structural and functional homolog of the human ataxia telangiectasia gene. Cell 1995; 82:815-821.

79. Cleaver JE. Defective repair replication in xeroderma pigmentosum. Nature 1968; 218:652-656.

80. Tung BS, McGregor WG, Wang YC et al. Comparison of the rate of excision of major UV photoproducts in the strands of the human *HPRT* gene of normal and xeroderma pigmentosum variant cells. Mutat Res 1996; 362:65-74.

81. Misra RR. Vos J-M H. Defective replication of psoralen adducts detected at the gene-specific level in xeroderma pigmentosum variant cells. Mol Cell Biol 1993; 13:1002-1012.

82. Maher VM, Ouellette LM, Curren RD. Frequency of ultraviolet light-induced mutations is higher in xeroderma pigmentosum variant cells than in normal human cells. Nature 1976; 261:593-595.

83. Myhr BC, Turnbull D, DiPaolo JA. Ultraviolet mutagenesis of normal and xeroderma pigmentosum variant human fibroblasts. Mutat Res 1979; 63:341-353.

84. Wang YC, Maher VM, McCormick JJ. Xeroderma pigmentosum variant cells are less likely than normal cells to incorporate dAMP opposite photoproducts during replication of UV-irradiated plasmids. Proc Natl Acad Sci USA 1991; 88:7810-7814.

85. Wang YC, Maher VM, Mitchell DL et al. Evidence from mutation spectra that the UV hypermutability of xeroderma pigmentosum variant cells reflects abnormal, error-prone replication on a template containing photoproducts. Mol Cell Biol 1993; 13:4276-4283.

86. Raha M, Wang G, Seidman MM et al. Mutagenesis by third-strand-directed psoralen adducts in repair-deficient cells: high frequency and altered spectrum in a xeroderma pigmentosum variant. Proc Natl Acad Sci USA 1996; 93:2941-2946.

87. Carty MP, Hauser J, Levine AS et al. Replication and mutagenesis of UV-damaged DNA templates in human and monkey cell extracts. Mol Cell Biol 1993; 13:533-542.

88. Thomas DC, Kunkel TA. Replication of UV-irradiated DNA in human cell extracts: evidence for mutagenic bypass of pyrimidine dimers. Proc Natl Acad Sci USA 1993; 90:7744-7748.

89. Carty MP, El-Saleh S, Zernik-Kobak M et al. Analysis of mutations induced by replication of UV-damaged plasmid DNA in HeLa cell extract. Environ Molecul Mutagen 1995; 26:139-146.

90. Thomas DC, Nguyen DC, Piegorsch WW et al. Relative probability of mutagenic translesion synthesis on the leading and lagging strands during replication of UV-irradiated DNA in a human cell extract.

Biochemistry 1993; 32:11476-11482.

91. Thomas DC, Svoboda DL, Vos JMH et al. Strand specificity of mutagenic bypass replication of DNA containing psoralen monoadducts in a human cell extract. Mol Cell Biol 1996; 16:2537-2544.

92. Thomas DC, Veaute X, Kunkel TA et al. Mutagenic replication in human cell extracts of DNA containing site-specific *N*-2-acetylaminofluorene adducts. Proc Natl Acad Sci USA 1994; 91:7752-7756

DNA DAMAGE RECOGNITION:
TOXICOLOGICAL AND MEDICAL PROSPECTS

There is mounting evidence that knowledge about the mechanisms by which DNA damage is recognized in mammalian cells may have major impacts in toxicological and medical sciences. This chapter is intended to provide a few very different examples of currently unresolved human health-related problems (low-dose risk assessment, anticancer therapy, antiviral agents), where DNA damage recognition processes play a prominent role.

As outlined in the previous chapters, the biological endpoint of genotoxic agents is governed by an intricate network of DNA damage response pathways, all of which are capable of modulating the cytotoxic, mutagenic and carcinogenic outcome of genotoxic insults (see scheme in Fig. 1.1). Whether a particular type of genetic lesion is processed in a mutagenic or nonmutagenic manner depends primarily on the prevailing cellular reactions which, in turn, are determined by the initial substrate recognition step of each pathway in the damage response network. In most cases, multiple damage processing mechanisms with largely diverging effects compete for the same substrate (for example a cyclobutane pyrimidine dimer or a carcinogen-DNA adduct), and DNA damage recognition is the primary biochemical step that regulates the differential activation of these cellular reactions. Some processes cooperate synergistically (for example DNA repair and cell cycle checkpoints) while other pathways act antagonistically (for example DNA repair and mutagenic translesion replication).

DNA DAMAGE RECOGNITION AND ITS CONSEQUENCES

The intracellular fate of a genetic lesion is dictated by the relative affinity of various damage recognition subunits for this particular lesion. In general, recognition by DNA repair systems (damage reversal, DNA excision repair) effectively reduces the cytotoxic, mutagenic and carcinogenic consequences of DNA damage (see chapters 1 and 2), although a few significant exceptions have been reported in the literature.[1-4] In addition, cell cycle checkpoint systems provide the means to arrest cell proliferation in response to DNA damage, thereby facilitating DNA repair and preventing the dangerous propagation of mutated or incomplete sets of the genetic information (chapters 1 and 2).[5,6] Cell cycle control circuits may also culminate in the activation of apoptotic death programs to eliminate genetically compromised cells.[7,8] On the other hand, there is evidence suggesting that mammalian and human cells possess specialized mechanisms that operate during S phase to facilitate error-

Mechanisms of DNA Damage Recognition in Mammalian Cells, by Hanspeter Naegeli.
© 1997 R.G. Landes Company.

free replication of damaged templates (chapter 10).[9-11]

These nonmutagenic pathways of the damage response network are opposed by several mechanisms that potentiate mutagenic reactions, for example by promoting error-prone modes of bypass during replication of damaged templates (chapter 10).[12-14] In mammalian cells, genetic instability following exposure to DNA damaging agents is often a product of illegitimate recombination, i.e., a genetic rearrangement mechanism that requires little or no sequence homology and results in deleterious chromosomal alterations.[15] Also, the activation of alternative replication origins in damaged mammalian chromosomes (discussed in chapter 10) may have disastrous consequences by promoting overreplication of critical genes involved in cell transformation and cancer.[16]

Finally, DNA-reactive agents also exert cytotoxic effects, mainly through the action of transcription or replication blocking lesions. This cytotoxic response is strongly modulated by the biochemical properties of RNA and DNA polymerases or other nucleic acid metabolizing enzymes. As discussed in chapter 9, RNA polymerase II is capable of mediating the preferential repair of transcribed strands thereby reducing cytotoxicity, whereas RNA polymerases I and III lack this ability for transcription-repair coupling.[17-19] Similarly, various mammalian DNA polymerases differ significantly in their sensitivity to damaged templates (see chapter 10).[20] Experimental work summarized in chapter 8 suggests that, in some cases, cytotoxicity may be enhanced by the fortuitous binding of cellular proteins to specific types of base adducts, resulting in their sequestration (or "hijacking") at improper sites in the genome.[21,22]

MOLECULAR MODEL SYSTEMS

The rate and efficiency of substrate recognition by each pathway of the damage response network depends on the chemical and structural properties of the offending lesion. For example, recognition of carcinogen-DNA adducts by nucleotide excision repair is thought to vary as a function of the site of covalent modification on the bases, the molecular composition of the adduct, its bulkiness and hydrophobicity, and the resulting alterations in the secondary structure of the DNA duplex. Additional factors such as nucleotide sequence context, interactions with transcription factors and the local chromatin structure also play a decisive role in DNA damage recognition.

To understand in detail these different levels of complexity, it is necessary to first examine the problem of DNA damage recognition using appropriate molecular model systems. In a second phase, knowledge generated in such model studies should be applicable to more relevant systems that mimic human exposure to genotoxic agents in the environment, in our homes or in the workplace. During the last decades, information on the chemistry, metabolism and reactivity of environmental or man-made carcinogens has been growing continuously (see for example refs. 23,24). In parallel, the chemical structure of the resulting DNA adducts has been elucidated (see for example refs. 25-27). More recently, the solution conformations or crystallographic structures of DNA molecules containing a particular type of carcinogen adduct have been determined at striking levels of resolution (see for example refs. 28,29). Critical target sequences in a number of oncogenes or tumor suppressor genes have also been identified. For example, exposure to solar light is correlated with $C \cdot G \rightarrow T \cdot A$ transition mutations at several dipyrimidine sites in the *p53* tumor suppressor gene.[30] Dietary aflatoxin B_1 exposure is correlated with $G \cdot C \rightarrow T \cdot A$ transversions that lead to serine substitutions at residue 249 of p53 protein in hepatocellular carcinomas.[31,32] Exposure to cigarette smoke is correlated with $G \cdot C \rightarrow T \cdot A$ transversions in the *p53* gene of lung, head and neck carcinoma cells, and these mutations have been associated with benzo[*a*]pyrene adduct formation.[33,34] This wealth of information should be further exploited to develop procedures that allow to predict mutagenic and carcinogenic responses in experimental animals and human populations.

PUBLIC POLICY

In the long term, knowledge of the mechanisms of DNA damage recognition may have a significant impact on the regulatory systems concerned with human exposure to carcinogens. Many types of cancer in humans are preventable since epidemiological studies reveal that diet, customs and certain known environmental factors have a predominant influence on tumor incidence. In many cases, such as cancer of the skin or lung, the causal relationship between exposure to a known carcinogen (sunlight or cigarette smoke) and tumor development is well established.[34,35] Instead of focussing our efforts on minimizing these known risk factors, enormous resources are spent to counteract low-dose exposure to a multitude of chemicals despite the fact that their true carcinogenic potential in humans is, in many cases, poorly understood.

Potential carcinogens are normally tested by administration to rodents at concentrations that exceed by several orders of magnitude those expected in human populations. However, there is a growing awareness that a more realistic assessment of human cancer risks can be achieved by an interdisciplinary approach that takes into account all intracellular reactions that modulate, by either attenuating or potentiating, the biological consequences of low levels of genotoxic agents. The final goal of this approach is to reexamine the postulated linear extrapolation for genotoxic chemicals and, therefore, allow more accurate assessments of the mutagenic and carcinogenic risks following low-dose exposures.

Several lines of investigations converge on the central concept that the mutagenic risk of a DNA lesion, and hence its carcinogenic potential, depends on the efficiency by which this lesion is recognized by DNA repair processes. A direct link between DNA repair and carcinogenesis was first established by the human genetic disorder xeroderma pigmentosum, in which the lack of

nucleotide excision repair activity predisposes the affected individuals to elevated rates of mutagenesis and carcinogenesis.[36-38] A strict requirement for DNA repair activities in preventing mutations and cancer was confirmed using transgenic mice models in which the nucleotide excision repair pathway was deleted.[39-41] More recent studies have unequivocally confirmed that poor repairability coincides with increased mutation rates in specific target sequences that are directly implicated in the carcinogenesis process (see below).

DNA lesions that are rapidly repaired have a smaller probability of developing their mutagenic potential as compared to those that are repaired more slowly. Conceivably, certain types of DNA lesions may escape detection by DNA repair, and a quantitatively minor lesion which is resistant to removal by DNA repair enzymes poses a greater threat to the genetic integrity than a quantitatively major product that is rapidly repaired. Hence, it may be possible that some rare lesions have unexpectedly large biological effects because of their inefficient recognition by DNA repair processes. Conversely, other lesions may exert mutagenic and carcinogenic effects less frequently than predicted from simple linear extrapolations. As a consequence, improved assessments of the impact of carcinogen exposure on human health requires knowledge of the ability of human tissues to repair carcinogen-induced DNA damage. In the future, valuable predictions on the kinetics and efficiency of DNA damage recognition and subsequent removal from critical target sequences will constitute an important basis for low-dose risk assessments.

Another aspect that is often not considered in current risk assessment studies is that human populations are always exposed to a multiplicity of different carcinogens which may exert additive effects on cellular defense systems. Conceivably, exposure to certain combinations of DNA damaging agents, although at very low concentrations, may eventually saturate the cellular DNA repair mechanisms. It is possible that, under these circumstances, the phenotypic consequences of DNA damage are severely enhanced because DNA repair proteins are diverted from their physiological function in processing DNA damage generated by endogenous agents (for example oxygen radicals) or ubiquitous environmental agents (for example UV-B radiation or natural food ingredients). Chapter 7 describes a set of experiments in which we developed a novel repair competition assay to compare the relative affinity of nucleotide excision repair recognition factors for various carcinogen-DNA adducts. Interestingly, we found that acetylaminofluorene-C^8-guanine adducts are able to attract the nucleotide excision repair system about 10-fold more efficiently than pyrimidine(6-4)pyrimidone photoproducts and about 100-fold more efficiently than cyclobutane pyrimidine dimers. Therefore, a low number of a particular carcinogen-DNA adduct (in this case acetylaminofluorene-C^8-guanine lesions) displays a strong inhibitory effect on excision repair of base damage inflicted by UV radiation. These results suggest that low-dose exposure to specific carcinogens may have dramatic and unexpected effects by potentiating the carcinogenic risk following UV-B uptake during sunning. In any case, studies on the mechanisms of DNA damage recognition are expected to facilitate the toxicological evaluation of complex carcinogen mixtures.

DNA REPAIR
AND MUTAGENESIS
AT NUCLEOTIDE RESOLUTION

Analysis of DNA repair and mutagenesis at nucleotide resolution in single genes demonstrated that DNA repair efficiency is a major determinant for the formation of mutational hotspots. It has been recognized in recent years that damage recognition is a highly heterogeneous process in the context of mammalian chromosomes. For example, the template strand of genes transcribed by RNA polymerase II is often repaired faster and more efficiently than its complementary coding strand or other nontranscribed

genomic sequences.[17,19] The rates of DNA damage excision also vary between different nontranscribed genes, as repressed genomic loci condensed in heterochromatin are highly refractory to DNA repair processes.[42,43]

Another degree of heterogeneity has emerged when DNA repair rates within single genes were analyzed at nucleotide resolution. These studies were made possible by a new technology based on two prokaryotic enzymes that convert cyclobutane pyrimidine dimers to ligatable DNA strand breaks.[44,45] Human skin fibroblasts were exposed to UV radiation, and excision repair along specific genes was monitored by treating genomic DNA with T4 endonuclease V to generate nicks at cyclobutane pyrimidine dimers; DNA photolyase in combination with visible light was then used to remove the resulting 5'-pyrimidine overhangs. After amplification by ligation-mediated polymerase chain reaction, the DNA fragments were analyzed on sequencing gels. On the resulting autoradiographs, the appearance of DNA breaks indicates the presence of cyclobutane pyrimidine dimers at specific positions of the gene, whereas the progressive loss of these enzymatically induced breaks after different time periods reflects excision repair of the dimers.

Tornaletti et al[44] used this sensitive method to assess the location, frequency and repair rate of cyclobutane dimers along the human *p53* tumor suppressor gene. This study was prompted by previous reports demonstrating that, in many cases, the *p53* gene of human skin cancer cells contains mutations (C→T or CC→TT double transitions) that are typically induced by the UV component of sunlight. These mutations are not evenly distributed along the *p53* gene but are rather clustered in hotspots such as those listed in Table 11.1. Interestingly, the repair rates varied dramatically with sequence position. Some positions were repaired with 70-95% efficiency within 24 h after irradiation, whereas other positions were less than 10% repaired in the same time period. On the transcribed strand, between 23% and 56% of the dimers remained at

Table 11.1. Hotspots for UV-induced mutations in the transcribed (T) or nontranscribed strand (NT) of p53 tumor suppressor gene[44]

Codon (strand)	Hotspot for damage	Slowspot for repair	Hotspot for mutation
151 (NT)	+	−	+
177 (NT)	−	+	+
196 (NT)	−	+	+
245 (T)	−	+	+
248 (T)	−	+	+
278 (NT)	+	+	+
286 (T)	+	+	+
294 (T)	+	+	+

codons 245, 248, 286 and 294 after 24 h of repair, whereas adjacent positions that are not at mutational hotspots were almost completely repaired within 24 h (Table 11.1). On the nontranscribed strand, cyclobutane dimers at codons 177, 196 and 278 were only 0-33% repaired after 24 h, whereas other positions were 70-90% repaired after 24 h. Table 11.1 shows that, in most cases, sites that are hotspots for UV-induced mutations in skin cancer cells are also sites of slow repair, strongly suggesting that inefficient repair is a primary determinant of cancer-associated mutations. This conclusion is supported by the fact that at least four sites in human p53 (codons 177, 196, 245 and 248), which are known hotspots for mutations and suffer from slow repair, are not among those most frequently damaged by UV radiation (Table 11.1). The existence of highly nonuniform repair rates within a single genomic locus was confirmed when excision of cyclobutane pyrimidine dimers from the human PGK1 (phosphoglycerate kinase 1) gene was analyzed.[45] These measurements showed that repair rates along a single gene may vary by more than one order of magnitude.

Ligation-mediated polymerase chain reaction was subsequently adapted to analyze the repair of bulky adducts formed by the polycyclic hydrocarbon $(\pm)$-7b,8a-dihydroxy-9a,10a-epoxy-7,8,9,10-tetra-hydrobenzo[a]pyrene (BPDE). Wei et al[46] measured the rates of removal of BPDE adducts from individual sites in exon 3 of the hypoxanthine phosphoribosyltransferase (HPRT) gene. In their experiments, strand breaks at sites of covalent modification were introduced by treatment of genomic DNA with Escherichia coli (A)BC excinuclease (see chapter 5). The reaction products were annealed with 5' biotinylated primers and extended with the Sequenase 2.0 DNA polymerase to generate a blunt end at the site of each cut. A linker was ligated to these blunt ends, and the DNA fragments were isolated from the rest of the genomic DNA with magnetic beads. After amplification by PCR, the fragments were resolved on sequencing gels. This method yielded large differences in the rate of repair at individual sites along the nontranscribed strand (Table 11.2). Repair was particularly slow at BPDE-induced mutational hotspots (nucleotides 208, 211, 212 and 229). Conversely, sites characterized

Table 11.2. Comparison between the number of BPDE-induced base substitution mutants in the nontranscribed strand of exon 3 of the HPRT gene and rates of repair of BPDE-guanine adducts in the same region[46]

Nucleotide position	Number of base substitution mutants (from a total of 20 mutants)	BPDE adducts remaining after 30 h of repair
197	0	8%
199	0	12%
201	0	18%
207	0	38%
208	1	50%
209	0	54%
211	2	63%
212	4	75%
226	0	24%
229	2	81%
238	0	68%

by very fast repair of BPDE-guanine adducts such as nucleotide positions 197, 199, 201, 207 or 226 are not mutational hotspots (Table 11.2). These results strengthen the hypothesis that poor damage recognition and, hence, inefficient DNA repair is a major determinant of damage-induced mutation rates.

NEW CANCER CHEMOTHERAPY

An important medical problem that remains to be solved is the development of new and more effective chemotherapeutic strategies against cancer. Nuclear DNA is a major pharmacological target of antitumor therapy, and many new DNA-reactive agents are being tested in clinical trials or developed for potential use as antineoplastics.[47,48] This approach to cancer treatment is based on the idea that damage to DNA interferes with chromosomal replication and causes arrest of rapidly dividing cell populations, primarily cancer cells.

Unfortunately, several problems are associated with conventional approaches such as treatment with ionizing radiation or DNA-reactive drugs. First, the efficacy of these radio- or chemotherapeutic agents depends on the number and distribution of unrepaired DNA lesions in cancer cells. Hence, DNA repair represents a possible mechanism of chemotherapy resistance.[49] Second, radio- and many types of chemotherapy induce mutagenic DNA lesions that may accelerate further genetic changes in cancer cells, thereby enhancing progression to tumor malignancy or therapy resistance. Also, drug-induced mutations are thought to be responsible for the development of secondary tumors in cancer patients.[50-52]

Thus, the search for highly cytotoxic DNA lesions that are poorly repaired and, additionally, display low mutagenicity is critical to the development of new and more effective anticancer strategies.

To explore a new chemotherapeutic strategy, we have recently tested a novel class of deoxyribose derivatives which were manipulated at their position C4' (see Fig. 7.5 in chapter 7). We found that DNA synthesis by a model DNA polymerase (Klenow fragment of prokaryotic DNA polymerase I) was effectively terminated at these C4' deoxyribose adducts.[53] Additionally, primer extension studies in the presence of single deoxyribonucleotides showed intact base pairing fidelity opposite the tested C4' variants regardless of whether the Klenow fragment or its proofreading-deficient mutant were tested.[53] These results imply that the inherent base pairing fidelity of the DNA polymerase was not detectably disturbed at altered C4' deoxyribose moieties. However, their strong capacity to impede DNA polymerase progression suggests a potential target for new anticancer chemotherapeutics that operate by a nonmutagenic mechanism. In a parallel study, we observed that modified C4' residues are unable to stimulate DNA excision repair processes in human cell extracts.[54] In summary, the combination of DNA synthesis arrest (indicating cytotoxicity), low mutagenicity and poor repairability (indicating biological persistence) suggests that C4' DNA backbone analogs may be further developed for chemotherapeutic purposes. The availability of crystallographic structures of mammalian polymerases complexed with template/primers and deoxyribonucleotide substrates should facilitate the design of appropriate C4' deoxyribose analogs.[55] In addition, the observation that certain C4'-modified TTP residues are used as a substrate for strand elongation by DNA polymerases (B. Giese and A. Marx, personal communication) suggests a possible pathway to target such novel deoxyribose modifications to the DNA of cancer cells. Of course, many alternative approaches are being taken to improve cancer treatment. Another line of experimentation, discussed in chapter 8, involves intense scrutiny of the mechanisms of action of the antitumor agent cisplatin. This strategy will hopefully result in the production of a new generation of platinum derivatives with enhanced clinical effectiveness.

DEVELOPMENT OF ANTIVIRAL AGENTS

Inhibitors of substrate recognition by DNA glycosylases may lead to the development of new pharmacological agents that are active against neuroinvasive herpes viruses. This class of viruses have a worldwide distribution and, because of their neuroinvasiveness, are among the most difficult pathogens to control. The human herpesviruses include herpes simplex types 1 and 2 (HSV-1 and HSV-2), varicella-zoster and Epstein-Barr viruses, cytomegalovirus, human herpesviruses types 6 and 7, and Kaposi's sarcoma herpes virus. These agents share the properties of latency and reactivation, i.e., they can cause productive lytic infections, in which infectious virus is produced and cells are killed, or nonproductive infections, in which cells survive and viral DNA persists.[56,57] During nonproductive (latent) infections viral DNA is not replicated and only limited parts of the genome are transcribed. After acute lytic infections, herpesviruses often persist in this latent form for years before being reactivated. The clinical features of neuroinvasive herpesvirus infections range from vesicles on the skin or mucous membranes in adults, to viremia in children. Infections of the fetus involves very high mortality. These herpesviruses also cause encephalitis, neuralgia, myelitis and facial paralysis.[58-60]

Although vaccines have been successful in preventing many viral diseases, there is still a pressing need for effective chemotherapeutic treatment of viral infections. Only with the recognition of viral enzymes and proteins that can serve as molecular targets for drugs, antiviral chemotherapy has become possible. Synthetic nucleoside analogs are currently the most important antiviral drug available for clinical use. These compounds result in inhibition of viral DNA replication but are not active against latent herpes viruses.[61,62] Also, problems associated with acquired drug resistance to nucleoside analogs have to be considered.[63-65]

As mentioned in chapter 4, several viruses encode their own uracil-DNA glycosylase. A completely new strategy for the development of drugs that are active against neuroinvasive herpesviruses is suggested by the report of Pyles and Thompson,[66] who showed that HSV-1 uracil-DNA glycosylase is required for both viral replication and reactivation in the murine nervous system. Presumably, the episomal viral genome accumulates large numbers of uracil residues. Upon reactivation, these mutagenic lesions are normally repaired by the virally encoded uracil-DNA glycosylase. Pyles and Thompson[66] constructed two HSV-1 strains lacking the gene for uracil-DNA glycosylase. Following direct intracranial inoculation in mice, both mutants exhibited a 10-fold reduction in neurovirulence compared to the parental strain. After inoculation at a peripheral site (footpads), the mutant strains were 100,000-fold less neuroinvasive than the control strain. Restoration of the uracil-DNA glycosylase locus in the two mutants resulted in full neurovirulence and neuroinvasiveness. Taken together, these results suggest that compounds that specifically inhibit viral uracil-DNA glycosylase may provide clinically useful drugs against neuroinvasive herpesvirus infections.[67] Although the substrate specificity of the human and HSV-1 uracil-DNA glycosylases is identical, there appear to be subtle differences in the uracil cleavage mechanism.[68,69] These differences may be exploited to design herpes-specific uracil-DNA glycosylase inhibitors. Important structural insights for the design of such inhibitors are provided by the crystal structure of the uracil-DNA glycosylase inhibitor from *Bacillus subtilis* bacteriophage PBS2 complexed with human uracil-DNA glycosylase.[70]

REFERENCES

1. Kat A, Thilly WG, Fang WH et al. An alkylation-tolerant, mutator human cell line is deficient in strand-specific mismatch repair. Proc Natl Acad Sci USA 1993; 90:6424-6428.

2. Branch P, Aquilina G, Bignami M et al. Defective mismatch binding and a mutator phenotype in cells tolerant to DNA damage. Nature 1993; 362:652-654.

3. Klein JC, Bleeker MJ, Roelen HCPF et al. Role of nucleotide excision repair in processing of O^4-alkylthymines in human cells. J Biol Chem 1994; 269:25521-25528.

4. Hays JB, Ackerman EJ, Pang QS. Rapid and apparently error-prone excision repair of nonreplicating UV-irradiated plasmids in *Xenopus laevis* oocytes. Mol Cell Biol 1990; 10:3505-3511.

5. Hartwell LH. Defects in a cell cycle checkpoint may be responsible for the genomic instability of cancer cells. Cell 1992; 71:543-546.

6. Kuerbitz SJ, Plunkett BS, Walsh WV et al. Wild-type p53 is a cell cycle checkpoint determinant following irradiation. Proc Natl Acad Sci USA 1992; 89:7491-7495.

7. Williams GT, Smith CA. Molecular regulation of apoptosis: genetic controls on cell death. Cell 1993; 74:777-779.

8. Wang XW, Vermeulen W, Coursen JD et al. The XPB and XPD DNA helicases are components of the p53-mediated apoptosis pathway. Genes Dev 1996; 10:1219-1232.

9. Myhr BC, Turnbull D, DiPaolo JA. Ultraviolet mutagenesis of normal and xeroderma pigmentosum variant human fibroblasts. Mutat Res 1979; 63:341-353.

10. Wang YC, Maher VM, McCormick JJ. Xeroderma pigmentosum variant cells are less likely than normal cells to incorporate dAMP opposite photoproducts during replication of UV-irradiated plasmids. Proc Natl Acad Sci USA 1991; 88:7810-7814.

11. Wang YC, Maher VM, Mitchell DL et al. Evidence from mutation spectra that the UV hypermutability of xeroderma pigmentosum variant cells reflects abnormal, error-prone replication on a template containing photoproducts. Mol Cell Biol 1993; 13:4276-4283.

12. Lawrence C. The RAD6 DNA repair pathway in *Saccharomyces cerevisiae*: what does it do, and how does it do it? BioEssays 1994; 16:253-257.

13. Nelson JR, Lawrence CW, Hinkle DC. Deoxycytidyl transferase activity of yeast REV1 protein. Nature 382; 382:729-731.

14. Nelson JR, Lawrence CW, Hinkle DC. Thymine-thymine dimer bypass by yeast DNA polymerase ζ. Science 1996; 272:1646-1649.

15. Würgler FE. Recombination and gene conversion. Mutat Res 1992; 284:3-14.

16. Schimke RT, Sherwood SW, Hill AB et al. Overreplication and recombination of DNA in higher eukaryotes: potential consequences and biological implications. Proc Natl Acad Sci USA 1986; 83:2157-2161.

17. Bohr VA, Smith CA, Okumoto DS et al. DNA repair in an active gene: removal of pyrimidine dimers from the *DHFR* gene of CHO cells is much more efficient than in the genome overall. Cell 1985; 40:359-369.

18. Christians FC, Hanawalt PC. Lack of transcription-coupled repair in mammalian ribosomal RNA genes. Biochemistry 1993; 32:10512-10518.

19. Friedberg EC. Relationships between DNA repair and transcription. Annu Rev Biochem 1996; 65:15-42.

20. Hoffmann J-S, Pillaire M-J, Maga G et al. DNA polymerase β bypasses in vitro a single d(GpG)-cisplatin adduct placed on codon 13 of the *HRAS* gene. Proc Natl Acad Sci USA 1995; 92:5356-5360

21. Treiber DK, Zhai X, Jantzen H-M et al. Cisplatin-DNA adducts are molecular decoys for the ribosomal RNA transcription factor hUBF (human upstream binding factor). Proc Natl Acad Sci USA 1994; 91:5672-5676.

22. MacLeod MC, Powell KL, Tran N. Binding of the transcription factor, Sp1, to non-target sites in DNA modified by benzo[*a*]pyrene diol epoxide. Carcinogenesis 1995; 16:975-983.

23. Miller JA, Miller EC. Metabolic activation of carcinogenic aromatic amines and amides via *N*-hydroxylation and *N*-hydroxy esterification and its relationship to ultimate carcinogens as electrophilic reactants. In: Bergmann ED, Pullman B, eds. The Jerusalem Symposia on Quantum Chemistry and Biochemistry, Physicochemical Mechanism of Carcinogenesis. Jerusalem: The Israel Academy of Sciences and Humanities, 1969;237-261.

24. Conney AH. Induction of microsomal enzymes by foreign chemicals and carcinogenesis by polycyclic aromatic hydrocarbons. Cancer Res 1982; 42:4875-4917.

25. Meehan T, Straub K, Calvin M. Benzo[*a*]pyrene diol epoxide covalently binds to deoxyguanosine and deoxyadenosine in DNA. Nature 1977; 269:725-727.

26. Mehan T, Straub K. Double-stranded DNA stereoselectively binds benzo[*a*]pyrene diol epoxides. Nature 1979; 277:410-412.

27. Cheng SC, Hilton BD, Roman JM et al. DNA adducts from carcinogenic and noncarcinogenic enantiomers of benzo[*a*]pyrene dihydrodiol epoxide. Chem Res Toxicol 1989; 2:334-340.

28. Cosman M, de los Santos C, Fiala R et al. Solution conformation of the (+)-*cis-anti*-[BP]dG adduct in a DNA duplex: intercalation of the covalently attached benzo[*a*]pyrenyl ring into the helix and displacement of the modified deoxyguanosine. Biochemistry 1993; 32:4146-4155.

29. Cosman M, Hingerty BE, Luneva N et al. Solution conformation of the (-)-*cis-anti*-benzo[*a*]pyrenyl-dG adduct opposite dC in a DNA duplex: intercalation of the covalently attached BP ring into the helix with base displacement of the modified deoxyguanosine into the major groove. Biochemistry 1996; 35:9850-9863.

30. Brash DE, Rudolph JA, Simon JA et al. A role for sunlight in skin cancer: UV-induced *p53* mutations in squamous cell carcinoma. Proc Natl Acad Sci USA 1991; 88:10124-10128.

31. Hsu IC, Metcalf RA, Sun T et al. Mutational hotspot in the *p53* gene in human hepatocellular carcinomas. Nature 1991; 350:427-428.

32. Bressac B, Kew M, Wands J et al. Selective G to T mutations of *p53* gene in hepatocellular carcinoma from southern

Africa. Nature 1991; 350:429-431.

33. Denissenko MF, Pao A, Tang M-S et al. Preferential formation of benzo[a]-pyrene adducts at lung cancer mutational hotspots in *P53*. Science 1996; 274:430-432.

34. Harris CC. p53: at the crossroads of molecular carcinogenesis and risk assessment. Science 1993; 262:1980-1981.

35. Ames BN, Shinegaga MK, Hagen TM. Oxidants, antioxidants, and the degenerative diseases of aging. Proc Natl Acad Sci USA 1993; 90:7915-7922.

36. Maher VM, McCormick JJ. Effect of DNA repair on the cytotoxicity and mutagenicity of UV irradiation and of chemical carcinogens in normal and xeroderma pigmentosum cells. In: Yuhas JM, Tennant RW, Regan JD, eds. Biology of Radiation Carcinogens. New York: Raven Press, 1976:129-145.

37. Maher VM, Ouellette LM, Curren RD et al. Frequency of ultraviolet light-induced mutations is higher in xeroderma pigmentosum variant cells than in normal human cells. Nature 1976; 261:593-595.

38. Kraemer KH, Lee M-M, Andrews AD et al. The role of sunlight and DNA repair in melanoma and nonmelanoma skin cancer. Arch Dermatol 1994; 130:1018-1021.

39. Nakane H, Takeuchi S, Yuba S et al. High incidence of ultraviolet-B- or chemical-carcinogen-induced skin tumours in mice lacking the xeroderma pigmentosum group A gene. Nature 1995; 377:165-168.

40. de Vries A, van Oostrom CTM, Hofhuis FMA et al. Increased susceptibility to ultraviolet-B and carcinogens of mice lacking the DNA excision repair gene XPA. Nature 1995; 377:169-173.

41. Sands AT, Abuin A, Sanchez A et al. High susceptibility to ultraviolet-induced carcinogenesis in mice lacking *XPC*. Nature 1995; 377:162-165.

42. van Hoffen A, Natarajan AT, Mayne LV et al. Deficient repair of the transcribed strand of active genes in Cockayne's syndrome cells. Nucleic Acids Res 1993; 21:5890-5895.

43. Kantor GJ, Bastin SA. Repair of some active genes in Cockayne syndrome cells is at the genome overall rate. Mutat Res 1995; 336:223-233.

44. Tornaletti S, Pfeifer GP. Slow repair of pyrimidine dimers at *p53* mutation hotspots in skin cancer. Science 1994; 263:1436-1438.

45. Gao S, Drouin R, Holmquist GP. DNA repair rates along the human *PGK1* gene at nucleotide resolution. Science 1994; 263:1438-1440.

46. Wei D, Maher VM, McCormick JJ. Site-specific rates of excision repair of benzo-[a]pyrene diol epoxide in the hypoxanthine phosphoribosyltransferase gene of human fibroblasts: correlation with mutation spectra. Proc Natl Acad Sci USA 1995; 92:2204-2208.

47. Hurley LH, Boyd FL. DNA as a target for drug action. Trends Pharmacol Sci 1988; 91:402-407.

48. Wang G, Glazer PM. Altered repair of targeted psoralen photoadducts in the context of an oligonucleotide-mediated triple helix. J Biol Chem 1995; 270:22595-22601.

49. Chu G. Cellular responses to cisplatin. J Biol Chem 1994; 269:787-790.

50. Bradley LJN, Yarema KJ, Lippard et al. Mutagenicity and genotoxicity of the major DNA adduct of the antitumor drug cis-diamminedichloroplatinum. Biochemistry 1993; 32:982-988.

51. Hennings H, Shores RA, Poirier MC et al. Enhanced malignant conversion of benign mouse skin tumors by cisplatin. J Natl Cancer Inst 1990; 82:836-840.

52. Greene MH. Is cisplatin a human carcinogen? J Natl Cancer Inst 1992; 84:306-312.

53. Hess MT, Schwitter U, Petretta M et al. DNA synthesis arrest at C4'-modified deoxyribose residues. Biochemistry 1997; in press.

54. Hess MT, Schwitter U, Petretta M et al. Site-specific DNA substrates for human excision repair: comparison between deoxyribose and base adducts. Chem Biol 1996; 3:121-128.

55. Pelletier H, Sawaya MR, Kumar A et al. Structures of ternary complexes of rat DNA polymerase β, a DNA template-primer, and ddCTP. Science 1994; 264:1891-1903.

56. Fawl RL, Roizman B. Induction of reactivation of herpes simplex virus in murine sensory ganglia in vivo by cadmium. J Virol 1993; 67:7025-7031.

57. Minson AC. Interactions of herpes simplex viruses with the host cell. Biochem Soc Trans 1994; 22:298-301.

58. O'Meara M, Ouvrier R. Viral encephalitis in children. Curr Opin Pediatr 1996; 8:11-15.

59. Whitley RJ, Lakeman F. Herpes simplex virus infections of the central nervous system: therapeutic and diagnostic considerations. Clin Infect Dis 1995; 20:414-420.

60. Bowsher D. Post-herpetic neuralgia in older patients. Incidence and optimal treatment. Drugs Aging 1994; 5:411-418.

61. Beutner KR. Valacyclovir: a review of its antiviral activity, pharmacokinetic properties, and clinical efficacy. Antiviral Res 1995; 28:281-290.

62. Wagstaff AF, Faulds D, Goa KL. Acyclovir. A reappraisal of its antiviral activity, pharmacokinetic properties and therapeutic efficacy. Drugs 1994; 47:153-205.

63. Coen DM. Acyclovir-resistant, pathogenic herpesviruses. Trends Microbiol 1994; 2:481-485.

64. Pottage JC, Kessler HA. Herpes simplex virus resistance to acyclovir: clinical relevance. Infect Agents Dis 1995; 4:115-124.

65. Laufer DS, Starr SE. Resistance to antivirals. Pediatr Clin North Am 1995; 42:583-599.

66. Pyles RB, Thompson RL. Evidence that the herpes simplex virus type 1 uracil DNA glycosylase is required for efficient viral replication and latency in the murine nervous system. J Virol 1994; 68:4963-4972.

67. Focher F, Verri A, Spadari S et al. Herpes simplex virus type 1 uracil-DNA glycosylase: isolation and selective inhibition by novel uracil derivatives. Biochem J 1993; 292:883-889.

68. Mol CD, Arvai AS, Slupphaug G et al. Crystal structure and mutational analysis of human uracil-DNA glycosylase: structural basis for specificity and catalysis. Cell 1995; 80:869-878.

69. Savva R, McAuley-Hecht K, Brown T et al. The structural basis of specific base-excision repair by uracil-DNA glycosylase. Nature 1995; 373:487-493.

70. Mol CD, Arvai AS, Sanderson RJ et al. Crystal structure of human uracil-DNA glycosylase in complex with a protein inhibitor: protein mimicry of DNA. Cell 1995; 82:701-708.

INDEX